THE SEMANTIC FOUNDATIONS OF LOGIC
VOLUME 1: PROPOSITIONAL LOGICS

Nijhoff International Philosophy Series

VOLUME 35

For a list of volumes in this series see final page of this volume.

Richard L. Epstein

The Semantic Foundations of Logic Volume 1: Propositional Logics

with the assistance and collaboration of

Walter A. Carnielli
Itala M.L. D'Ottaviano
Stanisław Krajewski
Roger D. Maddux

KLUWER ACADEMIC PUBLISHERS
DORDRECHT / BOSTON / LONDON

Library of Congress Cataloging in Publication Data

Epstein, Richard L., 1947-
The semantic foundations of logic / by Richard L. Epstein ; with the assistance and collaboration of Walter A. Carnielli ... [et al.].
p. cm. -- (Nijhoff international philosophy series ; 35-)
Includes bibliographical references.
Contents: v. 1. Propositional logics
ISBN 0-7923-0622-8 (v. 1)
1. Logic. 2. Logic, Symbolic and mathematical. 3. Semantics.
I. Title. II. Series: Nijhoff international philosophy series ; v. 35, etc.
BC71.E56 1990
160--dc20 --dc20 89-48568

ISBN 0-7923-0622-8

Published by Kluwer Academic Publishers,
P.O. Box 17, 3300 AA Dordrecht, The Netherlands.

Kluwer Academic Publishers incorporates
the publishing programmes of
D. Reidel, Martinus Nijhoff, Dr W. Junk and MTP Press.

Sold and distributed in the U.S.A. and Canada
by Kluwer Academic Publishers,
101 Philip Drive, Norwell, MA 02061, U.S.A.

In all other countries, sold and distributed
by Kluwer Academic Publishers Group,
P.O. Box 322, 3300 AH Dordrecht, The Netherlands.

Final pages for this book were
prepared for press by Richard L. Epstein

Printed on acid-free paper

Printed in The Netherlands

Dedicated to

Peter Eggenberger
Harold Mann
and
Benson Mates

with gratitude for their encouragement and guidance

Preface

This book grew out of my confusion. If logic is objective how can there be so many logics? Is there one right logic, or many right ones? Is there some underlying unity that connects them? What is the significance of the mathematical theorems about logic which I've learned if they have no connection to our everyday reasoning?

The answers I propose revolve around the perception that what one pays attention to in reasoning determines which logic is appropriate. The act of abstracting from our reasoning in our usual language is the stepping stone from reasoned argument to logic. We cannot take this step alone, for we reason together: logic is reasoning which has some objective value.

For you to understand my answers, or perhaps better, conjectures, I have retraced my steps: from the concrete to the abstract, from examples, to general theory, to further confirming examples, to reflections on the significance of the work. In doing so I have had to begin at the beginning: what is logic? what is a proposition? what is a connective? If much seems too well known to be of interest, then plunge ahead without a guide: the chapters for the most part can be read independently (that explains the occasional repetitions); the Introduction is a tour guide for the more experienced traveller. But the path I have chosen is not always the most familiar. At places I think I have found an easier way, though because it is new, or perhaps simply because I may not know it well, the way may seem more difficult.

I want to see where this path leads, whether the path, in the end, seems smoother and takes us to higher ground with a better view. So I may not always stop to argue each principle at length. It is the trip as a whole which I hope you will find refreshing.

> In the discussions of the wise there is found unrolling and rolling up, convincing and conceding; agreements and disagreements are reached. And in all that the wise suffer no disturbance.
>
> Nagasena

Come, let us reason together.

Preface

Acknowledgements

The story of this book began in Wellington, New Zealand. Working with the logic group there, Douglas Walton and I developed relatedness logic in 1977.

In 1978 I met Niels Egmont Christensen who led me to see that a slight variation in the work on relatedness logic could model his ideas on analytic implication. Later that year at Iowa State University I began to question what was the "right" logic. From then until I left I.S.U. Roger Maddux challenged me and helped me to technically clarify my intuitions. He and Donald Pigozzi introduced me to nonclassical logic and the algebras of them. In 1980 Roger Maddux, Douglas Walton, and I wrote a monograph which contained the basis of much of the technical work of Chapters V–VIII here. In 1980 and 1982 I gave lectures on propositional and predicate logics at I.S.U. where Howard Blair, William Robinson, and later Gary Iseminger challenged me to explain my philosophical assumptions. I am grateful that while at Iowa State University I was given ample time for research, due particularly to Dan Zaffarano, Dean of Research.

In 1981 I visited the University of Warsaw for six months on an exchange sponsored by the U.S. National Academy of Sciences and the Polish Academy of Sciences. There I met and began collaborating with Stanisław Krajewski whose insights led me to clarify the relationship between formal languages and the languages we speak. Part of our joint work is the chapter on translations between logics, which was influenced by discussions with L. Szczerba.

In 1982 I moved to Berkeley where it was my good fortune to meet Peter Eggenberger and Benson Mates. They are fine teachers: listening to my early inchoate ideas, reading my confused analyses, guiding my reading in philosophy they have helped me understand what I was trying to say. The shape of many of the discussions here comes from conversations I had with them.

In 1984 I lectured on what was then still a series of separate papers to a group of Brazilians at Berkeley. It was through the urgings of Walter Carnielli that I then made the decision to turn those papers into a book and, finally, to publish the work on propositional logics as a separate volume. Much of the form and outline of this volume was developed in discussions with him. He, Newton da Costa, and Itala D'Ottaviano read versions of several of the chapters, and later in 1986 Karl Henderscheid read a draft of the entire volume. Their questions and criticisms substantially improved the exposition.

In 1985 the Fundação de Amparo à Pesquisa do Estado do São Paulo provided me with a grant (number 84/1963–2) to visit Brazil and lecture at the VII Latin American Symposium on Mathematical Logic. That lecture was published in the proceedings of the conference as *Epstein, 1988*, and parts of it are reprinted in Chapter IV with permission of the American Mathematical Society. In 1987 the Fulbright Foundation awarded me a fellowship to lecture and do research at the Center for Logic, Epistemology and History of Science at the University of Campinas and at the Universidade Federal da Paraíba in Brazil. These visits gave me an opportunity to collaborate with Walter Carnielli and Itala D'Ottaviano and resulted in Chapter IX and Appendix 2 to Chapter IV. Part of Chapter IX appears in *Reports on Mathematical Logic, 22* and is reprinted here with permission.

In 1987 I visited the University of Auckland and met Stanisław Surma who gave me many useful suggestions for the book and, most importantly, found a home for it with Martinus Nijhof/Kluwer Academic Publishers.

And throughout, David Gross helped me master typing on a computer and also gave me useful suggestions for Chapter I.

To all these people, and any others I have inadvertently forgotten, I am most grateful. Much that is good in this book is due to them; the mistakes and confusions are mine alone. It is with great pleasure I thank them here.

> In spite of everything, a man was given a chance to get a little peace. He allotted himself a task, and, while performing it, realized that it was meaningless, that it was lost among a mass of human endeavors and strivings. But when a pen hung in air and there was a problem of interpretation or syntax to solve, all those who once, long ago had applied thought and used language were near us. You touched the delicate tracings warmed by their breath, and communion with them brought peace. Who could be so conceited as to be quite sure that he knew which actions were linked up and complementary; and which would recede into futility and be forgotten, forming no part of the common heritage? But was it not better, instead, to ponder the only important question: how a man could preserve himself from the taint of sadness and indifference.
>
> Czeslaw Milosz

And I am grateful to Harold Mann, too, who helped me ponder that question.

Contents

III Relatedness Logic: The Subject Matter of a Proposition – S and R –

IX A Paraconsistent Logic: $\mathbf{J_3}$

in collaboration with **Itala M. L. D'Ottaviano**

X Translations Between Logics

Introduction

Why are there so many logics? Is there no one right way to reason, no one notion of necessity, of objectivity?

This book is devoted to showing that there is a simple structural unity based on semantic assumptions common to many logics. In this Introduction I will briefly describe that unity and the contents of this book. In doing so I will occasionally have to draw on technical terms from the mathematics and philosophy of logic so that someone unfamiliar with logic may prefer to proceed directly to the text.

The proper place to begin is with classical propositional logic. There are good reasons why it's so widely accepted. Something is right about it. It's not the whole story, but it's a fundamental part, a standard of reference for all other logics.

Classical logic is fundamental because it's the simplest symbolic model of reasoning we can devise once certain assumptions are made about what logic is. A proposition in classical logic is abstracted to only its truth-value and its form relative to the propositional connectives. I set out those assumptions in Chapter I, and develop the classical logic in Chapter II.

Something like the division of propositions into the true and the false which is basic to classical logic seems basic to all reasoning. Every logician in the end divides propositions into those which are acceptable and those which are not. I will argue throughout that it is correct in each case to understand these divisions as the division into the true propositions and the false. We have no direct access to the world but only our uncertain perceptions of it, and no two of us can share exactly the same perception. Therefore, to call a sentence true is at best a hypothesis which we hope to share with others, and a realistic humility demands that we see the distinction between logical and pragmatic grounds for rejecting a proposition as a matter of degree and not of kind.

Moreover, even were we to agree that there is something in the world which objectively determines whether 'Ralph is a dog' is true or false, it seems to me that there is nothing in the world external to us which can determine in the same way the truth-value of 'If Ralph is a dog, then George is a duck.' The truth-value of that depends not only on whether Ralph is a dog and whether George is a duck, but on how we are to interpret 'if ... then ...'. For most of us, for most logics, the truth-value of an 'if ... then ...' sentence depends on more than just the truth-values of the antecedent and consequent. Modalities, tenses, accessibility to understanding, constructive mathematical content, or subject matters may enter in, and logics, including logics which seem to explicitly reject the true-false dichotomy of propositions, have been based on all these aspects as I show in Chapter III and Chapters V–IX.

A general form of semantics arises from the view that all these logics comprise a spectrum. Each, except for classical logic, incorporates into the semantics some aspect of propositions other than truth-value and form. As we vary the aspect we vary the logic. In essence, each logic analyzes an 'if ... then ...' proposition classically if the aspects of antecedent and consequent are appropriately connected, while rejecting the proposition otherwise. This overview is presented in Chapter IV; mathematicians interested primarily in the structural analysis of logic may prefer to begin with that. Thereafter it should be clear which sections are devoted to motivation or philosophical analysis and which are primarily technical.

To show that these semantics indeed yield a structural and conceptual overview of many logics, I present chapters on analytic implication, modal logics, intuitionistic logic, many-valued logics, and a paraconsistent logic. In each I first present an introduction to the assumptions of the logic along with the standard semantics in terms which I hope are a reasonably accurate reflection of how the logic is commonly understood. Then I show how the logic can be understood in terms of the overview, giving semantics within the general framework which I believe conform to and reflect the intuitions of the original practitioners of the logic. But I do not feel that I have to show that my reading of, for example, intuitionistic logic is in absolute accord with the intuitionists' understanding in order to justify it, only that it is a way to grasp their logical analyses strongly enough to give a projective knowledge of their work based on well-motivated semantic assumptions.

This general form of semantics is not intended to replace other semantics. For example, under certain assumptions possible world semantics are a good explanation of the ideas of modal logics. Providing uniform semantics which are in reasonable conformity with the ideas on which various logics are based allows for comparisons and gives us a uniform way in which to approach the sometimes overwhelming multiplicity of logics.

It is equally important that this general form of semantics provides a simple tool for incorporating into logic many different aspects of propositions which until now have generally been treated informally. We now have a framework in which to formalize, discuss, and compare different notions as they affect our reasoning. For example, in Chapter V I present a case study of how to use the general framework to develop a logic which incorporates a notion of referential content of propositions.

The referential content of a proposition is not, however, a primitive notion. It depends on the referential content of the predicates and names of which the proposition is composed. The internal structure of propositions matters, and predicate logic is a further test of the aptness of these ideas. Building on the work of Chapter IV, it is possible to extend the structural framework I've given to apply to the predicate calculus. A general form of semantics for predicate logics is based on the perception that when predicates and names are primitive then it is the aspects of these in addition to their extensions which determine the truth of a proposition: which aspect we deem significant determines the logic. Many more assumptions about reasoning and language must be made to establish a logic which deals with the internal structure of propositions; it is unnecessary to introduce those here, jeopardizing the general agreements about reasoning with propositions as wholes which I hope we can reach (or uncover). Therefore, I have deferred to Volume 2 my investigations with Stanisław Krajewski on predicate logic.

The semantic framework which I set out in Chapter IV may be proposed as a very weak general form of logic, a general form which becomes usable only upon the choice of which aspect of propositions we deem to be significant. But then is logic relative to the logician? Or does a notion of necessary truth lie in this very general framework? Briefly, I believe that it is our agreements about how we will reason together which determine our notion of objectivity. I do not mean by this only active, explict agreements, but also implicit ones, what may be said to be our common background. Most of our agreements are implicit, and not necessarily freely made: lack of disagreement I understand as agreement. I discuss this in Chapters I and XI.

Throughout I have tried to find and make explict fundamental assumptions or agreements on which our reasoning and logic are based. I have repeated the statement of certain of these assumptions in a number of different places, partly because I want the chapters to be as self-contained as possible, but also because it is important to see these assumptions and agreements in different contexts and applied differently to be able to grasp their plausibility and pervasiveness.

What I am doing could be seen as founding logic in natural language and reasoning. When nonconstructive, platonist assumptions are used to do either the mathematics of logic, that is, prove theorems about our formalizations, or to apply logic to a particular subject matter, such as arithmetic or geometry, we can see precisely where they are needed. Those assumptions I treat as abstractions from experience, for that is how I understand abstract things. However, they need not be viewed that way and I have attempted to provide alternative readings of the technical work based on the assumption that intangible, nonsensible abstract things are as real or more real than the objects we daily encounter. The primary discussion of these matters can be found in Chapter I and in the development of classical logic in Chapter II. In Chapter IV I point out specific nonconstructive, infinitistic abstractions of the semantics which we usually make in pursuing metalogical investigations.

In Chapter II I also present a Hilbert-style formalization of the notion of proof and syntactic deduction which will be used throughout the book. For most logics in this book the metalogical investigations which I concentrate on concern the relation between the semantic and syntactic notions of consequence and whether or how those can be represented in terms of theorems or valid formulas by means of a Deduction Theorem.

The many examples of logics presented in this book allow us to consider the extent to which one logic or way of seeing the world can be reduced to another by a translation. In Chapter X I present a general theory of translations which I have developed with Stanisław Krajewski.

There are many important subjects in the study of propositional logics which I do not deal with in this book. I have not discussed the algebraic analyses of propositional logics; for that you can consult *Rasiowa, 1974* and *Blok and Pigozzi, 1989.* I have made no attempt to connect this work with the categorial interpretation of logic, for which *Goldblatt, 1979* is a good reference. Nor have I dealt with other approaches to the notion of proof in propositional logics. But a lack which I feel most strongly is that there are many other well-known propositional logics I have not discussed here, many of which are surveyed in *Marciszewski, 1981*, and in *Haack, 1974,* who also discusses the philosophical issues surrounding them. I believe those, too, can be developed within the general overview. But the provocative study of predicate logic beckons.

This is not the story of all propositional logics. But I hope to have presented enough to convice you that it is a good story of many logics which brings a kind of unity to them.

I The Basic Assumptions of Propositional Logic

A. What Is Logic?

Logic is concerned with how to reason, how to deduce from hypotheses, how to demonstrate truths. As presented here, logic is concerned with providing symbolic models of acceptable reasoning.

What can we mean by 'acceptable'? Is logic concerned only with the psychology of how people reason, with setting out pragmatic standards? I, or you and I together, can reflect on our rules for reasoning, but those cover only very simple cases. We are led, therefore, to formal systems, devised to reflect, model, guide, and/or abstract from our native ability to reason. These formal systems are based on our understanding of certain notions such as truth and reference, and those seem to be dependent on i. how we understand the world, and ii. how the world really is.

Is there any difference between (i) and (ii)? And if so, is it a difference which we can perceive and take into account? These are questions that must be raised in doing logic, for they concern how we will account for objectivity in logic and to what extent we will see our work as prescriptive, not just a model of what *is* done, but what *should* be done.

B. Propositions

Let us begin by asking ourselves what objects, what things we are going to study in logic.

1. Sentences, propositions, and truth

When we argue, when we prove, we do so in a language. And we seem to be able to confine ourselves to declarative sentences and to certain ways of forming complex sentences from these.

If you ask me what a sentence is I could perhaps direct you to a grammar book. But I believe that the notion of a sentence and a declarative sentence are sufficiently well understood by us to be taken as primitive here. Disagreements about some particular examples may arise and need to be resolved by us, but I do not think that a general theory of sentences would make our work more "scientific".

So we begin with sentences, written (or uttered) concatenations of inscriptions (or sounds). To study these we may abstract away from them certain aspects, for example what language they are in, or who said them. When we abstract away enough properties from some of these linguistic acts we seem to find common features, the most important of which, for logic, we call 'truth' and 'falsity' .

Which declarative sentences are true or false, that is, have a *truth-value*? Some it would seem are too ambiguous, such as 'I am half-seated', or nonsensical, such as '7 is divisible by lightbulbs'. But if only sentences which are completely objective and unambiguous are true or false then 'Strawberries are red' can be neither true nor false—for which strawberries are meant? what hue of red? measured by what instrument or person? And then we couldn't analyze:

(1) If strawberries are red, then some color blind people cannot see strawberries among their leaves.
Strawberries are red.
Therefore:
Some color blind people cannot see strawberries among their leaves.

Surely this is an example of valid reasoning; and it is reasoning which is important for us to formalize, for this is reasoning as we actually do it. And yet, I believe, any attempt to make these sentences fully precise will fail. At best we can redefine terms, using others which may be *less* ambiguous; but always we have to rely on our common understanding. (Rosser and Turquette, *1952*, pp. 2–8, give a good sample of the hopelessness of expunging all ambiguity from a sentence.) What we need in order to justify the sentences in (1) as a valid argument is that we may treat 'Strawberries are red' and the other two sentences there *as if* they had truth-values, not that they are unambiguous. All declarative sentences, except perhaps those in highly technical work such as mathematics, have ambiguity. This ambiguity is an essential component of communication, I believe, for no two persons can have exactly the same thoughts or perceptions and hence must understand every linguistic act somewhat differently.

It is sufficient for our logical purposes to ask whether we can agree that a particular sentence, or class of sentences as in a formal language, is declarative and whether it is suitable to hypothesize a truth-value for it. If we cannot agree on certain cases, such as 'The King of France is bald', then we cannot reason together about them. That does not then mean that we do different logics or that logic is psychological; it only means that we differ on borderline cases. I believe that the assumption that we agree that a sentence has a truth-value, that the ambiguities are inessential, is always implicit.

The word 'agree' may be misleading. Agreements needn't and usually aren't explicit and may be forced on us by convention, or for physiological, psychological, or metaphysical reasons. Or they may be made explicitly for just the course of a short discussion. They may depend on how we understand the world, or how the world really is. In Chapter XI I will discuss further this notion of agreement and how it relates to an explanation of the objectivity of logic.

I have not explained the notions of truth and falsity. I must assume that for the most part you understand them and that you know for a simple sentence such as 'Ralph is a dog' what it means for it to be taken as true or false. Basically, truth will be taken as a primitive notion for such simple sentences, a notion we all understand how to use in most applications, while falsity will be understood as the opposite of truth, the not-true. It is these, truth and falsity, which we will try to formalize in more complex and controversial situations.

To summarize, *the basic object of study of logic is a proposition.*

Propositions A *proposition* is a written or uttered sentence which is declarative and which we agree to view as being either true or false, but not both.

From now on I will be less careful and often say that a proposition *has* a truth-value since we've agreed to view it as if it does.

Suppose, now, that we are having a discussion. An implicit assumption that underlies it is that we will consistently use words in the same way, or if you prefer, that the meanings and references of the words we use don't vary. This assumption is so embedded in our use of language that it's hard to even think of a word except as a *type*, that is, as a representative of all inscriptions that look the same and utterances which sound the same. There is no way for me to make precise what I mean by 'look the same': if there were then we could program a computer to read handwriting. But we know well enough in writing and conversation what it means for two inscriptions or utterances to be *equiform*. And so we can make the following agreement.

Words Are Types We will assume that in any discussion equiform words will have the same meaning and reference and in general the same properties of interest to logic. We therefore identify them and treat them as the same word. Briefly, *a word is a type*.

Lest this assumption lead us astray we will be careful in our reasoning to avoid ambiguous words and words like 'I' or 'now' which depend on context. Where we can't we will need to distinguish equiform utterances by labels of some sort. For example, though we understand 'Rose rose and picked a rose' well enough, we must distinguish the three occurrences of the equiform inscriptions if we are to be unambiguous enough to use this sentence in logic. We use some device such as 'Rose$_1$ rose$_2$ and picked a rose$_3$' or 'Rose$_{\text{person}}$ rose$_{\text{verb}}$ and picked a rose$_{\text{flower}}$'. The assumption that words are types is a goad to us to get rid of the grossest ambiguities in our speech when we do logic.

Suppose in a discussion I write down a sentence which we take to be a proposition:

> Socrates was Athenian.

Later I want to use that sentence in an argument, say:

> If Socrates was Athenian, then Socrates was Greek.
> Socrates was Athenian.
> Therefore: ...

But we have different sentences, since sentences are inscriptions. How are we to proceed?

Since words are types we argue that these two equiform sentences should both be true or both false. We don't care where they're placed on the paper, or who said them, or when they were uttered. Their properties for logic depend only on what words appear in them in what order. Any property which differentiates them isn't of concern to logic and reasoning. In this sense these two propositions are timeless.

This sort of argument can't be made in general. If first I say, 'I am over 6 feet tall' and then you say 'I am over 6 feet tall', we would be completely unjustified in assuming that these two utterances of the same words in the same order have all the same properties of concern to logic. Even avoiding the use of words like 'I' we still encounter difficulties with formalized versions of self-referential sentences such as 'This sentence is false', as I describe in Epstein, *1985* and *198?*. In this book, however, we will try to avoid such problem sentences. We make the following assumption.

Propositions Are Types In the course of any discussion in which we use logic we will consider a sentence to be a proposition only if any other sentence or phrase which is composed of the same words in the same order can be assumed to have the same properties of concern to logic during that discussion. We therefore identify equiform sentences or phrases and treat them as the same sentence. Briefly, *a proposition is a type.*

It is important to identify both sentences and phrases, because, for example, in (1) we want to be able to identify the phrase 'strawberries are red' in the first sentence with the second sentence.

The device I just used of putting *single quotation marks* around a word or phrase is a way of naming that word or phrase, or any linguistic unit. We need it because sometimes a confusion can arise if it has not been made clear whether a word or phrase is being referred to *as* a word or phrase as when I say 'The Taj Mahal has eleven letters', where I don't mean that the building has eleven letters but that the phrase does. When we refer to a word or phrase through the use of this device we'll say that we have given it a *quotation name* and that we *mention* the word or phrase. Otherwise we simply *use* the word or phrase with it's normal meaning. We are justified in using quotation names because words and propositions are to be understood as types.

I use this device as well as italicizing for mentioning words and phrases with some reluctance because there is not always a clear distinction between using a word and mentioning it. Moreover, when we write 'and' do we mean a string of symbols or the word with all its aspects? If the word, then when we write 'Ralph is a dog' do we mean those words in that order, or do we mean the proposition? The linguistic unit intended must be inferred from the context, and sometimes it's not even clear to the user of the convention.

But more importantly, use-mention confusions can often be fruitful, indicating that our intuitions may be in conflict with distinctions we have made. Such, I believe, is the case surrounding the notion of implication, as I discuss particularly in §II.E and §VI.B.2 (I use the symbol '§' to indicate a section of this book).

I will also use single quotation marks in the usual manner for quoting direct speech.

The device of enclosing a word or phrase in *double quotation marks* is equivalent to a wink or a nod in conversation, a nudge in the ribs indicating that I'm not to be taken literally, or that I don't really ascribe to what I'm saying. Double quotes are called *scare quotes* and they allow me to get away with "murder".

I want to make one additional restriction concerning the propositions we'll consider in this book. It seems to me that it will be hard for us to agree that a particular sentence is a proposition if we are speaking different languages. Therefore, throughout this book I will deal only with propositions in English. I believe that most if not all of what I'll say in this volume is applicable to reasoning in general in all languages; in Chapter XI I will discuss that claim.

2. Another view: abstract propositions

Many logicians start with a different analysis of the basic objects of study of logic. They postulate or argue for such a thing as an abstract proposition which is supposed to be the common element of many utterances all of which "say the same thing." Thus

(2) It is raining.
Pada deszcz.
Il pleut.

if uttered at the same time and place all express or stand for the same abstract proposition. It is argued that the word 'true' can only be properly applied to these things which cannot be seen, heard, or touched. Sentences are understood to "express" or "represent" or "participate in" such propositions.

These timeless and eternal objects then serve to explain the objectivity of truth and logic. It is held that an abstract proposition *is* true or false, and is completely objective and unambiguous. Those who take abstract propositions as the basis of logic argue that we cannot give unambiguous answers to the questions: What is a sentence? What constitutes a use of a sentence? When has one been used assertively, or even put forward for discussion? These questions, they say, can and should be avoided by taking things inflexible, rigid, timeless as propositions. But that only pushes back these same problems to: How do we use logic? What is the relation of these formal theories of mathematical symbols to our arguments, discussions, and search for truth? How can we tell if this utterance is an instance of that abstract proposition? It's not that taking utterances of sentences as propositions brings up questions which can be avoided. For example, were we to confine logic to the study of abstract propositions, argument (1) would be defective: the sentences there could not be taken to express propositions because of their ambiguity.

Abstract propositions seem apt in dealing with technical subjects such as mathematics. But I believe that it is our common reasoning which comes first and that mathematical proofs, to be convincing, must be shown to conform to that in their essentials. Confining logic to only the study of the "timeless and eternal truths" of mathematics and perhaps science seems to me not only too restrictive, but a reversal of the proper development of logic.

In summary, I do not believe that the sentences in (2) are all "saying the same thing." Rather, they do so only on the convention that there is some such thing as meaning which we can abstract, since they mean differently if only at the level of being in different languages and having different linguistic structures. The process of abstracting from these objects of our experience is what I take to be important. I reject abstractions but I embrace the use of abstracting.

Nonetheless, I hope that much of what follows can be of use to those who hold this other view of propositions by understanding the sentences which I discuss as representing or expressing abstract propositions, and the properties of those sentences as properties of the abstract propositions. In several places I have tried to give such an alternative reading (perhaps not always sympathetically), most notably in the discussion of the logical form of a proposition in Chapter II.E. At those times I refer to the advocates of abstract objects, and in particular abstract propositions, as 'platonists'.

There are other views of what kind of thing a proposition is. Williamson, *1968*, compares various of these from a viewpoint similar to mine. Most notably, though, Frege has taken the thought of a sentence to be what is true or false. I find it difficult to understand how two people can have the same thought which is in any case not a material thing, so I will direct you to Frege, *1918*, for his explanation.

C. Form and Content

We begin with propositions which are sentences. There are two features of them which contribute to our reasonings and proofs: their *syntax*, by which we mean their form or grammar, and their *semantics*, by which we mean their meaning or content. These are inextricably linked: the choice of what forms of propositions we'll study will lead to what and how we can mean, and the meaning of the forms will lead to which of those are acceptable.

Often forms are chosen as primary, as when a logic is presented solely as a collection of forms of sentences which are acceptable and ways to syntactically manipulate those. It often seems easier to gain agreement on some few acceptable forms than on questions of content. That is because forms can be exhibited and we can, each of us, invest these with our own meanings. That is, forms are (comparatively) objective. Consider:

All men are mortal.
Socrates is a man.
Therefore:
Socrates is mortal.

Is this an example of valid reasoning acceptable on the basis of it's form only? It's often said so. But why that form? How can we distinguish as valid the form of that argument from the form of:

All men are mortal.
Socrates is mortal.
Therefore:
Socrates is a man.

Only by reference to the notions of truth and meaning.

I'll emphasize these notions in creating models of reasoning. It's in the semantics above all, I believe, that agreement must be reached. Without exposing the assumptions that lead to our choice of acceptable forms the objectivity of those forms is based on implicit misunderstandings between us. My goal is to make explicit the disagreements as well as the agreements. Moreover, by studying what the forms mean before asking which are acceptable I hope to make the formal systems easier to understand.

D. Propositional Logic and the Basic Connectives

There are so many properties of propositions which could affect logic that we must begin by restricting our attention to only some of them. In this volume we will consider only the properties of propositions as wholes and ways to connect propositions to form new ones. *We will ignore the internal structure of propositions except insofar as they are built from other propositions in specified ways.* This is what is called the study of *propositional logic*.

There are many, many ways to connect propositions to form a new proposition. Some are easy to recognize and use. For example, 'Ralph is a dog and dogs bark' can be viewed as two sentences joined by the connective 'and'. Note that to view 'and' as a connective of sentences we need to assume that, for example, 'Dogs bark.' and 'dogs bark' are equiform.

Some other common connectives are: 'but', 'or', 'although', 'while', 'if ... then ...', 'only if', 'neither ... nor ...', and so on. We want to strike a balance between choosing as few to concentrate on and hence simplifying our semantic analyses and as many as possible so that our analyses will be broadly applicable.

Our starting point will be the four traditional basic connectives of logic: 'and', 'or', 'if ... then ...', and 'not'. We must be a bit careful with 'not' as it is to operate on propositions. In English it can occur in many different ways in a sentence, so let's agree that we'll study it as the connective which precedes a sentence as in 'it's not the case that ...'. These four connectives will give us a rich enough grammatical basis to begin our logical investigation. But whether they will be enough, or the most suitable connectives, are questions we will have to raise in relation to each particular type of semantic analysis.

These English connectives have many connotations and properties, some of which may be of no concern to us in logic or in a particular type of semantic analysis; for example, in American English 'not' is usually said more loudly than the surrounding words in the sentence. Therefore, we are going to replace these connectives with formal symbols to which we will give fairly explicit and precise meaning in each semantic analysis, based on our understanding of the English ones.

symbol	*what it will be an abstraction of*
$\wedge$	'and'
$\vee$	'or'
$\neg$	'it's not the case that'
$\rightarrow$	'if ... then ...'

Thus a complex sentence we might study is 'Ralph is a dog $\wedge$ dogs bark', corresponding to the earlier example.

A further formal device is important in reducing ambiguity: parentheses. In ordinary speech we might say 'If George is a duck then Ralph is a dog and Dusty is a horse.' It is not clear whether we should formalize this proposition as $A \rightarrow (B \wedge C)$ or $(A \rightarrow B) \wedge C$. Parentheses avoid the ambiguity and enforce on us one of the two readings.

To lessen the proliferation of quotation marks in naming linguistic items I will assume that *a formal symbol names itself* when it seems no serious confusion will arise. Thus I might say that $\wedge$ is a connective.

Here is some terminology that goes with these symbols:

The sentence formed by joining two sentences by $\wedge$ is called the *conjunction* of the two; we say we *conjoin* them and each is a *conjunct.*

The sentence formed by putting $\vee$ between two sentences is called a *disjunction* and each sentence is called a *disjunct*. Sometimes a disjunction is called an *alternation.*

The sentence formed by putting $\neg$ in front of another is called a *negation.* Thus the negation of 'Ralph is a dog' is '$\neg$ (Ralph is a dog)' where parentheses are useful to indicate the phrasing.

We use the terms *antecedent* and *consequent* to refer to the former and latter sentences that are joined between by an $\rightarrow$. The $\rightarrow$ is sometimes called the *conditional* and the operation associated with it *conditionalization*, though in some instances discussed below it's called *implication*.

We say that $\wedge$, $\vee$, $\rightarrow$ are *binary* connectives because they join two propositions to form a new one; $\neg$ is a *unary* connective operating on one proposition to form a new one.

Before we begin our first semantic analysis of these connectives in the next chapter I want once more to stress that though I believe the views I present here are the best way to motivate and understand the technical work that follows, they are not the only way. The assumptions I have laid out in this chapter will seem clearer and more reasonable as they are used and argued for in many different contexts in the following chapters. But I will try to present other views as the occasion demands.

II Classical Propositional Logic – PC –

In this chapter I will present the classical propositional logic which, I believe, is the simplest logic that can be developed from the assumptions of Chapter I. In this setting I present the notions of a formal language, a model, the logical form of a proposition, proof, consequence, and the notion of a logic.

I have developed these with considerable detail and motivation here so that we can understand in what way they are tied to classical logic and how they can be generalized. In §G I recapitulate their definitions in mathematical format after first discussing the role of mathematics and infinitistic assumptions in the study of logic.

More experienced readers may wish to skip ahead and use this chapter for reference only, though the analyses here are important for motivating the general framework in Chapter IV and I will assume familiarity with them.

A. The Classical Abstraction and the Fregean Assumption

The logic that most mathematicians endorse, that's probably the simplest we can study, is the classical propositional logic. We simplify as much as possible, abstracting away all properties of a proposition until there's only one left in addition to its form, the feature of a sentence that makes it a proposition: it has a truth-value.

Explicitly, we make the following assumption.

The Classical Abstraction The only properties of a proposition which matter to logic are its form and its truth-value.

The forms with which we will begin our investigation are those we discussed in §I.D.

If the only things that matter about a proposition are its truth-value and its form, then the truth-value of a complex proposition can depend only on the connectives appearing in it and the truth-values of its parts. If this were not the case then the truth-values of 'Ralph is a dog' and of 'cats are nasty' wouldn't determine the truth-value of 'Ralph is a dog $\wedge$ cats are nasty'. But if the truth-values don't, and we've agreed that no other property of these propositions matters, what does determine the truth-value of the complex proposition? If there is something nonfunctional, transcendent, that occurs when, say, $\wedge$ connects two sentences, then how are we to reason? From the truth of one proposition how could we deduce the truth of another? Without some regularity reasoning cannot take place. This is a basic assumption.

> The truth-value of a complex proposition is determined by its form and the properties of its constituents.

Here 'constituent' means proper constituent, a part and not the whole.

I call this the *Fregean Assumption* after Frege who so strongly emphasized it and carefully made it explicit. It acts as a kind of simplicity constraint on our models. If propositions using 'and', 'or', 'not', and 'if... then...' can only be understood within the context of all of language, then we'll just have to make do with our formal abstractions if we want to give a model of reasoning.

B. Truth-Functions and the Extensionality Consideration

The simplest propositions are those which contain no formal connectives, for example 'Ralph is a dog' or 'Every bird sings'. They are *atomic*. Since they have no form of significance to propositional logic, the Classical Abstraction limits us to considering only their truth-values. The Fregean Assumption then tells us that if we join two atomic propositions with $\wedge$, $\vee$, or $\rightarrow$, or place $\neg$ in front of one, the truth-value of the resulting proposition must depend on only the truth-value of those atomic propositions. That is, the connectives must operate semantically as functions of the truth-values of the constituent atomic propositions, they are *truth-functions*.

Which truth-functions correspond to our connectives? I'll let you convince yourself that the only reasonable choices for $\neg$ and $\wedge$ are given by the following tables, where I use p and q to stand for atomic propositions, and 'T' to stand for 'true', 'F' for 'false'.

p	¬p
T	F
F	T

p	q	p∧q
T	T	T
T	F	F
F	T	F
F	F	F

That is, if p is T, then $\neg$p is F; if p is F, then $\neg$p is T. And p$\wedge$q is T if both p and q are T; otherwise it is F.

For $\vee$ there are two choices, corresponding to an inclusive or exclusive reading of 'or' in English. The choice is arbitrary: it's customary now to use the inclusive version, as in 'p or q or both'.

p	q	p∨q
T	T	T
T	F	T
F	T	T
F	F	F

In §E.4 I'll discuss the exclusive use of 'or' with examples.

The connective which has always generated the most debate is 'if ... then ...'. But with the Classical Abstraction and the Fregean Assumption there is really little choice for its formalization. We use

p	q	p→q
T	T	T
T	F	F
F	T	T
F	F	T

The first two lines seem to be the very essence of 'if ... then ...'. For the last two lines, if we were to take them both as F then we'd have that $p \rightarrow q$ would be the same as $p \wedge q$. If we were to take the third line as T and the last line as F, then $p \rightarrow q$ would have the same truth-value as q; were the third line F and the last T, then $p \rightarrow q$ would be the same as $q \rightarrow p$. Those choices are counterintuitive. So our table is the most generous one in assigning T to conditional propositions: $p \rightarrow q$ is T unless p is T and q is F.

An illustration from mathematics will show why it is useful to classify a conditional with false antecedent as true. Consider the following theorem of arithmetic: if a and b are odd integers, then $a + b$ is an even integer. For this to be true every instance of it must be true. Yet were we to take either of the last two lines of the table for '→' to be F then the formalization of this theorem would be false: $4 + 8 = 12$ which is even, and $4 + 7 = 11$ which is not even. Our formalization of 'if ... then ...' propositions allows us to deal with cases where the antecedent "does not apply" by treating them as vacuously true.

Thus given any atomic propositions p and q, we have a semantic analysis of $\neg p$, $p \wedge q$, $p \vee q$, and $p \rightarrow q$. But what about '(Ralph is a dog ∧ dogs bark) → Ralph barks'? Here the antecedent is not a formless entity which is simply true or false: it contains a formal connective which we must take into account. If we let A, B, C, ... stand for any propositions whatever, whether containing formal connectives or not, we need to ask in what way the forms of A and B matter in determining the truth-values of $A \wedge B$, $A \vee B$, $A \rightarrow B$, and $\neg A$.

How much does the form of a proposition matter? Certainly $p \wedge q$ and $p \rightarrow q$ are to be evaluated by different methods. But if both are evaluated as true or both false is there anything else that can semantically distinguish them? The only semantic value we attribute to a proposition in classical logic is a truth-value. So if two propositions have the same truth-value, they are semantically indistinguishable.

Thus to evaluate '(Ralph is a dog ∧ dogs bark) → Ralph barks' we first determine the truth-value of 'Ralph is a dog ∧ dogs bark', and then only that truth-value and the truth-value of 'Ralph barks' matter in determining the truth-value of the whole. We proceed in exactly the same manner in determining the truth-value of '(Ralph is a dog ∨ dogs bark) → Ralph barks', so that if 'Ralph is a dog ∧ dogs bark' and 'Ralph is a dog ∨ dogs bark' have the same truth-value, then the two conditionals will have the same truth-value.

In summary, we make the following assumption, which is, I believe, the only one consonant with our earlier ones.

The Extensionality Consideration If two propositions have the same semantic properties then they are indistinguishable in any semantic analysis, regardless of their forms. In particular, in classical logic if A is part of C, then the truth-value of C depends only on the truth-value and not the form of A, except insofar as the form of A determines the truth-value of A.

Thus the connectives operate semantically as truth-functions in all circumstances, and we have already decided which truth-functions those are.

(1)

A	¬A
T	F
F	T

(2)

A	B	A ∧ B
T	T	T
T	F	F
F	T	F
F	F	F

(2)

A	B	A ∨ B
T	T	T
T	F	T
F	T	T
F	F	F

(4)

A	B	A → B
T	T	T
T	F	F
F	T	T
F	F	T

If later I say that a *connective is classical* I will mean that it's one of ¬, →, ∧, ∨ interpreted according to these *truth-tables*.

Note that I've used 'and', 'or', 'not', and 'if ... then ...' to present the meaning of the formal connectives: you need them to read the tables. This isn't circular: we are not defining or giving meaning to 'and', 'or', 'not', 'if ... then ...' but to ∧, ∨, ¬, →. I must assume that you understand the ordinary English connectives.

Have we really restricted ourselves by looking at only these four connectives? What about others which may be important to reasoning and the search for true propositions? Our assumptions tell us that they must be treated as truth-functions. In §G.3 below we'll see that it's possible to define every truth-functional connective of propositions from just these four formal connectives once we've established the role of mathematics in proving such facts.

C. The Formal Language and Models for It

We have no dictionary, no list of all propositions in English, nor do we have a method for generating all propositions, for English is not a fixed, formal, static language. But we can introduce generality and a rigid formal language to investigate propositions by using variables.

Let p_0, p_1, ... be *propositional variables*. These can stand for any propositions, but the intention is that they'll be the ones whose internal structure won't be under consideration. Our formal language is built from these using the connectives $\neg$, $\rightarrow$, $\wedge$, $\vee$ and parentheses. This will be the *formal object language*. But to be able to talk about *that* language we'll need to use variables, say p, q, q_0, q_1, ... , to stand for any of the p_i's, and other variables A, B, ... , A_0, A_1, ... , to range over all the p_i's and complex expressions formed from the p_i's in the formal language. These are the *metavariables*. The analogue to a sentence in English in the formal language is a *well-formed-formula* (*wff*). Here's how we generate the wffs:

1. Each (p_i) is a wff for $i = 0, 1, 2, \ldots$.
2. If A, B are wffs, then so are $(\neg A)$, $(A \rightarrow B)$, $(A \wedge B)$, $(A \vee B)$.
3. These are the only ways to generate wffs.

I'll denote this formal language as '$L(p_0, p_1, \ldots \neg, \rightarrow, \wedge, \vee)$', often abbreviated to '$L(\neg, \rightarrow, \wedge, \vee)$'.

Though it may seem obvious that parentheses ensure that the reading of each wff is unambiguous, that requires a demonstration. In §G.1 below I will show that given any sequence of symbols in the formal language, if it's a wff then there's only one way to read it. That is, there's no wff which can be parsed as both, say, $A \wedge B$ and $C \rightarrow D$.

Excessive parentheses can make it difficult to read a formal wff. I will sometimes use an informal convention in presenting wffs which will reduce the need for parentheses: $\neg$ binds more strongly than $\wedge$ and $\vee$, which bind more strongly than $\rightarrow$. Thus $\neg A \rightarrow B$ is understood as $((\neg A) \rightarrow B)$. I'll often dispense with the outermost parentheses, too, and the ones surrounding a variable p_i . It's also useful to have other types of parentheses to use informally, such as '{ }' and '[ ]'. I'll also continue the convention of letting any formal symbol name itself (cf. §I.D).

$L(p_0, p_1, \ldots \neg, \rightarrow, \wedge, \vee)$ is the formal language. Clearly no wff such as $p_0 \wedge \neg p_1$ is true or false. It's only when we make an assignment of propositions to the variables, such as p_0 stands for 'Ralph is a dog' and p_1 stands for 'Four otters are sitting on a log', that we have a semi-formal proposition 'Ralph is a dog $\wedge$ $\neg$(four otters are sitting on a log)', one which we agree can be viewed as having a truth-value. We may read this as 'Ralph is a dog and it's not the case that four otters are sitting on a log' so long as we remember that we've agreed that all that 'and' and 'it's not the case that' mean is captured by the tables for $\wedge$ and $\neg$.

We may not know the truth-value of, say, 'Ralph is a dog' and wish to test the hypothesis that it's true. After we specify an assignment of propositions to the variables we then have to specify which of the propositions are to be taken as true and which as false. Compound propositions can then be evaluated using the truth tables. These assignments and evaluation are what we call a *model*. Taking p_0, p_1, ... to stand for the actual English propositions which have been assigned to the variables p_0, p_1, ... , respectively, a model can be presented schematically.

I

$$L(p_0, p_1, \ldots \neg, \rightarrow, \wedge, \vee)$$

$$\downarrow \quad \Big\} \text{ realization}$$

$$\{p_0, p_1, \ldots, \text{complex propositions formed from these using } \neg, \rightarrow, \wedge, \vee\}$$

$$\downarrow \quad \Big\} \; v \text{ plus truth-tables}$$

$$\{T, F\}$$

Here the second level is what I call *the semi-formal object language*, or *formalized English*: these are the propositions which we agree to view as being true or false. There are two kinds of propositions in this level. First there are those propositions p_0, p_1, ... assigned to the variables p_0, p_1, ... These are called the *atomic propositions*: their internal structure will not be under consideration. There are also *compound* or *complex propositions* which can be formed from these using the propositional connectives. We've already given the formation rules for wffs and we can use those to make explicit the propositions in the semi-formal language: given any formal wff A we call the *realization of* A in this model the result of replacing each propositional variable in A by the proposition assigned to that variable. The propositions of the semi-formal language are the realizations of the wffs of the formal language.

Note that we have thus agreed to view any such semi-formal sentence as a proposition. We do this because we have agreed on how to understand the formal connectives. From now on I will use the *metavariables* $A, B, \ldots A_0, B_0, \ldots$ to range over wffs in the formal language, sentences in the semi-formal language, or English sentences taken as propositions, trusting to the context to make it clear which is meant in those cases where the distinction matters.

In the diagram v is a *valuation*, a way we've agreed to assign T or F to those propositions assigned to the propositional variables (v is the lowercase Roman letter and should be distinguished from the connective $\vee$). Then v plus the truth-tables uniquely determine the truth-value of any complex proposition of the semi-formal language because there is only one way to parse each of those. We view the truth-tables as a method to extend v to all propositions of the semi-formal language and write $\mathsf{v}(A) = \mathsf{T}$ or $\mathsf{v}(A) = \mathsf{F}$ to indicate that the proposition A is true or false in this model.

When, in later chapters, there may be a question of which semantic analysis is under discussion I will call these *classical models*.

D. Validity and Semantic Consequence

1. Validity

Some propositions we may take to be obviously true from our experience, such as 'Dogs bark'. The truth of others, however, follows solely from the logical agreements we have made. The study of those will illuminate our assumptions and help us to develop a stock of propositions which we may use in any of our deductions. For instance, we might argue that relative to the classical interpretation of the connectives 'Ralph is a dog or Ralph is not a dog' is true due solely to its form because we would formalize it as '(Ralph is a dog) $\vee\neg$(Ralph is a dog)' which can be taken as a realization of, e.g., $p_1 \vee \neg p_1$. In any model the realization of that wff will be true. Hence it is the form of the original proposition that ensures its truth. To state this procedure in a general way we need some definitions. To begin with, I will use M to range over models.

For a model M and *semi-formal* wff A, if $\mathsf{v}(A) = \mathsf{T}$ we'll say that A is *true in* M, or M *validates* A, written $\mathsf{M} \models A$. We'll then say that *a wff* A *of the formal language is true* (or *valid*) *in* M *if the realization of* A *is true in* M. If A is false in M we write $\mathsf{M} \not\models A$. Thus $p_1 \wedge \neg p_2$ is false in the model which assigns 'Grass is green' to p_1 and 'Every dog is a canine' to p_2. Whenever I give an example like this I'm assuming that we take the obvious commonly agreed upon truth-values for the propositions assigned to the variables.

We say that M is a *model for a collection of wffs of the formal language* if every wff in the collection is true in M. Thus the last assignment I gave is a model for, e.g., $\{p_1 \to p_2, \neg(\neg p_1 \wedge \neg p_2), p_1 \vee p_2\}$.

Some wffs are true in every model: their truth-value is independent of any particular realization. One example is $p_1 \vee \neg p_1$. Such a wff is a *tautology* and we call it *valid*. We say that a *semi-formal proposition is a tautology* or *valid* if it is the realization of a valid wff. *Valid propositions are those which are true due to their (propositional) form only* (relative to our semantics). Valid wffs exemplify those forms. In the following chapters when we want to stress that this notion of validity is for the classical interpretation of the connectives we'll use the terms *classical tautology* or *classically valid*.

We sometimes say that a sentence in our ordinary language, English, is valid if there is a straightforward formalization of it into semi-formal English on which we feel certain we'll all agree and which is valid. Thus 'If George is a duck, then George is a duck' is valid and hence true due to its form only.

Other wffs are false in every model, they are false due to their form only. An example is $p_1 \wedge \neg p_1$. Such a wff is called a *contradiction* or *anti-tautology*. Note that in classical logic if A is a tautology then $\neg$A is a contradiction, and if A is a contradiction then $\neg$A is a tautology.

Formal wffs exemplify the forms of propositions. To talk about the forms of wffs of the formal language we use *schemas*: formal wffs with the propositional variables replaced by metavariables. Thus to say that *a schema*, such as $A \vee \neg A$, *is valid* means that every result of replacing the metavariables by wffs is a valid wff. Here the replacement must be uniform; for example, $A \rightarrow (B \rightarrow A)$ is a (classically) valid schema but only if when we replace one A by a wff we use that same wff to replace the other A.

The Classical Propositional Calculus, **PC**, is the collection of all formal wffs which are classical tautologies. We also call this 'Classical (Propositional) Logic'. It is a formal logic in that it contains only formal wffs. **PC** is the collection of all valid forms of propositions in classical logic.

I'll sometimes use the name **PC** to refer not only to the tautologies but to all the work we've done in establishing them: the assumptions, the semantics, etc. Also, when discussing other logics I will often refer to the notions of this section by prefacing them with ' **PC**-' as in **PC**-tautologies, **PC**-models, **PC**-validity.

The name 'calculus' is apt because we can calculate whether any particular wff is a tautology: for each of the finite number of ways of assigning T or F to the propositions assigned to its variables we can check mechanically, using the truth-tables, whether it comes out true. If it always does then it's a tautology, otherwise not. I'll discuss this decision method in §G.5 below.

2. Semantic consequence

One of the main tasks of logic is to say when one proposition follows from another. For instance, we might argue that 'Ralph is a dog' follows from '(Ralph is a dog or George is a duck) and George is not a duck'. That's because, at least relative to classical logic, if the latter is formalized in the obvious way then whenever it is true so is 'Ralph is a dog'.

The notion we are discussing here is a metalogical one. To be precise, we'll say that *a wff* B *is a semantic consequence* (in classical logic) *of a wff* A, written $A \vDash B$, if in every model in which A is true, so is B. We can extend this definition to collections of wffs, which I denote using the *metavariables* Γ, Δ, Σ. We say that B *is a semantic consequence of* Γ, written $\Gamma \models B$, if for every model in which all the wffs in Γ are true, so is B. And similarly we define $\Gamma \models \Delta$ if for every model in which all the wffs in Γ are true, so are all those in Δ. We write $\models A$ if A *is true in all models*, that is, A is a tautology. If other logics are also under consideration I'll use '$\models_{PC}$' for '$\models$'.

These definitions are then extended to semi-formal and ordinary language propositions on the pattern of our definition of validity above.

The notion of semantic consequence will be taken as one possible formalization of valid deduction. We say that Γ *therefore* B is a *valid deduction* if $\Gamma \models B$. Such a deduction is sometimes called a *valid argument*. If we have $A \models B$ we sometimes say that A *implies* B, or A *entails* B (in classical logic). For example, 'Ralph is a dog $\wedge$ George is a duck' implies 'Ralph is a dog' in classical logic.

We may, however, think of one proposition, B, following from another, A, just in case $A \rightarrow B$ is true. For what does 'if ... then ...' mean in English if not that? Still, we can distinguish that case from B *following logically* from A: for that we'd want that $A \rightarrow B$ is a logical tautology, not just true. If we hold this view we might use 'implies that' interchangeably with 'if ... then ...', thus: Ralph is a dog and George is a duck implies that Ralph is a dog.

Do we have two distinct notions of 'follows from'? In **PC** there's only one, for we have: $A \models B$ if and only if $\models A \rightarrow B$. That's easy to see, for if whenever A is true we have that B is true, then $A \rightarrow B$ is always true in our **PC**-models; and if $A \rightarrow B$ is always true then if A is true, so is B.

The question remains, however: do we have two distinct pre-formal notions, implication and the validity of an 'if ... then ...' proposition, or one? It seems to me that we have only one pre-formal notion of one proposition following from another and that we formalize it in several ways. I will argue that point throughout this volume. However, it is now common to distinguish these two notions, reserving 'implies' and 'implication' for the metalogical one, and 'if...then' and 'conditional' for the notion formalized by '$\rightarrow$'. Nonetheless, some logics, such as the modal logics of Chapter VI, are designed to formalize implication as '$\rightarrow$'. Whether you think that is a use-mention confusion will depend, I believe, on how you understand the English 'if ... then ...'. Fortunately, for **PC** the question is moot.

Finally, we say that *two propositions* (or *wffs*) *are semantically equivalent* if each is a semantic consequence of the other: in every model they have the same truth-value. Thus in classical logic '$\neg$(Ralph is a dog $\wedge$ $\neg$(George is a duck))' is semantically equivalent to 'Ralph is a dog $\rightarrow$ George is a duck', as you can show. Semantic equivalence is *one* formalization of the informal notion of two propositions meaning the same thing for all our logical purposes.

Let's review what we've done. We made the assumption that the only aspects of a proposition that are of concern to logic are its truth-value and its form as built up from sentence connectives. We set up a formal language to help us abstract the logical form of propositions. We based that formalization on the four connectives : 'and', 'or', 'it's not the case that', and 'if ... then ...'. It was the semi-formal language in which we were primarily interested: English with the formal propositional connectives $\neg$, $\rightarrow$, $\wedge$, $\vee$.

We then gave an analysis of the truth of a complex proposition in terms of it's parts by using the truth-tables. Using that we could explain by reference to the formal language what it means for a proposition to be true due to its (propositional) form only, and what it means for one proposition to follow from another.

It's time now to see how to apply some of these ideas to examples in our ordinary language.

E. The Logical Form of a Proposition

1. On logical form

We have chosen to be very stingy with the grammatical tools we allow ourselves in parsing the propositional structure of a proposition. If we confine ourselves to formalizing only 'and', 'or', 'not', 'if ... then ...', then how are we to analyze 'Ralph barks but George is a duck' or 'Horses eat grass because grass is green'?

In this section I want to discuss some agreements and methods which allow us to rewrite a wide class of propositions in such a way that the new English sentence is equivalent to the original for all our logical purposes and yet can be easily formalized. Here 'for all our logical purposes' means relative to the assumptions of classical propositional logic. Similar methods and remarks will apply to other assumptions for other logics.

We will have:

English proposition
$\Updownarrow$
useful rewriting of English
$\Updownarrow$
proposition of semi-formal English

though sometimes we will go from the English sentence directly to the semi-formal one. The rewriting in English will exhibit how we've agreed to treat the proposition in logic. And the formal equivalent is called its *(propositional) logical form.*

I can't give rules for rewriting every proposition which we'll want to use in logic. And there's often more than one way to rewrite a sentence, leading to equally reasonable choices for its logical form. Many discussions in the literature about what is the logical form of a particular sentence or type of sentence seem to me to be analyses of how our assumptions about logic apply to the sentence.

I should point out that for a platonist there is no question of what the logical form of a proposition is. Propositions are abstract objects which have logical form and the difficulty, instead, is how to determine which proposition is being expressed by a particular sentence.

2. Atomic propositions

In §C I gave an example of an assignment to p_0 and p_1, respectively 'Ralph is a dog' and 'Four otters are sitting on a log'. Another might be to let p_0 stand for '2+2 = 4' and p_1 stand for 'George is a duck'. And yet another could be to let p_0 stand for 'Stars shine and mathematics is a difficult subject', and p_1 for 'It is not possible that Ralph is a dog'. You might object to this last assignment for p_0 saying that 'Stars shine and mathematics is a difficult subject' contains the connective 'and' which we were formalizing by $\wedge$. If we assign that proposition to p_0 then our model will be insensitive to any relationship between it and the two propositions: 'Stars shine', 'Mathematics is a difficult subject'.

The propositions assigned to the propositional variables in a model are taken by us to be *atomic*: they are indivisible in that model, their internal structure does not matter. In classical logic each is simply taken to be true or false. It is of course better if the atomic propositions do not contain the English connectives we're formalizing, especially if we wish to give colloquial readings to the formal connectives. But it is a matter of choice which propositions are taken to be atomic.

3. Ambiguity

Our ordinary speech is full of ambiguity. How are we to understand 'If the play is sold out then Don will be here and Sam will meet us tonight'? Does this have the form $A \rightarrow (B \wedge C)$ or $(A \rightarrow B) \wedge C$? Neither logic nor any conventions can resolve this. We must inquire of the utterer, or rely on the context or an educated guess. All that our formal tools can do is indicate that there is an ambiguity which the precision of our logical work cannot tolerate.

Similar remarks apply to cross-referencing of pronouns and other types of ambiguity, some of which I discuss in detail in Volume 2.

In classical logic, however, two kinds of ambiguity are harmless. Consider 'Ralph is a dog and George is a duck and Dusty is a horse'. Are we to take this as $(A \wedge B) \wedge C$ or as $A \wedge (B \wedge C)$? For our semantic analysis it doesn't matter, for in every model these two readings will have the same truth-value. So on occasion I will write $A_1 \wedge A_2 \wedge A_3$ to indifferently stand for $A_1 \wedge (A_2 \wedge A_3)$ or $(A_1 \wedge A_2) \wedge A_3$. Similarly, the grouping of propositions joined solely by $\vee$ does not matter to our semantics. Though such groupings may be of significance in certain syntactic analyses which we'll make in §J and §K below, we have a precision here that is not observed or needed in ordinary speech.

4. Other propositional connectives

Some propositions contain sentence connectives other than those we are formalizing. For instance,

(5) Ralph barks but George is a duck.

We have several choices in dealing with propositions such as these. Since they don't contain a connective which we're formalizing, we may treat them as atomic. This, however, isn't suitable if we are also investigating the propositions from which they're comprised. Thus if our model also contains 'Ralph barks' and 'George is a duck', we're left with three distinct atomic propositions and can deduce nothing about their logical relations from their form.

A second choice is to formalize the connective using one of $\neg$, $\rightarrow$, $\wedge$, $\vee$ or some combination of these. For instance, (5) could be formalized as 'Ralph barks $\wedge$ George is a duck'. In that case we'd argue that all the *logical* properties of 'but' in (5) are captured by $\wedge$: the word 'but' only indicates some surprise or contrary movement of thought over and above conjoining the propositions. Similarly we may formalize 'Neither horses bark nor dogs crow' as '$\neg$(horses bark) $\wedge$ $\neg$(dogs crow)'.

We will often formalize connectives other than 'if ... then ...', 'it's not the case that', 'and', 'or' (and their obvious equivalents) by some combination of our formal connectives. But each time we do, we must either invoke some general agreement concerning such formalizations or else give an argument why such formalization is appropriate relative to our semantics. We have to say what it is about the connective which makes us feel justified in ignoring or changing it.

Since in classical logic any formalization of a connective must be truth-functional, we should be ready to argue that the connective is indeed truth-functional. Thus, it seems to me, our only choices for 'Horses eat grass because grass is green' are to take it as atomic or else say that it is not suitable to be dealt with in classical logic, for 'because' is not truth-functional (if you think that 'because' could be rendered by '$\rightarrow$', then consider 'Horses don't eat grass because grass is green'). We should not be afraid of this last choice: our formalism can't model everything.

Some general conventions are worth listing here, though I won't stop to argue for each.

(6)

English connective	*Formalization*
since A, B	$A \rightarrow B$
given A, then B	$A \rightarrow B$
A only if B	$A \rightarrow B$
B if A	$A \rightarrow B$
if A, B	$A \rightarrow B$
B in case A	$A \rightarrow B$

B provided that A	$A \rightarrow B$
not A unless B	$A \rightarrow B$
when A, B	$A \rightarrow B$
A, so B	$A \rightarrow B$
A if and only if B	$(A \rightarrow B) \wedge (B \rightarrow A)$
both A, B	$A \wedge B$
neither A nor B	$\neg A \wedge \neg B$
not both A and B	$\neg(A \wedge B)$
A or B (exclusive 'or', 'or else')	$(A \vee B) \wedge \neg(A \wedge B)$

'A if and only if B' is usually abbreviated 'A iff B', and we introduce a convenient abbreviation for its formalization above: $A \leftrightarrow B$. We call '$\leftrightarrow$' the *biconditional*. Often I'll call 'if A then B' the 'left to right direction' of 'A iff B', and 'if B then A' the 'right to left direction'. $B \rightarrow A$ is called the *converse* of $A \rightarrow B$.

A warning, however: we need to be careful about formalizing cases of even those connectives we've agreed upon. For instance, consider: 'If Robin Hood had been French then he would have spoken German'. Here is our connective 'if... then...'. Yet it would be inappropriate to formalize it as the truth-functional '$\rightarrow$', for in that case we would get a semi-formal proposition which is true, and I daresay that most of us would argue for the falsity of this proposition. This is an example of a *counterfactual*, a conditional whose antecedent is false yet whose truth-value does not depend only on that (I will discuss such propositions further in §IV.B.1 and §VI.C.1). The moral of this is that we should be ready to justify a truth-functional reading of any use of a connective if we formalize it in classical logic.

A particular difficulty we have in English is whether to read 'or' as inclusive or exclusive. Thus are we to take 'John will go to the movies or John will go out for ice cream' as being true if John both goes to the movies and gets ice cream, as we would have to if we formalized it as $A \vee B$? Or are we to understand 'or' as exclusive alternation, the proposition being true if exactly one of its parts is true? In that case we'd formalize it as in table (6). English does not distinguish between these two readings except by context, and where that fails us let's agree to read 'or' as $\vee$. Some languages do distinguish these two readings by using different words; for instance, in Latin *vel* is used for inclusive 'or' and *aut* for exclusive 'or'. It's from the former that we get the symbol $\vee$.

5. Other ways to incorporate one proposition within another

Consider the following propositions:

(7) John wonders whether Ralph is a dog.

(8) John, you know, eats, sleeps, and snores.

(9) Jack Sprat is lean; his wife Jane is fat.

For each of these we have the three choices described in the last section: we may treat it as atomic; we can somehow formalize it using one or a combination of the formal connectives, arguing why that is suitable relative to the semantics; or we can classify it as unsuitable to deal with in this propositional logic.

The last choice is what we do with (7) in classical logic. It would be inappropriate to treat it as atomic since it contains another proposition, 'Ralph is a dog' as a part. Yet the truth of the proposition as a whole does not depend on the truth or falsity of that part, so there is no suitable way to model it using truth-functional connectives.

The use of commas, semicolons, and colons is a common way to join propositions in English. Many of their uses can be formalized in classical logic. In (8) we first argue that the phrase 'you know' is only for emphasis and doesn't affect how we'll reason with the sentence. So we rewrite it as 'John eats, sleeps, and snores'. Then we argue that the subject of each verb is John, so making that explicit we have 'John eats, John sleeps, and John snores'. Now it seems reasonable to formalize the first comma as $\wedge$.

In (9) let's consider the second part, 'his wife, Jane, is fat'. First, 'his' refers to Jack Sprat, so being explict we have 'Jack Sprat's wife, Jane, is fat'. Who is Jane? The last rewrite indicates she is Jack Sprat's wife, and also that she is fat. So rewriting we have 'Jane is Jack Sprat's wife and Jane is fat'. Finally, the semicolon is acting as a conjunction when we want to conjoin two nearly unrelated sentences. So we can formalize (9) as

(10) Jack Sprat is lean $\wedge$ (Jane is Jack Sprat's wife $\wedge$ Jane is fat).

There is a great temptation to say that (10) "is what (9) says" and then to think of (10) as somehow lurking behind (9) when (9) is uttered. Or to say that (9) and (10) "say the same thing", as if there were some one abstract proposition expressed by both (9) and (10). To me that abstract proposition is nothing more than the fact that we identify (9) and (10) for all purposes of classical logic. We may wish to make an abstraction and treat all identified propositions as if they were one; the abstraction comes from the identification. But to postulate an abstract proposition which if expressed by two distinct sentences leads us to make an identification seems to me to leave us hopelessly groping for the intangible with no guides for how to hook our language into it.

There are no rules for rewriting sentences such as (9) into (10). Much of it is "common sense" as we reflect on our initial assumptions about logic. Exposure to many examples and practice formalizing English sentences are really the only ways to become familiar with the conventions and to learn how to "speak logic."

F. Further Abstractions: The Role of Mathematics in Logic

In classical propositional logic every atomic proposition is abstracted to just its truth-value. Nothing else matters. Thus if we have two models such that for each propositional variable p_i the one model assigns to p_i a proposition which is true if and only if the other model assigns one that is true, then the two models are indistinguishable for the purposes of classical logic. Of course the propositions in one may be about mathematics and in the other about animal husbandry, but that cannot enter into our deliberations if we are using classical logic.

It is thus appropriate to ignore those differences between models of type **I** which do not matter to classical logic and simplify the models to:

II

$$L(p_0, p_1, \ldots \neg, \to, \wedge, \vee) \xrightarrow{\;\upsilon,\ \text{truth-tables}\;} \{T, F\}$$

Here υ assigns truth-values to the variables $p_0, p_1, \ldots$ and is extended to all formal wffs by the truth-tables (1), (2), (3), and (4). I'll often name this abstracted form of a model by its valuation, υ.

Our language and semantics have now become very formal and abstract. This is a virtue, for with these abstract systems we can analyze many propositions which would have been nearly intractable before because of their complexity. Simplification and abstraction are equally important because they allow us to use mathematics to establish general results about our logic. But these results will have significance for us only if we remember that models of type **II** come from models of type **I**.

I now want to discuss certain abstractions and idealizations that will allow us to use mathematics more fruitfully in our logic. But first we need to consider the relation between mathematics and logic.

At the beginning of this century Whitehead and Russell, *1910–1913*, developed a great deal of mathematics from a basis of formal logic. That was part of a program to justify the view that mathematics just is a *part* of logic. That view, called the *logicist* conception of mathematics is still an important current in the philosophy of mathematics.

On the other hand there were those such as Tarski, *1934*, who believed that we are justified in using all of mathematics to develop logic. No work could be called logical if it were not sufficiently mathematicized, made precise by mathematical definitions and axiomatizations. Thus before Tarski defined a formal language he gave a mathematical theory of concatenations of symbols. It would almost seem from this point of view that logic is part of mathematics.

I believe that there is more of a symbiosis between mathematics and logic. In any advanced parts of mathematics we want to be able to use logic in establishing theorems. For instance, one could argue that you need formal logic in order to give an axiomatization of the theory of symbol concatenation. And we want to be able to claim that the reasoning we do in mathematics is of the same sort as we've formalized in logic, perhaps supplemented by a few special forms of reasoning such as mathematical induction. Yet to develop logic to any degree useful to mathematics and the sciences and to establish any really general results about logic we need to make it more mathematical and use mathematics on it. There is no clear division between the disciplines. To my way of thinking both are involved in abstracting to a high degree from experience.

What we must do, I believe, is use our common sense, our ordinary reasoning and understanding of language and the world, in establishing formal logic. That same reasoning is what is used in mathematics. An informal mathematics will then be useful in further abstractions and methods of analysis of logic. It is, however, important to see where those further mathematical abstractions and applications are made so that one who disagrees with them can go with us up to that point and no further.

Propositions as I've presented them are written or uttered. There can only be a finite, though perhaps potentially infinite, number of them. We can view any fragment of the formal or semi-formal language as a completed whole, say all wffs using fewer than 47 symbols made up from $p_0, p_1, \ldots, p_{13}$ or the realizations of those in a particular model. We could achieve more generality by establishing results for all wffs using fewer than m symbols made up from $p_0, p_1, \ldots, p_n$ where m and n are any natural numbers 1, 2, 3, … . Already this is an abstraction. Are we to suppose that '¬¬¬¬¬¬¬¬¬¬¬¬¬(Ralph is a dog)' is of significance in reasoning? We would never normally use such a proposition, or if you think we would then come up with your own example of a proposition preceded by so many negations that it seems preposterous to conceive of reasoning with it. Yet by including all such propositions and well-formed-formulas in our logical analyses we may simplify our work considerably. And who is to say with certainty that some mechanical procedure for analyzing the truth-value of a relatively simple compound proposition may never need such a complex proposition? Therefore, we will consider the *collection of propositional variables,* denoted 'PV', as a completed whole, to be treated mathematically as a set. And we will further abstract from experience to treat the formal and semi-formal languages as completed infinite wholes, not simply as generated according to some rules. These are the syntactic abstractions we make.

Semantically, we now treat a model of type **II** as a function $v: \text{PV} \rightarrow \{\text{T}, \text{F}\}$ which is extended inductively to all wffs by the truth-tables. Finally we make a very powerful generalization for classical logic.

The Fully General Classical Abstraction Any function $v: \text{PV} \rightarrow \{\text{T}, \text{F}\}$ which is extended to all wffs by the truth-tables is a model.

Thus not only does it not matter how a model of type **II** arises: we assume any mathematically possible one can arise. In particular, we assume that we can independently assign truth-values to all the variables in PV.

All the definitions in §D are now to be understood in terms of these models and assumptions.

The fundamental mathematical tool I will use in this book is induction. I will assume that you are familiar with that proof method (see *Epstein and Carnielli,* Chapter 5). On occasion I will refer to a collection as being *recursive*. This is a technical term which you may choose to understand informally as meaning 'given by an inductive definition' (see *Epstein and Carnielli*). I will enclose by parentheses of the sort '{ }' either several things or a description of things to indicate that those things are to be taken as a collection, that is, mathematically as a set.

These assumptions are not about logic but about how to do the mathematics of metalogic. They apply to models of type **II**, not type **I**. The great simplicity and generality we get by making these assumptions is justified, I believe, so long as we don't arrive at any contradiction with our previous more fundamental assumptions and intuitions when we apply a metalogical result to actual propositions.

The question of whether we are now committed to believing in the existence of infinite totalities and whether we could prove our theorems about logic without such a commitment depend in the first place on how we view the notion of a collection or set which I discuss in Volume 2. It seems to me possible that what we want from these idealizations could be accomplished with a finitistic approach to mathematics, albeit at the cost of greater complexity. But this is not a book about the foundations of mathematics, so I will content myself with pointing out where these assumptions are made.

Someone who holds a platonist conception of logic and views propositions as abstract things would most likely take as basic facts about the world what I have called assumptions, idealizations, and generalizations. Words and sentences are understood as abstract objects which are represented in inscriptions or utterances, not equivalences we impose on the phenomena of our experience. All of them exist whether they are ever uttered or not. The formal and informal language and any model are abstract objects complete in themselves, composed of an infinity of things. Though the distinction between a model of type **I** and **II** can be observed by a platonist, it does not mater whether a model is ever actually expressed or brought to our attention. It simply exists and hence the Fully General Classical Abstraction is no abstraction but an observation about the world.

G. A Mathematical Presentation of PC

In this section I'll collect together the definitions of the formal logic we've developed and prove some facts about it, freely using the abstractions of the last section. This will be important as a reference for later chapters.

1. The formal language

We have the following primitives:

binary connectives	$\rightarrow, \wedge, \vee$
unary connective	$\neg$
parentheses	(,)
variables	$p_0, p_1, \ldots$

We will use p, q, q_0, q_1, ... as metavariables ranging over PV = $\{ p_0, p_1, \ldots \}$.

The collection of well-formed-formulas, *Wffs*, is defined inductively using A, B, ... as metavariables:

1. (p_i) is a wff for each i = 0, 1, 2, ...
2. If A, B are wffs, then so are $(\neg A)$, $(A \rightarrow B)$, $(A \wedge B)$, $(A \vee B)$.
3. No other concatenations of symbols are wffs.

We may define a formal language using fewer of these connectives by deleting the appropriate clauses of (2). We may also define a language with any other n-ary connective, γ, by adding to (2): $\gamma(A_1, \ldots, A_n)$ is a wff. All the definitions and results of this section (§G.1) can be generalized to any such language in a straightforward way, and I will assume those generalizations below. Sometimes I will refer to a formal language by the primitive connectives it is based on, e.g., the language of $\neg$ and $\wedge$.

We need to show that there is only one way to read each wff.

***Theorem 1* (*Unique Readability of Wffs*)**

There is one and only one way to parse each wff.

Proof: If A is a wff, then there is at least one way to read it since it has a definition. To show that there is only one way to read it I'll establish that no initial segment of a wff is a wff.

The idea is that if we begin at the left of a wff and subtract 1 for every left parenthesis and add 1 for every right parenthesis, then we will sum up to 0 only at the end of the wff. More precisely, define a function f from any concatenation of primitive symbols $\lambda_1, \ldots, \lambda_n$ of our formal language to the integers by:

$f(\neg)=0$	$f(\wedge)=0$	$f(\vee)=0$	$f(\rightarrow)=0$
$f(p_i)=0$	$f(\,(\,)=-1$	$f(\,)\,)=+1$	

and $f(\lambda_1, \ldots, \lambda_n) = f(\lambda_1) + \cdots + f(\lambda_n)$. To show that for every wff A, $f(A)=0$, we proceed by induction on the number of symbols in A. The wffs with fewest symbols are (p_i), i = 0, 1, 2, ... and for them it is immediate. So now suppose it is true for all wffs with fewer symbols than A. We then have four cases, which we cannot yet assume are distinct:

Case i. A arises as $(\neg B)$. Then B has fewer symbols than A, so by induction $f(B)=0$, and so $f(A)=0$.
Case ii. A arises as $(B \wedge C)$. Then B and C have fewer symbols than A, so $f(B)=f(C)=0$ and hence $f(A)=0$.
Case iii. A arises as $(B \vee C)$ } are done similarly
Case iv. A arises as $(B \rightarrow C)$ }

I'll leave to you to show by the same method that any initial part B′ of a wff B must have $f(B') < 0$. Hence no initial part of a wff is a wff.

Now suppose that we are given a wff which can be parsed as both $(A \wedge B)$ and $(C \rightarrow D)$. Then $A \wedge B)$ is identical to $C \rightarrow D)$. Hence A is identical to C for otherwise one would be an initial segment of the other, contradicting what we have just proved. But then $\wedge B)$ is identical to $\rightarrow D)$, which is a contradiction. The other cases are virtually the same. ■

Hence we may define the *principal* or *main connective* of $(A \rightarrow B)$ to be '$\rightarrow$' and the *immediate constituents* to be A and B, and similarly for other wffs. We sometimes refer to a wff which has, for example, '$\rightarrow$' as it's main connective as an '$\rightarrow$-wff'.

By taking account of the way that wffs are built we may use induction more effectively to prove facts about them. By Theorem 1 we may define *the length of a wff*:

1. The length of (p_i) is 1.
2. If the length of A is n then the length of $(\neg A)$ is $n+1$.
3. If the maximum of the length of A and the length of B is n, then each of $(A \rightarrow B)$, $(A \wedge B)$, $(A \vee B)$ has length $n+1$.

Thus to show that all wffs have a property we can use *induction on the length of wffs*: first show that the atomic wffs have the property and then show that if all wffs of length n have it, so do all wffs of length $n+1$. A good example is the proof of Theorem 8 below. In *Epstein and Carnielli* you can find a discussion of the general notion of induction on "inductively defined classes."

A further aid in using induction on wffs is an *ordering of wffs*:

1. The primitive symbols are alphabetized in the order:
$\neg, \rightarrow, \wedge, \vee, (,), p_0, p_1, p_2, \ldots$
2. If the length of B is greater than the length of A, then A comes before B.
3. If A and B have the same length and reading left to right at the first place at which they have different symbols A has the prior one in the alphabetization above, then A comes before B.

Finally, we define what we mean by A is a *subformula* of B:

1. A is a subformula of A.

2. If A is a subformula of C, then A is a subformula of $(\neg C)$, and for every D, A is a subformula of $(C \rightarrow D)$, $(D \rightarrow C)$, $(C \wedge D)$, $(D \wedge C)$, $(C \vee D)$, and $(D \vee C)$.
3. A is a subformula of B only if it is classified as such by repeated application of (1) and (2).

We say that p_i *appears in* A if (p_i) is a subformula of A, and denote by 'PV(A)' the set $\{p_i : p_i \text{ appears in } A\}$. When I write A(q) I mean that q appears in A; C(A) means that A is a subformula of C.

2. Models and the semantic consequence relation

A *model* for classical logic is a function $\upsilon: PV \rightarrow \{T, F\}$ which is extended to all wffs by the tables (1), (2), (3), and (4). The extension is unique by Theorem 1. When it is necessary to distinguish a model of this logic from ones for other logics I will call υ a **PC**-model.

We say that a *wff* A *is true or valid in model* υ if $\upsilon(A) = T$ and notate that $\upsilon \models A$, often read 'υ validates A'. We say that a wff A *is false in* υ if $\upsilon(A) = F$, and write $\upsilon \not\models A$.

Letting Γ, Σ, Δ stand for collections of wffs, we write $\Gamma \models A$, read as Γ *validates* A or A *is a semantic consequence of* Γ, if given any model υ which validates all wffs in Γ we have that $\upsilon(A) = T$. We write $A \models B$ for $\{A\} \models B$, and $\models A$ for $\varnothing \models A$, that is, A is true in every model ('$\varnothing$' denotes the empty collection). If $\models A$ we say that A is a (**PC**)-*tautology* or is **PC**-*valid*. We say that υ is a *model of* Γ, written $\upsilon \models \Gamma$, if $\upsilon \models A$ for every A in Γ.

We denote as **PC** the collection of wffs $\{A : \models A\}$. The relation $\models$ is called *the semantic consequence relation of* **PC** and is often denoted $\models_{\mathbf{PC}}$. For reference I will list some of its important properties.

Theorem 2 (Properties of the Semantic Consequence Relation)

a. $A \models A$

b. If $\models A$, then $\Gamma \models A$.

c. If $A \in \Gamma$, then $\Gamma \models A$.

d. If $\Gamma \models A$ and $\Gamma \subseteq \Delta$, then $\Delta \models A$.

e. (***Transitivity***)
If $\Gamma \models A$ and $A \models B$, then $\Gamma \models B$.

f. (***Transitivity, the Cut Rule***)
If $\Gamma \models A$ and $\Delta, A \models B$, then $\Gamma \cup \Delta \models B$.

g. If $\Gamma \cup \{A_1, \dots, A_n\} \models B$ and $\Gamma \models A_i$ for $i = 1, \dots, n$, then $\Gamma \models B$.

(All of the above will apply to the semantic consequence relation of any logic we study in this book, whereas those that follow may not.)

h. $\models A$ iff for every nonempty Γ, $\Gamma \models A$.

j. (***The Semantic Deduction Theorem for* PC**)
$\Gamma, A \models B$ iff $\Gamma \models A \rightarrow B$.

k. (***The Semantic Deduction Theorem for* PC *for Finite Consequences***)
$\Gamma \cup \{A_1, \ldots, A_n\} \models B$ iff $\Gamma \models A_1 \rightarrow (A_2 \rightarrow (\cdots \rightarrow (A_n \rightarrow B) \cdots)$.

l. (***Alternate version of* (k)**)
$\Gamma \cup \{A_1, \ldots, A_n\} \models B$ iff $\Gamma \models (A_1 \wedge \cdots \wedge A_n) \rightarrow B$.

m. (***Substitution***)
If $\models A(p)$ then $\models A(B)$, where $A(B)$ is the result of substituting B uniformly for p in A (i.e., B replaces every occurrence of p in A).

Proof: I'll leave all of these to you except for (h) and (m).

h. The left to right direction follows from (b). For the other direction, if every nonempty Γ validates A, then in particular $\neg A \models A$. So by (j), $\models \neg A \rightarrow A$ and hence $\neg A \rightarrow A$ is true in every model. Therefore A is true in every model.

m. The only place that p can occur in A(B) is within B itself. Hence given a model v, once we have evaluated $v(B)$ the evaluation of $v(A(B))$ proceeds as if we were evaluating A(p) in a model w, where w and v agree on all propositional variables except p and $w(p) = v(B)$. Since A(p) is true in every model, $w(A(p)) = T$ and so $v(A(B)) = T$. Since v is an arbitrary model, $\models A(B)$. ∎

Part (m) says that propositional variables are really variables which can stand for any proposition, a consequence of the Extensionality Consideration.

3. The truth-functional completeness of the connectives

Suppose we add to the language a new connective $\gamma(A_1, \ldots, A_n)$ which we wish to interpret as a truth-function. Then we must specify its interpretation by a truth-table which for every sequence of n T's and F's assigns either T or F. If we do this we'll say that we have added a *truth-functional connective* to the language.

I will show in this section that every truth-functional connective can be defined in terms of $\neg, \rightarrow, \wedge, \vee$ in classical logic, that is, $\{\neg, \rightarrow, \wedge, \vee\}$ is *truth-functionally complete*. More precisely, given a connective $*$ as above, we can find a schema $S(A_1, \ldots, A_n)$ which uses only $\neg, \rightarrow, \wedge, \vee$ such that for all wffs $A_1, \ldots, A_n$ we have that $S(A_1, \ldots, A_n)$ is semantically equivalent to $\gamma(A_1, \ldots, A_n)$. Thus except for the convenience of abbreviation, further truth-functional connectives add nothing of significance to our language and semantics.

Theorem 3 In **PC** every truth-functional connective can be defined in terms of $\neg, \rightarrow, \wedge, \vee$.

Proof: Let γ be a truth-functional connective as above. If γ takes only the value F then $\gamma(A_1, \ldots, A_n)$ is semantically equivalent to $(A_1 \wedge \neg A_1) \vee \cdots \vee (A_n \wedge \neg A_n)$. Otherwise let

$$\alpha_1 = (\alpha_{11}, \ldots, \alpha_{1n})$$
$$\vdots$$
$$\alpha_k = (\alpha_{k1}, \ldots, \alpha_{kn})$$

be those sequences of T's and F's which are assigned T by the table for γ. Define for $i = 1, \ldots, k$ and $j = 1, \ldots, n$,

$$B_{ij} = \begin{cases} A_{ij} & \text{if } \alpha_{ij} = \mathsf{T} \\ \neg A_{ij} & \text{if } \alpha_{ij} = \mathsf{F} \end{cases}$$

and $D_i = B_{i1} \wedge \cdots \wedge B_{in}$, associating to the right (this corresponds to the ith row of the table). Then $D_1 \vee D_2 \vee \cdots \vee D_n$, associating to the right, is semantically equivalent to $\gamma(A_1, \ldots, A_n)$. ■

As an example of this procedure, suppose we wish to define in **PC** a ternary connective γ which has the following table :

A_1	A_2	A_3	$\gamma(A_1, A_2, A_3)$
T	T	T	T
T	T	F	F
T	F	T	F
T	F	F	T
F	T	T	T
F	T	F	T
F	F	T	F
F	F	F	F

Then

$$\alpha_1 = (\mathsf{T}, \mathsf{T}, \mathsf{T}) \quad \alpha_2 = (\mathsf{T}, \mathsf{F}, \mathsf{F}) \quad \alpha_3 = (\mathsf{F}, \mathsf{T}, \mathsf{T}) \quad \alpha_4 = (\mathsf{F}, \mathsf{T}, \mathsf{F})$$

and

$$D_1 = A_1 \wedge (A_2 \wedge A_3) \qquad D_2 = A_1 \wedge (\neg A_2 \wedge \neg A_3)$$
$$D_3 = \neg A_1 \wedge (A_2 \wedge A_3) \quad D_4 = \neg A_1 \wedge (A_2 \wedge \neg A_3)$$

So $D_1 \vee (D_2 \vee (D_3 \vee D_4))$ is semantically equivalent to $\gamma(A_1, A_2, A_3)$.

Our proof of Theorem 3 actually shows that $\{\neg, \wedge, \vee\}$ is truth-functionally complete. By noting the following *semantic equivalences* we can show that *each of* $\{\neg, \rightarrow\}$, $\{\neg, \wedge\}$, $\{\neg, \vee\}$ *is truth-functionally complete*:

(11) *Equivalences in* **PC**

$A \to B$ and $\neg(A \wedge \neg B)$
$A \to B$ and $\neg A \vee B$
$A \wedge B$ and $\neg(\neg A \vee \neg B)$
$A \wedge B$ and $\neg(A \to \neg B)$
$A \vee B$ and $\neg(\neg A \wedge \neg B)$
$A \vee B$ and $\neg A \to B$

However, $\{\to, \wedge, \vee\}$ *is not truth-functionally complete* because any wff made from these connectives using only p_0 will take value T if p_0 has value T, so $\neg$ cannot be defined.

There are two truth-functional connectives each of which by itself is truth-functionally complete. The first is the *Sheffer stroke* ('nand') written $A | B$ which is T iff not both A and B are T. Then $\neg A$ is semantically equivalent to $A | A$, and $A \to B$ to $A | \neg B$. The other connective formalizes 'neither ... nor ...' and is symbolized $A \downarrow B$. It is true iff both A and B are false. It is truth-functionally complete because $\neg A$ is semantically equivalent to $A \downarrow A$, and $A \wedge B$ to $\neg A \downarrow \neg B$.

4. The choice of language for PC

Since every truth-functional connective is definable in **PC** from $\neg$ and $\to$, the connectives $\wedge$ and $\vee$ are in some sense superfluous. We could take $L(p_0, p_1, \ldots \neg, \to)$ as the language of classical logic and then define connectives $\wedge$ and $\vee$ as in table (11). We feel that we have the same logic because we have a translation from the language $L(p_0, p_1, \ldots \neg, \to, \wedge, \vee)$ to $L(p_0, p_1, \ldots \neg, \to)$:

$$(p_i)^\dagger = p_i$$
$$(\neg A)^\dagger = \neg(A^\dagger)$$
$$(A \to B)^\dagger = A^\dagger \to B^\dagger$$
$$(A \wedge B)^\dagger = \neg(A^\dagger \to \neg B^\dagger)$$
$$(A \vee B)^\dagger = \neg(A^\dagger) \to B^\dagger$$

This translation is faithful to our models: given any $v: PV \to \{T, F\}$ if we first extend it to all wffs of $L(p_0, p_1, \ldots \neg, \to, \wedge, \vee)$ and call that v_1, and then extend it to $L(p_0, p_1, \ldots \neg, \to)$ and call that v_2, we have that $v_1 \models A$ iff $v_2 \models A^\dagger$.

Similarly, we feel we have the same logic if we formalize it using $\{\neg, \wedge\}$ or $\{\neg, \vee\}$ or any other truth-functionally complete set of connectives. We recognize this by saying, loosely, that **PC** *may be formalized in any of several languages*. And we may speak of *a formalization of* **PC** *in the language of* $L(p_0, p_1, \ldots \neg, \to)$. We need to be careful speaking like this lest we begin to think that there is some formless body of truths comprising **PC**. These phrases are meant only to refer to the translations and equivalences we've established. Similar comments will apply to any other logic we study in which some of the primitive connectives may be defined in terms of the others.

5. The decidability of tautologies

It may seem obvious that we can effectively decide whether a given wff is a tautology or not, but it's worth pointing out exactly what that depends on. In doing so we'll see that our infinitistic assumptions have not fouled up the obvious.

a. In any model, the truth-value of any wff depends only on the truth-values of the propositional variables appearing in it.

b. Given any assignment of truth-values to a finite number of propositional variables, the assignment can be extended to all variables (e.g. take F to be the value assigned to all other propositions).

c. Therefore, a wff is not a tautology iff there is an assignment of truth-values to the variables appearing in it for which it takes the value F. For if there is such a one, it can be extended to be a model in which the wff is false. And if there is none then the wff must be a tautology.

d. For any assignment of truth-values to the variables appearing in a wff we can effectively calculate the truth-value of the wff by repeatedly using the tables (1), (2), (3), and (4).

Thus we can decide whether any wff is a **PC**-tautology by listing all possible truth-value assignments to the variables appearing in it—if there are n variables then there are 2^n assignments—and for each assignment mechanically checking the resulting truth-value for the wff. The wff is a tautology iff it always receives the value T. See *Epstein and Carnielli,* Chapter 19, for a formal presentation of this procedure showing that it is mechanically computable.

Note that this decision procedure for the validity of wffs also allows us to decide whether any two given wffs are semantically equivalent: A is semantically equivalent to B iff $A \leftrightarrow B$ is a tautology.

This decision procedure for validity though effective is practically unusable for wffs containing, say, 40 or more propositional variables, for we would have to check some 2^{40} assignments. There are many shortcuts we can take, quite a few of which are spelled out in *Quine, 1950* . But to date no one has come up with a method of checking validity which is substantially faster and could be run on a computer in less than (roughly) exponential time relative to the number of variables appearing in the wff; it has been conjectured that there is no faster method.

One particular way of using (a)–(d) above is worth noting: we attempt to falsify the wff. If we can it's not a tautology, otherwise it is. This is particularly useful for $\rightarrow$-wffs because only one line of the table yields F: $A \rightarrow B$ can be falsified iff there is an assignment that makes both A true and B false. Generally we consider schemas rather than particular wffs. For example

$$\neg(A \wedge B) \rightarrow (B \rightarrow A)$$

	$\neg(A \wedge B)$	$(B \rightarrow A)$	
is false iff	T	F	
which is iff	$A \wedge B$	B	A
is	F	T	F

and that is the falsifying assignment. Hence the schema is not tautological.

Similarly,

$$((A \wedge B) \rightarrow C) \rightarrow (A \rightarrow (B \rightarrow C))$$

	$(A \wedge B) \rightarrow C$	$A \rightarrow (B \rightarrow C)$		
is false iff	T	F		
iff		A	$B \rightarrow C$	
is		T	F	
iff			B	C
is			T	F

But if A is T, B is T, and C is F, then $(A \wedge B) \rightarrow C$ is F so there is no way to falsify the schema. Hence it is a tautology.

6. Some PC-tautologies

As a reference with respect to other logics I'm going to list some **PC**-tautologies in the form of schemas. These schemas can be seen as the syntactic codification of many of our assumptions, especially since we may identify the validity of an $\rightarrow$ -schema with an assertion about semantic consequence in **PC**. It is good practice in the methods of **PC** to establish those schemas which are not obvious.

1. $A \vee \neg A$ *Excluded Middle*, *tertium non datur* (a third way is not given)
2. $\neg(A \wedge \neg A)$ the law of *noncontradiction*
3. $\neg\neg A \rightarrow A$ } the laws of *double negation*
4. $A \rightarrow \neg\neg A$ }
5. $(A \wedge B) \rightarrow A$ } the laws of *simplification for conjunction*
   $(A \wedge B) \rightarrow B$ }
6. $(A \wedge A) \leftrightarrow A$ the principle of *tautology for conjunction*
7. $(A \wedge B) \rightarrow (B \wedge A)$ *commutativity of conjunction*
8. $((A \wedge B) \wedge C) \leftrightarrow (A \wedge (B \wedge C))$ *associativity of conjunction*
9. $A \rightarrow (A \vee B)$ } the laws of *addition for disjunction*
   $B \rightarrow (A \vee B)$ }
10. $(A \vee A) \leftrightarrow A$ the principle of *tautology for disjunction*
11. $(A \vee B) \rightarrow (B \vee A)$ *commutativity of disjunction*
12. $((A \vee B) \vee C) \leftrightarrow (A \vee (B \vee C))$ *associativity of disjunction*

13. $A \vee (B \wedge C) \leftrightarrow [(A \vee B) \wedge (A \vee C)]$ } *Distribution laws for conjunction and disjunction*
14. $A \wedge (B \vee C) \leftrightarrow [(A \wedge B) \vee (A \wedge C)]$
15. $\neg(A \wedge B) \leftrightarrow (\neg A \vee \neg B)$ } *De Morgan's laws*
16. $\neg(A \vee B) \leftrightarrow (\neg A \wedge \neg B)$
17. $[(A \vee B) \wedge \neg A] \rightarrow B$ a propositional form of the *Disjunctive Syllogism*
18. $A \rightarrow A$ *Identity*
19. $(\neg A \rightarrow A) \rightarrow A$ *Clavius' Law*
20. $A \rightarrow (B \rightarrow A)$
21. $\neg A \rightarrow (B \rightarrow A)$
22. $A \rightarrow (B \rightarrow B)$
23. $(A \wedge \neg A) \rightarrow B$

Numbers (20) and (21) are sometimes called the *paradoxes of material implication*, and (22) and (23) the *paradoxes of strict implication*. I'll discuss them in §L.

24. $[(A \rightarrow B) \wedge (B \rightarrow C)] \rightarrow (A \rightarrow C)$ } two expressions of the *transitivity of* $\rightarrow$
25. $(A \rightarrow B) \rightarrow [(B \rightarrow C) \rightarrow (A \rightarrow C)]$
26. $[A \wedge (A \rightarrow B)] \rightarrow B$ } two expressions of the principle of *detachment*
27. $A \rightarrow [(A \rightarrow B) \rightarrow B]$
28. $[\neg B \wedge (A \rightarrow B)] \rightarrow \neg A$ } two expressions of the principle of *modus tollens*
29. $\neg B \rightarrow [(A \rightarrow B) \rightarrow \neg A]$
30. $(A \rightarrow B) \rightarrow [(A \rightarrow \neg B) \rightarrow \neg A]$ a version of the principle of *reductio ad absurdum*
31. $[(A \wedge B) \rightarrow C] \rightarrow [A \rightarrow (B \rightarrow C)]$ *Exportation*
32. $[A \rightarrow (B \rightarrow C)] \rightarrow [(A \wedge B) \rightarrow C]$ *Importation*
33. $[A \rightarrow (B \rightarrow C)] \rightarrow [B \rightarrow (A \rightarrow C)]$ *Interchange of Premisses*
34. $(A \rightarrow B) \leftrightarrow (\neg B \rightarrow \neg A)$ *Transposition*

    $\neg B \rightarrow \neg A$ is called the *contrapositive* of $A \rightarrow B$
35. $(A \leftrightarrow B) \leftrightarrow (\neg A \leftrightarrow \neg B)$
36. $(A \rightarrow B) \leftrightarrow \neg(A \wedge \neg B)$
37. $[A \rightarrow (B \wedge C)] \rightarrow [(A \rightarrow B) \wedge (A \rightarrow C)]$
38. $[A \rightarrow (B \vee C)] \rightarrow [(A \rightarrow B) \vee (A \rightarrow C)]$
39. $[A \rightarrow (B \rightarrow C)] \rightarrow [(A \rightarrow B) \rightarrow (A \rightarrow C)]$

7. Normal forms

Given any wff we can find another wff which has a particularly simple form and which is semantically equivalent to it by using the proof of Theorem 3.

We say that a wff is in *disjunctive normal form* if it is a disjunction of conjunctions of variables or negations of variables; it is in *conjunctive normal form* if it is a conjunction of disjunctions of variables or negations of variables.

By the principles of tautology for conjunction and disjunction ((6) and (10) of the last section), we can consider either of (p) or $(\neg p)$ to be a disjunction or a conjunction. Thus a wff A is in disjunctive normal form if A is $B_1 \vee B_2 \vee \cdots \vee B_n$ where each B_i is, for some m, of the form $C_i \wedge \cdots \wedge C_m$, where each C_i is either a propositional variable or the negation of one. Because of the associativity of conjunction and disjunction ((8) and (12) of the last section) the conjuncts or disjuncts may be associated in any way.

Theorem 4 (***The Normal Form Theorem for*** **PC**) Given any wff A there is an effective procedure for finding a wff B which is in disjunctive normal form and contains exactly the same propositional variables as A and is semantically equivalent to A in **PC**; similarly, there is an equivalent wff in conjunctive normal form which contains the same propositional variables.

Proof: Any wff A determines a truth-function: if $PV(A) = \{q_1, \ldots, q_n\}$ then we can make up a table with 2^n rows corresponding to the ways we can assign T or F to each of the variables, and we can calculate the truth value of A under each assignment. The proof of Theorem 3 then establishes a disjunctive normal form for A. By repeated use of the distribution laws for conjunction and disjunction ((13) and (14) of the preceding section) we can obtain a conjunctive normal form for A. ■

8. The principle of duality for PC

Consider the truth-conditions for $\wedge$ and $\vee$:

$A \wedge B$ is T iff both A, B are T; otherwise F
$A \vee B$ is F iff both A, B are F; otherwise T

They are dual to each other in the sense that one arises by reversing the roles of T and F in the other. It's no accident that the symbol for conjunction is an upside down $\vee$.

Let A be any wff which contains no occurrence of '$\rightarrow$'. The *dual of* A, which I'll notate as A*, is the wff we get by replacing every occurrence of $\wedge$ by $\vee$ in A and every occurrence of $\vee$ by $\wedge$. For instance, the dual of $\neg(A \vee B)$ is $\neg(A \wedge B)$; the dual of $\neg A \wedge \neg B$ is $\neg A \vee \neg B$. The evaluation of A* proceeds exactly as for A except that the roles of T and F are reversed. If A has the same truth-value as B in all models of **PC**, then A* will have the same truth-value as B* in all models of **PC**. Recalling the definition of '$\leftrightarrow$' from §E.4, we see that $\vDash_{PC} A \leftrightarrow B$ iff $\vDash_{PC} A^* \leftrightarrow B^*$. This is called the *principle of duality for* **PC**. Using it we can deduce, for example, the second De Morgan Law, (16) in §6 above, from the first.

H. Formalizing the Notion of Proof

Though the material in this part is motivated by our work on **PC**, it is not tied to classical logic and will apply to all the other logics we'll study in this volume. *Nothing in this section presupposes any of the mathematical abstractions discussed in* §F: all the notions here make perfectly good sense in a finitistic, constructive context, though unless I state otherwise I will follow the abstractions of §F whenever I use any of these notions.

1. Reasons for formalizing

Historically, logic proceeded from observations that there were propositions true due to their form only, to an investigation of those forms, then to symbolization and generality, and finally to laying out a few forms from which all other logically acceptable ones could be mechanically derived. That approach dates back at least to the Stoics, and to Aristotle too, though he was concerned more with the form of logically acceptable arguments than valid wffs. The modern versions came to fruition in the work of Frege, *1879,* and Whitehead and Russell, *1910 –13.*

Logic in this tradition was seen as one language for all reasoning. It was only in the 19th Century with the use of models for non-euclidean geometry that mathematicians considered the idea that a set of formal propositions could have more than one interpretation. It is revealing to read the correspondence between Hilbert, who had combined the notion of models for geometry with rigorous formal logic, and Frege (*Frege, 1980,* pp. 31–52) to see how novel the idea of many models for one language appeared at that time.

Still, the tables for the classical connectives had been implicit in the writings of logicians since antiquity:

> Philo said that the conditional is true whenever it is not the case that its antecedent is true and its consequent is false; so that according to him, the conditional is true in three cases and false in one case. ... they say that a conjunction holds when all the conjuncts are true, but is false when it has at least one false conjunct ...
>
> Sextus Empiricus, *Against the Mathematicians,* VIII, 112, 125
> translation from Mates, *1953,* pp. 97–98

In 1921 Post showed that the propositional wffs derivable as theorems in Whitehead and Russell's system of logic coincide with the classical tautologies (see §K.1 below). Since we now know that we can semantically decide whether a wff is a **PC**-tautology, why should we bother to show that we can derive the **PC**-tautologies from a few acceptable forms?

For one thing, the decision procedure takes an unrealistic amount of time, and so for all practical purposes we have no decision procedure for wffs of even moderate length. Worse, when we investigate the internal form of propositions in Volume 2 we will find that classical predicate logic, which deals with quantification and is based on **PC**, has no decision method (see *Epstein and Carnielli*). In that case, and also for other propositional logics for which we have no decision procedure, a way to derive tautologies syntactically is essential.

Moreover, formalizing the notion of derivability, of when one proposition can be proved from others, is itself important not only for logic but for mathematics and science. The semantics of **PC** gave us a notion of semantic consequence. We would like a syntactic counterpart of that based on formalizing the notion of a proof.

Finally, by approaching logic in terms of the forms and not the meanings of wffs we will have another way to isolate our assumptions. We start with an informal notion of validity and take as fundamental the validity of a few simple wffs and the fact that some few basic rules lead always from valid wffs to valid wffs. Then we can say: if you accept these you will accept our logic. Unless, that is, you accept other wffs which are not derivable in our system. To convince you that there are no other acceptable wffs we must return to our formalization of the meaning of wffs and prove that the valid wffs, the tautologies, can all be derived within our system. That proof which links the syntax and semantics will always be a central point of our investigations of a logic.

My emphasis in this book is on semantics for logic; I will present only one approach to the theory of proofs, usually called Hilbert-style proof theory, which I will use throughout. For a general survey of formalizations of the notion of proof see Chapter IX of *Kneale and Kneale, 1962*.

2. The notions of proof, syntactic consequence, and theory

We have the formal language $L(p_0, p_1, \ldots \neg, \rightarrow, \wedge, \vee)$. Whenever I say 'proposition' below I mean one in the semi-formal object language of a model of L.

A *rule* is a direct consequence relation: given a collection of propositions of specified form, called the hypotheses, another proposition is taken to be a consequence. Since it's the forms that are significant, rules are given in the formal language and are usually presented as schemas,

$$\frac{A_1, \ldots, A_n}{B}$$

where any collection of wffs of the form $\{A_1, \ldots, A_n\}$ can be taken as hypotheses. For instance, we have

modus ponens $\quad \dfrac{A, A \rightarrow B}{B}$

the rule of adjunction $\quad \dfrac{A, B}{A \wedge B}$

When we choose a rule we believe that it's self-evident that it has the property that if the hypotheses are instantiated with true propositions then the conclusion will be true.

The notion of proof is formalized as follows. We begin with a collection of propositions which we take to be self-evidently true, called the *axioms*. We then set out a collection of rules which together with the axioms comprise an *axiom system.* There is a *proof* or *derivation of* A *from the axioms* if there is a sequence $A_1, \dots, A_n$ such that A_n is A and each A_i is either an axiom or is a direct consequence of some of the preceding A_j's by one of the rules. In that case we say that A is a *theorem* of that system, and write '$\vdash$A'. (We usually give a name to the system and subscript '$\vdash$' with that.)

At this point we have a choice whether to let a nonconstructive element enter into our formalization. We may decide that 'there is a proof of A' means that we can actually produce one, as is sometimes suggested by constructivists, or we may take it nonconstructively. Usually the choice depends on the underlying semantic assumptions motivating the development of the logic. I'll discuss the difference with respect to classical logic in §J.2, 3 and §K.1.

In axiomatizing logic itself we'll want to take as axioms propositions which we believe are self-evidently true due to their form only. The best way to describe those is by using formal wffs or schemas. The axiom system is then entirely in the formal language.

Once we have an axiom system for our logic we may wish to add further hypotheses specific to some discipline, say geometry or physics. A *proof of a proposition* A *from a collection of propositions* Σ is a sequence just as above except that we are also allowed to use the propositions of Σ as hypotheses in the rules. That is, each A_i in a proof either is an axiom, is in Σ, or is derivable by a rule from some of the preceding A_j's. If there is a proof of A from Σ then we say that A *is a syntactic consequence of* Σ and write $\Sigma \vdash A$. We sometimes read this as 'A is deducible from Σ'. We will write $A \vdash B$ for $\{A\} \vdash B$, and $\Sigma, A \vdash B$ for $\Sigma \cup \{A\} \vdash B$. Note that $\varnothing \vdash A$ is the same as $\vdash A$. A convenient notation for 'there is not a proof of A from Σ' is $\Sigma \nvdash A$.

With this definition of consequence and theorem, once we prove a theorem we can use it to prove further theorems. I'll let you convince yourself that if Σ is a collection of theorems and $\Sigma \vdash A$, then $\vdash A$.

The relation '$\vdash$' is our syntactic formalization of the notion of a proposition following from one or more propositions. In this we are again primarily interested in the syntactic forms so we will generally take Σ to be a collection of formal wffs and then pass back and forth between wffs and propositions as we did for the notions of truth and validity in §D.2. So I'll let Σ, Γ, Δ *and subscripted versions of these stand variously for collections of wffs, semi-formal propositions, or propositions* depending on the context.

Don't get confused and think that I have defined the notions of proof and consequence here. What I've done, relative to any particular axiom system, is give a definition of '$\vdash$' which we colloquially read as 'is a theorem' or 'has as a consequence' or 'proves'. I must assume that you already have an idea of what it means to prove something and it's that which we are formalizing as I need to use the informal notion in proving theorems about the formal or semi-formal object languages. I prove those (informal) theorems in the language of this book, the *metalanguage,* which is English supplemented with various technical notions.

We may take any particular syntactic consequence, or more usually schema of consequences $\{A_1, \ldots, A_n\} \vdash B$, and use it as a further *derived rule* in proofs. If we can prove A from Σ using a derived rule as well as the original rules, then we also have $\Sigma \vdash A$; hence, derived rules may shorten proofs but don't allow any new consequences (I'll demonstrate this in Theorem 5 below). For example, if we have $\{\neg(A \wedge \neg B), A\} \vdash B$ then we may use the derived rule

$$\frac{A, \neg(A \wedge \neg B)}{B}$$

We say that Σ is *closed under* a rule if whenever the hypotheses of the rule are in Σ then so is the consequence. The *closure* of Σ under a rule is the smallest collection of wffs Δ such that $\Delta \supseteq \Sigma$ and Δ is closed under the rule.

We call Σ a *theory* if it is closed under syntactic deduction, that is, if $\Sigma \vdash A$ then A is in Σ. This, roughly, is how we think of scientific theories as not just their hypotheses but their consequences too. In general, we call the collection of all syntactic consequences of Σ the *theory of* Σ, denoted $\text{Th}(\Sigma)$. That is, $\text{Th}(\Sigma) = \{A : \Sigma \vdash A\}$, and Σ is a theory if $\text{Th}(\Sigma) = \Sigma$. We sometimes say that Δ is the *closure of* Γ *under a rule* to mean that $\Delta = \text{Th}(\Gamma)$ where that rule is the only one used in derivations.

Some theories are better than others, as any scientist will tell you. As logicians we are especially interested in theories which are *consistent* , that is, contain no contradiction. A collection Σ which is not consistent we call *inconsistent*. The notion of contradiction which we'll employ will depend on the formal or informal semantics we are using, how we understand truth. For instance, in **PC** a consistent theory is one which for no A contains both A and $\neg A$. For a 3-valued logic we'll study in §VIII.C.1 the notion of a consistent theory is one which for no A contains all three of A, $\neg A$, $A \leftrightarrow \neg A$.

A *complete theory* is one which is as full a description as possible of "the way the world is" relative to the atomic propositions we've assumed and the semantics, formal and informal, which we are using. It might be inconsistent; if not it will contain as many complex propositions as possible relative to those atomic ones while still being consistent. So a complete and consistent theory corresponds to a possible description of the world, relative to our semantic intuitions and choice of atomic propositions. Therefore, *if we have a formal semantics we will want the notion of a complete and consistent theory to correspond to the notion of the set of propositions true in a model.*

3. Some properties of syntactic consequence relations

***Theorem 5* (*Properties of Syntactic Consequence Relations*)**

a. $A \vdash A$

b. If $\vdash A$, then $\Gamma \vdash A$.

c. If $A \in \Gamma$, then $\Gamma \vdash A$.

d. If $\Gamma \vdash A$ and $\Gamma \subseteq \Delta$, then $\Delta \vdash A$.

e. (***Transitivity***)
If $\Gamma \vdash A$ and $A \vdash B$, then $\Gamma \vdash B$.

f. (***Transitivity, the Cut Rule***)
If $\Gamma \vdash A$ and $\Delta, A \vdash B$, then $\Gamma \cup \Delta \vdash B$.

g. If $\Gamma \cup \{A_1, \ldots, A_n\} \vdash B$ and $\Gamma \vdash A_i$ for $i = 1, \ldots, n$, then $\Gamma \vDash B$.

h. (***Compactness***)
$\Gamma \vdash A$ iff there is some finite collection $\Delta \subseteq \Gamma$ such that $\Delta \vdash A$.

j. If there is a sequence $A_1, \ldots, A_n$ where each A_i is either an axiom, in Σ, or a consequence of some of the previous A_n's using a rule of the system or a derived rule, and $A_n = A$, then $\Sigma \vdash A$.

Proof: Most of these are relatively straightforward to establish so I'll only prove (j). Given such a sequence, each time a derived rule is employed, say to obtain A_k from $A_{i_1}, \ldots, A_{i_m}$, we can insert into the proof the derivation of A_k from those propositions (wffs) using only the axioms, rules of the system, and propositions (wffs) in Σ, because the use of a derived rule to obtain A_k is just a shorthand way of saying that A_k is a consequence of $\{A_{i_1}, \ldots, A_{i_m}\}$. In this way we can obtain a (much longer) proof of A using no derived rules. ■

Parts (e) and (f) are generalizations of the remark I made above that once we prove a proposition we may use it as if it were an axiom. Part (h) is of significance only if we make the assumption that we are dealing with completed infinite totalities of propositions.

We could, if we wish, rephrase several parts of the previous theorem and other facts about '$\vdash$' by talking about theories, and this is what many texts do. For example (c) can be stated as: $\Gamma \subseteq Th(\Gamma)$, and part (d) as: if $\Gamma \subseteq \Delta$, then $Th(\Gamma) \subseteq Th(\Delta)$.

4. The notion of a logic

Note that the first seven properties of syntactic consequence relations correspond to the first seven properties for semantic consequence relations listed in Theorem 2.

Given any logic we would like the semantic and syntactic consequence relations to coincide so that we have only one formal metalogical notion of 'follows from'. This is not always the case so we distinguish various possibilities.

There are several ways to present what might be called *a logic*:

1. Define a notion of model and truth in a model for a formal language. The logic is then the collection of wffs true in all models.

2. Define a notion of model and truth in a model for a formal language. Semantic consequence is then defined in the manner it was for **PC**, and the logic is considered to be the semantic consequence relation.

3. Present an axiom system in a formal language. Then the collection of all theorems is the logic.

4. Present an axiom system in a formal language. The logic is then considered to be the syntactic consequence relation as defined above.

I will consider any one of these to be *a presentation of a logic*; examples of each of them can be found in what follows and in the literature of modern logic. I'll say a logic is *presented semantically* if it is given by (1) or (2) and *presented syntactically* if it is given by (3) or (4). Sometimes we say that (3) or (4) is the *syntax of the logic*.

Given a semantics and a syntax for the same language, if every theorem is a tautology then we say that *the axiomatization is sound* or that we have *provided sound semantics for the axiomatization*, depending on which was presented first. In that case, for all A, if $\vdash A$ then $\models A$. This is the minimal condition we impose on any pair of semantics and syntax which we want to claim characterize the same pre-formal logical notions.

Given a sound axiomatization of a semantics if we also have that every tautology is a theorem, that is for all A, if $\models A$ then $\vdash A$, we say that *the axiomatization is complete for the semantics* or that *we have provided complete semantics for the axiomatization.* Sometimes we say simply that *the (axiomatized) logic has a semantics.* A proof of this fact is called a *completeness theorem.*

The strongest correlation we can have is when the semantic and syntactic consequence relations are the same. That is, for all Γ and A, $\Gamma \models A$ iff $\Gamma \vdash A$. In that case we say that we have *strongly complete semantics for the axiomatization*, or a *strongly complete axiomatization of the semantics.* A proof of the equivalence is called a *strong completeness theorem* for the logic.

If we can prove a strong completeness theorem for a logic then we can conclude by Theorem 5.h that $\Gamma \models A$ iff there is some finite collection of wffs $\Delta \subseteq \Gamma$ such that $\Delta \models A$. For most logics this compactness property cannot be proved without nonconstructive infinitistic assumptions (cf. §J.3 and §K.1) that may contradict the semantic ideas upon which the logic is based (cf. intuitionistic logic,

§VII.B.2). The practitioners of the logic may, therefore, choose to prove only a *finite* strong completeness theorem: for any finite Γ and any A, Γ⊨A iff Γ⊢A.

The notions of this section will become much clearer when we use them in the next section to prove a strong completeness theorem for **PC**.

J. An Axiomatization of PC

1. The axiom system

How are we to come up with axioms for **PC**? Generally we have some idea, some strategy for how to prove an axiom system complete or strongly complete. So we begin a proof with no axioms in hand and probably only one rule, say *modus ponens*. Then each time we need an axiom (schema) or an additional rule to make the proof work, we add it to the list. The list might get a bit long that way, but we can be sure it's complete if we've added enough axioms to make the proof work. We can always go back and try to simplify the list by showing that one of the axioms is superfluous by deducing it from the others . This is how I got the axiomatization for **PC** in the language L(¬, →) which I present in this section. I'll highlight uses of the axioms by printing them in boldface so you can see where they are needed. In §K I'll discuss other axiomatizations.

The syntactic consequence relation (see p.41) for **PC** will be defined with respect to the following axiomatization.

PC
in L(¬,→)

axiom schemas

Every formal wff which is an instance of one of the following schemas is an axiom.

1. ¬A → (A→B)
2. B → (A→B)
3. (A→B) → ((¬A→B) → B)
4. (A → (B→C)) → ((A→B) → (A→C))

rule $\dfrac{\text{A, A} \rightarrow \text{B}}{\text{B}}$ (*modus ponens*)

Here are two examples of derivations, both of which we will need in the completeness proof below.

Lemma 6 ***a.*** ⊢ A→A

b. {A,¬A}⊢B

Proof: On the right-hand side I give a justification for each step of the derivation.

a. 1. $\vdash A \to ((A \to A) \to A)$ — an instance of **axiom 2**
2. $\vdash A \to (A \to A)$ — an instance of axiom 2
3. $\vdash (A \to ((A \to A) \to A)) \to ((A \to (A \to A)) \to (A \to A))$ — an instance of **axiom 4**
4. $\vdash (A \to (A \to A)) \to (A \to A)$ — *modus ponens* using (1) and (3)
5. $\vdash A \to A$ — *modus ponens* using (2) and (4).

b. 1. $\vdash \neg A \to (A \to B)$ — **axiom 1**
2. $\neg A$ — premise
3. $(A \to B)$ — *modus ponens* using (1) and (2)
4. A — premise
5. B — *modus ponens* using (3) and (4). ■

We define: Γ is *consistent* if for every A *at most* one of $A, \neg A$ is a consequence of Γ. And Γ is *complete* if for every A, *at least* one of $A, \neg A$ is in Γ. As before, the *theory of* $\Gamma = \{A: \Gamma \vdash A\} = \text{Th}(\Gamma)$. And Γ is a *theory* if $\Gamma = \text{Th}(\Gamma)$. Note that if Γ is a theory, then Γ is consistent iff for every A at most one of $A, \neg A$ is in Γ.

Here are some facts about these formal notions. Not all of them are needed in the proof of the completeness theorem, but it is useful to list them in one place as a reference for comparing other notions of completeness and consistency for other logics.

Lemma 7

a. If Γ is a theory, then every axiom is in Γ.

b. If Γ is consistent, then for every A, $\Gamma \nvdash A$ or $\Gamma \nvdash \neg A$.

c. Γ is consistent iff every finite subset of Γ is consistent.

d. If Γ is consistent and $\Delta \subseteq \Gamma$, then Δ is consistent.

e. Γ is consistent iff there is some B such that $\Gamma \nvdash B$.

f. If Γ is complete, then for every A, $\Gamma \vdash A$ or $\Gamma \vdash \neg A$.

g. If Γ is complete and consistent, then for any $A \notin \Gamma$, $\Gamma, A \vdash B$ for every B.

h. If Γ is complete and consistent, then Γ is a theory.

Proof: (a) and (b) are immediate.

I'll prove the negated version of (c): Γ is inconsistent iff there is some finite subset of Γ which is inconsistent. Suppose Γ is inconsistent. Then there is some A such that $\Gamma \vdash A$ and $\Gamma \vdash \neg A$. So by Theorem 5.h there is some finite $\Sigma \subseteq \Gamma$ and finite $\Delta \subseteq \Gamma$ such that $\Sigma \vdash A$ and $\Delta \vdash \neg A$. So $\Delta \cup \Sigma$ is finite and inconsistent and contained in Γ. The other direction is immediate. Part (d) follows similarly from Theorem 5.d.

The negated version of (e), Γ is inconsistent iff for all B, $\Gamma \vdash B$, follows from (b) and Lemma 6.b.

Part (f) is immediate, and part (g) follows by Lemma 6.b.

Finally for (h), suppose Γ is complete and consistent, and yet is not a theory. Then for some A, $\Gamma \vdash A$ and $A \notin \Gamma$. But then $Th(\Gamma \cup \{A\}) = Th(\Gamma)$, which by (g) and (e) contradicts the consistency of Γ. So Γ is a theory. ■

Note: In some texts Γ is *consistent* is taken to mean that for every A at most one of $A, \neg A$ is actually in Γ. This has the advantage of not being dependent on any particular axiom system. If we were to use that terminology in this development then where we spoke of consistent sets of wffs we'd need to talk of consistent theories. And we would prove our lemmas about $Th(\Gamma)$, rather than Γ, being consistent or inconsistent.

Next we show that 'B follows syntactically from A' is equivalent to '$A \rightarrow B$ is a theorem'.

Theorem 8 ***a.*** ***(The Syntactic Deduction Theorem)***

$\Gamma, A \vdash B$ iff $\Gamma \vdash A \rightarrow B$

b. ***(The Syntactic Deduction Theorem for Finite Consequences)***

$\Gamma \cup \{A_1, \ldots, A_n\} \vdash B$ iff $\Gamma \vdash A_1 \rightarrow (A_2 \rightarrow (\cdots \rightarrow (A_n \rightarrow B) \cdots)$

Proof: a. $\Leftarrow$ Immediate.

(By '$\Leftarrow$' I mean the right-to-left direction of the iff-proposition. Analogously '$\Rightarrow$' means the left-to-right direction.)

$\Rightarrow$ Suppose $B_1, \ldots, B_n$ is a proof of B from $\Gamma \cup \{A\}$. I'll show by induction that for each i, $\Gamma \vdash A \rightarrow B_i$.

Either $B_1 \in \Gamma$ or B_1 is an axiom, or B_1 is A. In the first two cases the result follows by using **axiom 2**. If B is A we could just add $A \rightarrow A$ to our list of axioms. But I showed in Lemma 6.a that $\vdash A \rightarrow A$ using **axioms 2** and **4**.

Now suppose that for all $k < i$, $\vdash A \rightarrow B_k$. If B_i is an axiom, or $B_i \in \Gamma$, or B_i is A, then we have $\vdash A \rightarrow B_i$ as before. The only other case is when B_i is a consequence of B_m and $B_j = B_m \rightarrow B_i$ for some $m, j < i$ by *modus ponens*. Then by induction $\Gamma \vdash A \rightarrow (B_m \rightarrow B_i)$ and $\Gamma \vdash A \rightarrow B_m$, so by **axiom 4**, $\Gamma \vdash A \rightarrow B_i$.

Part (b) is an immediate corollary of part (a). ■

Note that this proof of the Deduction Theorem depends on axioms involving only the one connective '$\rightarrow$'.

2. A completeness proof

Lemma 9 Γ is complete and consistent iff there is a **PC** model v such that $\Gamma = \{A : v(A) = T\}$.

Proof: It's easy to show that the set of wffs true in a **PC** model is complete and consistent.

Suppose now that Γ is complete and consistent. Consider v: Wffs $\to \{T, F\}$ defined by $v(A) = T$ if $A \in \Gamma$, and $v(A) = F$ if $A \notin \Gamma$. We will show that v is a model by showing that it evaluates the connectives correctly.

If $v(\neg A) = T$ then $\neg A \in \Gamma$,
so $A \notin \Gamma$ by consistency,
so $v(A) = F$.

If $v(A) = F$ then $A \notin \Gamma$,
so $\neg A \in \Gamma$ by completeness,
so $v(\neg A) = T$.

Suppose $v(A \to B) = T$. Then $A \to B \in \Gamma$. Now note that Γ is a theory by Lemma 7.h . So if $v(A) = T$, we have $A \in \Gamma$ and by using the rule, $B \in \Gamma$ so $v(B) = T$. Conversely, if the conditions hold for $A \to B$ to be evaluated as T, then either $v(A) = F$ or $v(B) = T$. If the former then $A \notin \Gamma$ so $\neg A \in \Gamma$ and by **axiom 1,** $A \to B \in \Gamma$, so $v(A \to B) = T$. If the latter then $B \in \Gamma$ and as Γ is a theory, by **axiom 2,** $A \to B \in \Gamma$. ■

Lemma 10 ***a.*** If Γ is consistent then $\Gamma \cup \{A\}$ or $\Gamma \cup \{\neg A\}$ is consistent.
b. $\Gamma \nvdash A$ iff $\Gamma \cup \{\neg A\}$ is consistent.

Proof: a. For the contrapositive, suppose $\Gamma \cup \{A\}$ and $\Gamma \cup \{\neg A\}$ are both inconsistent. Then for any B, by Lemma 7.e, $\Gamma \cup \{A\} \vdash B$, and hence by Theorem 8, $\Gamma \vdash A \to B$. Similarly, $\Gamma \vdash \neg A \to B$. Hence by using **axiom 3,** $\Gamma \vdash B$, so Γ is inconsistent.

b. I'll prove the negated version. If $\Gamma \vdash A$ then $\Gamma \cup \{\neg A\}$ is inconsistent. If $\Gamma \cup \{\neg A\}$ is inconsistent, then by Lemma 6.b, $\Gamma \cup \{\neg A\} \vdash A$, hence, $\Gamma \vdash \neg A \to A$ by Theorem 8. By Lemma 6.a, $\vdash A \to A$. Hence using **axiom 3,** $\Gamma \vdash A$. Part (c) is proved analogously. ■

The next lemma is the crucial nonconstructive step in this method of proving completeness.

Lemma 11 If $\nvdash D$ then there is some complete and consistent collection of wffs Γ such that $\neg D \in \Gamma$.

Proof: Let $A_0, A_1, \ldots$ be a numbering of the wffs of the formal language (cf. the ordering of §G.1). By Lemma 10.b, $\{\neg D\}$ is consistent. Define

$$\Gamma_0 = \{\neg D\}$$

$$\Gamma_{n+1} = \begin{cases} \Gamma_n \cup \{A_n\} & \text{if this is consistent} \\ \Gamma_n \cup \{\neg A_n\} & \text{otherwise} \end{cases}$$

and

$$\Gamma = \bigcup_n \Gamma_n$$

Using Lemma 10.a and induction, each Γ_n is consistent. Hence Γ is consistent, for if not then some finite $\Delta \subseteq \Gamma$ is inconsistent by Lemma 7.c, and Δ being finite, $\Delta \subseteq \Gamma_n$ for some n. In that case Γ_n would be inconsistent, which would be a contradiction.

Finally, Γ is complete by construction. ■

Charles Silver, *1978*, has shown that we can use p_n in place of A_n in the definition of Γ above.

Theorem 12 *a.* **(*Completeness of the axiomatization of* PC)**
$\vdash A$ iff $\vDash A$.
b. **(*Finite Strong Completeness*)**
For finite Γ, $\Gamma \vdash A$ iff $\Gamma \vDash A$.

Proof: a. First, the axiomatization is sound: every axiom is a tautology, as you can check, and if A and $A \rightarrow B$ are tautologies then so is B. Hence if $\vdash A$ then A is a tautology.

Now suppose that $\nvdash A$. We will show that A is not a tautology. Since $\nvdash A$, by Lemma 11 there is a complete and consistent Γ such that $\neg A \in \Gamma$. So $A \notin \Gamma$. By Lemma 9 there is some **PC**-model v such that Γ is the set of wffs true in v. So $v(A) = \mathsf{F}$ and A is not a tautology. Thus if A is a tautology it is a theorem.

b. Suppose that $\Gamma = \{A_1, \ldots, A_n\}$. If $\{A_1, \ldots, A_n\} \vdash A$ then by the Syntactic Deduction Theorem for Finite Consequences (Theorem 8.b),
$\vdash A_1 \rightarrow (A_2 \rightarrow (\cdots \rightarrow (A_n \rightarrow B) \cdots))$. Hence by part (a),
$\vDash A_1 \rightarrow (A_2 \rightarrow (\cdots \rightarrow (A_n \rightarrow B) \cdots))$, and so $\{A_1, \ldots, A_n\} \vDash A$ (Theorem 2.k). ■

We have shown that for **PC** the syntactic and semantic formalizations of 'follows from' result in the same relation on wffs and hence on propositions. Moreover, these metalogical notions can be identified with the theoremhood or validity of the corresponding conditional.

Corollary 13 The following are equivalent:

a. $A \vDash B$ *b.* $A \vdash B$
c. $\vDash A \rightarrow B$ *d.* $\vdash A \rightarrow B$

Proof: The equivalence of (a) and (b) is Theorem 12.a. I remarked earlier on the equivalence of (c) and (a) (Theorem 2.j). The equivalence of (d) and (b) is the Syntactic Deduction Theorem. ■

The nonconstructive nature of this proof of completeness has a serious disadvantage: given a tautology such as $(p_1 \to (p_2 \to p_3)) \to (p_2 \to (p_1 \to p_3))$, Theorem 12 tells us that a derivation for it exists in our system, but it doesn't tell us how to produce it. In §K.2 I will give a constructive proof of completeness that will show how to produce the derivations, but the results of the next section, §J.3, depend heavily on the nonconstructive proof we've given above.

3. The Strong Completeness Theorem for PC

It is not possible to give a constructive proof of the strong completeness of **PC** (Henkin, *1954,* has shown that it is equivalent to the Boolean Prime Ideal Theorem). Therefore, a version of Lemma 11 is essential.

Lemma 14

a. If Σ is consistent and $\Sigma \nvdash D$, then there is a complete and consistent set Γ such that $\Sigma \cup \{\neg D\} \subseteq \Gamma$.

b. Every consistent collection of wffs has a model.

Proof: Part (a) follows just as for Lemma 11 except that we take Γ_0 to be $\Sigma \cup \{\neg D\}$. Part (b) then follows by Lemma 9. ■

Theorem 15 **(*Strong Completeness of the axiomatization of* PC)**

$\Sigma \vdash A$ iff $\Sigma \vDash A$.

Proof: If $\Gamma \vdash A$, then given any model of Γ all the wffs in Γ are true, as are the axioms. Since the rule is valid, A must be true in the model, too. Hence $\Gamma \vDash A$. Suppose $\Gamma \nvdash A$. Then proceed as in the proof of Theorem 12 using Lemma 14. ■

Here is a striking consequence of these infinitistic methods.

Theorem 16 **(*Compactness of the semantic consequence relation*)**

a. $\Gamma \vDash A$ iff there is some finite collection $\Delta \subseteq \Gamma$ such that $\Delta \vDash A$.

b. Γ has a model iff every finite subset of Γ has a model.

Proof: (a) follows from Theorem 15 and Lemma 5.h.

For (b), if Γ has a model then so does every finite subset of it. In the other direction, if every finite subset of Γ has a model, then every finite subset of Γ is consistent by the soundness part of Theorem 15. Hence by Lemma 7.c, Γ is consistent; so by Lemma 14 it has a model. ■

4. Derived rules: substitution

Derived rules were discussed in §H.2. By the Finite Strong Completeness Theorem we may prove that a schema

$$\frac{A_1, \ldots, A_n}{B}$$

is a derived rule by showing that $\{A_1, \ldots, A_n\} \vDash A$. Two derived rules for **PC** which concern how we may substitute into wffs are worth noting. First define $A \leftrightarrow B$ as $(A \rightarrow B) \wedge (B \rightarrow A)$, where $\wedge$ is defined from $\neg$ and $\rightarrow$ as in table (11), p. 34. In every model $\upsilon(A \leftrightarrow B) = T$ iff $\upsilon(A) = \upsilon(B)$.

Corollary 17 The following are derived rules in **PC**:

a. (***Substitution***) $\dfrac{\vdash A(p)}{\vdash A(B)}$

where B is any wff which is substituted uniformly for p (i.e., B replaces every occurrence of p).

b. (***Substitution of Logical Equivalents***) $\dfrac{A \leftrightarrow B}{C(A) \leftrightarrow C(B)}$

where C(B) is the result of substituting B for some but not necessarily all occurrences of the subformula A in C.

Proof: Part (a) follows from its semantic version, Theorem 2.m (p. 32).

For part (b), given evaluations of $\upsilon(A)$ and $\upsilon(B)$, the evaluations of $\upsilon(C(A))$ and $\upsilon(C(B))$ can differ only if $\upsilon(A) \neq \upsilon(B)$. So if $\upsilon \vDash A \leftrightarrow B$, then $\upsilon \vDash C(A) \leftrightarrow C(B)$. Hence, $A \leftrightarrow B \vDash C(A) \leftrightarrow C(B)$, and the corollary follows as remarked above. ∎

Thus propositional variables are really variables. If $\vdash A(p)$, then no matter what proposition 'p' is to refer to, simple or complex, we have a theorem. For this reason, axiomatizations are sometimes given using the rule of substitution rather than schemas, as I discuss further in §K.3.

K. Other Axiomatizations and Proofs of Completeness of PC

In this section I'll present some other ways to axiomatize **PC** and survey alternate proofs of the completeness theorem.

1. History and Post's proof

The first axiomatization of classical logic in terms that are recognizably the same as ours was Whitehead and Russell's system in *Principia Mathematica* (*1910 –13*), though they drew upon an earlier axiomatization by Frege, *1879*. For them the question of completeness simply did not arise. Their goal was to show that all logic, and indeed all mathematics, could be developed *within* their system, and hence did not look outside it for justification in terms of meanings of formulas. Dreben and van Heijenoort, *1986,* pp. 44–47, and Goldfarb, *1979,* discuss their views.

The first use of semantic notions to justify Whitehead and Russell's system by giving a completeness proof was by Bernays, though that was not published until *1926* (see *Dreben and van Heijenoort, 1986*). The first published proof, and long the most influential, was Post's, *1921*. He showed that every wff is semantically equivalent to one in disjunctive normal form. He then supplemented the system of *Principia Mathematica* with further axioms so that given any tautology A he could show how to produce a derivation of the equivalent disjunctive normal form of A, say B, and a derivation of $B \rightarrow A$. Joining those derivations he then had a proof of A. Only later was it shown that his further axioms could be proved in the system of *Principia Mathematica*.

Post's proof has a very clear advantage over the proof I gave above (§J.2): given a tautology it shows how to produce a derivation for it in the formal system. However, Post's proof has two disadvantages: it cannot be easily generalized to other logics, nor does it make clear how the completeness theorem is related to the notions of completeness and consistency. Moreover, the strong completeness of **PC** cannot be deduced from it, since the proof of that requires substantial infinitistic nonconstructive assumptions (see *Henkin, 1954*) as in the proof I gave (which is due to Łos, *1951*, based on the work of Lindenbaum—see Tarski, *1930*, Theorem 12). There are a number of other constructive proofs of completeness which Surma, *1973 B,* surveys, all of which share the same advantage and disadvantages. In the next section I present one.

2. A constructive proof of the completeness of PC

The constructive proof of the completeness of **PC** which I will give here is due to Kalmár, *1935*. In it I will use Lemma 6 and Theorem 8 from §J.1, since the proofs of those show how to produce the derivations which they assert exist.

Lemma 18 *a.* $\vdash A \rightarrow \neg\neg A$

b. $\{A, \neg B\} \vdash \neg(A \rightarrow B)$

Proof: a. As before, I'll put the justification to the right of each step of the derivation.

1. $(\neg A \rightarrow (A \rightarrow \neg\neg A)) \rightarrow [((\neg\neg A \rightarrow (A \rightarrow \neg\neg A)) \rightarrow (A \rightarrow \neg\neg A)]$ an instance of axiom 3
2. $\neg A \rightarrow (A \rightarrow \neg\neg A)$ an instance of axiom 1
3. $(\neg\neg A \rightarrow (A \rightarrow \neg\neg A)) \rightarrow (A \rightarrow \neg\neg A)$ *modus ponens* on (1) and (2)
4. $\neg\neg A \rightarrow (A \rightarrow \neg\neg A)$ an instance of axiom 2
5. $A \rightarrow \neg\neg A$ *modus ponens* on (3) and (4).

b. It's immediate that $\{A, \neg B, A \rightarrow B\} \vdash \neg B$ and $\{A, \neg B, A \rightarrow B\} \vdash B$. Hence by Lemma 6.b, $\{A, \neg B, A \rightarrow B\} \vdash \neg(A \rightarrow B)$. So by the Syntactic Deduction

Theorem (Theorem 8), $\{A, \neg B\} \vdash (A \to B) \to \neg(A \to B)$. By Lemma 6.a and Theorem 5.b, $\{A, \neg B\} \vdash \neg(A \to B) \to \neg(A \to B)$. An instance of axiom 3 is $[(A \to B) \to \neg(A \to B)] \to [(\neg(A \to B) \to \neg(A \to B)) \to \neg(A \to B)]$. Hence by using *modus ponens* twice, we get $\{A, \neg B\} \vdash \neg(A \to B)$. ■

In the proof of part (b) I have shown that our previous work allows us to conclude that there is such a derivation. The proofs of the earlier lemmas on which that depends show how to produce one.

Lemma 19 Let C be any wff and $q_1, \ldots, q_n$ the propositional variables appearing in it. Let υ be any valuation. Define, for all $i \leq n$,

$$Q_i = \begin{cases} q_i & \text{if } \upsilon(q_i) = \mathsf{T} \\ \neg q_i & \text{if } \upsilon(q_i) = \mathsf{F} \end{cases}$$

and define $\Delta = \{Q_1, \ldots, Q_n\}$. Then

i. If $\upsilon(C) = \mathsf{T}$, then $\Delta \vdash C$.

ii. If $\upsilon(C) = \mathsf{F}$, then $\Delta \vdash \neg C$.

Proof: The proof is by induction on the length of C. If C is (p_i) then we need to show that $\vdash p_i \to p_i$ and $\vdash \neg p_i \to \neg p_i$, which we did in Lemma 6.a. Now suppose it is true for all wffs shorter than C.

If C is $\neg A$, suppose $\upsilon(C) = \mathsf{T}$. Then $\upsilon(A) = \mathsf{F}$. So by induction $\Gamma \vdash \neg A$. If $\upsilon(C) = \mathsf{F}$, then $\upsilon(A) = \mathsf{T}$, so by induction $\Gamma \vdash A$. By the previous lemma and Lemma 5.b, $\Gamma \vdash A \to \neg\neg A$, so using *modus ponens*, $\Gamma \vdash \neg\neg A$ as we wish.

If C is $A \to B$ first suppose $\upsilon(C) = \mathsf{T}$. Then $\upsilon(A) = \mathsf{F}$ or $\upsilon(B) = \mathsf{T}$. If $\upsilon(A) = \mathsf{F}$ then $\Gamma \vdash \neg A$, so by axiom 1, $\Gamma \vdash A \to B$. If $\upsilon(B) = \mathsf{T}$, then axiom 2 allows us to conclude that $\Gamma \vdash A \to B$. Finally, if $\upsilon(C) = \mathsf{F}$, then $\upsilon(A) = \mathsf{T}$ and $\upsilon(B) = \mathsf{F}$. So $\Gamma \vdash A$ and $\Gamma \vdash \neg B$. Hence by the previous lemma, $\Gamma \vdash \neg(A \to B)$. ■

Theorem 20 ***a.*** **(*Completeness of the axiomatization of* PC)**

$\vdash A$ iff $\models A$.

b. **(*Finite Strong Completeness*)**

For finite Γ, $\Gamma \vdash A$ iff $\Gamma \models A$.

Proof: a. As in the previous proof (Theorem 12), if $\vdash A$ then $\models A$. For the converse, suppose A is valid and $q_1, \ldots, q_n$ are the propositional variables appearing in A. For every valuation υ, $\upsilon(A) = \mathsf{T}$. Let υ_1 assign T to all propositional variables and υ_2 assign T to all except q_n to which it assigns F. Then by the previous lemma,

$$\{q_1, \ldots, q_{n-1}, q_n\} \vdash A$$

and

$\{q_1, \ldots, q_{n-1}, \neg q_n\} \vdash A$

So by the Syntactic Deduction Theorem (Theorem 8), $\{q_1, \ldots, q_{n-1}\} \vdash q_n \to A$ and $\{q_1, \ldots, q_{n-1}\} \vdash \neg q_n \to A$. Using axiom 3 we then have that $\{q_1, \ldots, q_{n-1}\} \vdash A$. Repeating this procedure $n-1$ times we have that $\vdash A$.

b. This follows by the Syntactic Deduction Theorem as in the proof of Theroem 12.b. ■

In the proof of Theorem 20 I didn't exhibit the derivation of the given tautology, but you can trace through the earlier proofs on which it depends to produce one.

3. Schemas vs. the rule of substitution

One well-known axiomatization of **PC** is due to Łukasiewicz. It uses the rule of substitution rather than schemas.

Łukasiewicz's axiomatization of **PC**
in $L(\neg, \to)$

axioms

$$(p_1 \to p_2) \to ((p_2 \to p_3) \to (p_1 \to p_3))$$

$$(\neg p_1 \to p_1) \to p_1$$

$$p_1 \to (\neg p_1 \to p_2)$$

rules *modus ponens* substitution

$$\frac{A,\ A \to B}{B} \qquad \frac{\vdash A(p)}{\vdash A(B)}$$

Łukasiewicz and Tarski, *1930*, claim this axiomatization is complete. We could give a proof of that starting from scratch, perhaps using Łukasiewicz's methods (see *Surma, 1973 B*.) Or we could rely on our previous proof by showing that there is a proof in this system of each axiom schema of the system I gave in §J.

Any axiom system which uses the rule of substitution can be presented using schemas instead of that rule. However, only if substitution is a derived rule can an axiom system which uses schemas be converted to one using the rule of substitution. There is a drawback to using the rule of substitution as a proof method: we must formulate it to apply only to the derivation of theorems and not to the syntactic consequence relation. That is, we take the rule to be

$$\frac{\vdash A(p)}{\vdash A(B)}$$

and not

$$\frac{A(p)}{A(B)}$$

If we were to use the latter every wff would be a syntactic consequence of $\{ p_1 \}$, which we certainly do not want. Moreover, the rule of substitution only makes sense for the formal language and not for the semi-formal language of propositions. For these reasons I've presented every axiom system in this book using schemas rather than the rule of substitution, even if the originators of the logic did otherwise.

4. Independent axiom systems

Given an axiom system we'd like to know if any of the axioms are superfluous, that is, can some be proved from others? We want the simplest, most perspicuous system possible, for then the syntactic agreements necessary to establish the logic are as persuasive and as few as possible.

Given an axiom system we say that an *axiom is independent* of the others if we can't prove it from the system that results by deleting that axiom. An *axiom schema is independent* if no instance of that schema can be proved if the schema is deleted. *An axiom system is independent* if each axiom (schema) is independent. I will often say that an axiom is independent when I mean that the schema is.

In §VIII.G I will show that the axiomatization I gave of **PC** in §J is independent. Łukasiewicz and Tarski, *1930*, claim that Łukasiewicz's axiomatization is also independent.

5. Proofs using only rules

By using the Deduction Theorem and the Strong Completeness Theorem we can obtain a derived rule from each of our axioms: take the antecedent as hypothesis and consequent as conclusion. Thus we have, for example, the derived rule

$$\frac{\neg A}{A \rightarrow B}$$

Some of the tautologies of §G.7 are more familiar in the form of rules, for example,

modus tollens $$\frac{\neg B,\ A \rightarrow B}{\neg A}$$

and

reductio ad absurdum $$\frac{A \rightarrow B,\ A \rightarrow \neg B}{B}$$

With the notion of proof that we're using it's customary to minimize the number of rules that are taken as primitive at the expense of additional axioms. However, there are formalizations of the notion of proof that use only rules and no axioms, and in some cases these give clearer derivations. Kneale and Kneale, *1962*, Chapter IX §3, give a general history and discussion of this method, and Kleene,

1952, presents a formalization of both classical and intuitionistic logic (Chapter VII below) using only rules.

6. Axiomatizations of PC in other languages

We originally defined **PC** in $L(\neg, \rightarrow, \wedge, \vee)$ and it behooves us to give an axiomatization of it in that language. This isn't hard, for we can use our earlier proof from §J. The only place the other connectives will enter is in the proof of Lemma 9: we need to add axioms to ensure that the connectives $\wedge$ and $\vee$ are evaluated correctly.

PC
in $L(\neg, \rightarrow, \wedge, \vee)$

axiom schemas

Every instance of the following schemas is an axiom.

1. $\neg A \rightarrow (A \rightarrow B)$
2. $B \rightarrow (A \rightarrow B)$
3. $(A \rightarrow B) \rightarrow ((\neg A \rightarrow B) \rightarrow B)$
4. $(A \rightarrow (B \rightarrow C)) \rightarrow ((A \rightarrow B) \rightarrow (A \rightarrow C))$
5. $A \rightarrow (B \rightarrow (A \wedge B))$
6. $(A \wedge B) \rightarrow A$
7. $(A \wedge B) \rightarrow B$
8. $A \rightarrow (A \vee B)$
9. $B \rightarrow (A \vee B)$
10. $((A \vee B) \wedge \neg A) \rightarrow B$

rule $\dfrac{A,\ A \rightarrow B}{B}$ (*modus ponens*)

Theorem 21 This axiomatization is strongly complete for **PC**.

Proof: The proof follows as in §J for the axiomatization in $L(\neg, \rightarrow)$ except that we need to supplement the proof of Lemma 9 as follows:

$$\begin{aligned} \upsilon(A \wedge B) = \mathsf{T} \ &\text{iff} \ \upsilon(A) = \mathsf{T} \text{ and } \upsilon(B) = \mathsf{T} \\ &\text{iff} \ A \in \Gamma \text{ and } B \in \Gamma \\ &\text{iff} \ (A \wedge B) \in \Gamma \text{ using axioms 5, 6 and 7.} \end{aligned}$$

For disjunction, we have that if $A \in \Gamma$ or $B \in \Gamma$ then $(A \vee B) \in \Gamma$ by axioms 8 and 9. If $(A \vee B) \in \Gamma$ then either $A \in \Gamma$ or not. If not, then $\neg A \in \Gamma$, hence $B \in \Gamma$ by axioms 10 and 5. So

$$\begin{aligned} v(A \vee B) = \mathsf{T} \ &\text{iff}\ v(A) = \mathsf{T} \text{ or } v(B) = \mathsf{T} \\ &\text{iff}\ A \in \Gamma \text{ or } B \in \Gamma \\ &\text{iff}\ (A \vee B) \in \Gamma . \end{aligned}$$ ■

The conjunctive form of the Syntactic Deduction Theorem,

$$\Gamma \cup \{A_1, \ldots, A_n\} \vdash B \ \text{ iff } \ \Gamma \vdash (A_1 \wedge \cdots \wedge A_n) \rightarrow B$$

is a corollary to Theorem 21 .

For future reference, note that this proof also shows that *if we delete axioms* 8, 9, *and* 10 *then we have a strongly complete axiomatization of* **PC** *in the language* $L(\neg, \rightarrow, \wedge)$.

Later on we will need an axiomatization of **PC** in the language of $\neg$ and $\wedge$. It would be good practice for you to develop one using the methods of §J. For reference I'll give one here.

First, we use '$A \supset B$' as an abbreviation of $\neg(A \wedge \neg B)$. This connective is called *material implication* whenever $\neg$ and $\wedge$ are interpreted classically. This will be the only way I will use the symbol '$\supset$' as a connective anywhere in this book. I call by the name *material detachment* the rule $\dfrac{A, A \supset B}{B}$.

PC
in $L(\neg, \wedge)$
axiom schemas

$A \supset (A \wedge A)$

$(A \wedge B) \supset A$

$(A \supset B) \supset (\neg(B \wedge C) \supset \neg(C \wedge A))$

rule $\dfrac{A, A \supset B}{B}$ (material detachment)

This system is due to Rosser, *1953*, who proves that it is strongly complete.

L. The Reasonableness of PC

1. Why classical logic is classical

When Whitehead and Russell wrote *Principia Mathematica* (*1910 –13*) they codified what they saw as the laws of logic in order to use them as the basis for the foundations of mathematics. When Post, *1921*, showed that the propositional logic they set out axiomatically was characterized by the truth-tables of **PC** there was overwhelming reason for logicians and mathematicians to accept it: they could feel confident in using it because they could check their work with truth-tables, and this (semantically) simple logic seemed to be all that was needed to reason about propositions as wholes (without quantification), since so much mathematics had been developed from it. It was simple, easy to use, and applicable in science and mathematics. It soon came to be called 'classical,' though till then the truth-functional reading of the connectives had been only one of several competing logical traditions dating back to the ancient Greeks, as you can read in *Bochenski, 1970,* or *Mates, 1953*.

Here are some of the major features of classical propositional logic which recommend it to us as a good, useful model of reasoning.

1. It's simple. Indeed, given the general assumptions of propositional logic of Chapter I, I believe that it has the simplest semantics we can devise.
2. It's easy to use both formally and in applications because we can check our work with the truth-tables. And this simplicity is essential since we want to use our logic in reasoning on any subject. If a logic were harder to use and understand than, say, mathematics we'd hardly want to use it when we did mathematics.
3. It has broad scope: it can be used on any atomic propositions, any sentences which we agree to view as having truth-values.
4. We can give it a simple axiomatization and the resulting syntactic consequence relation is equivalent to the semantic consequence relation.

What more could there be to propositional logic? We know from §E.4 that there are propositional connectives we can't deal with in **PC**, but so much the worse for them we might argue. Logic is only concerned with truth and consequence. But does **PC** do a good job of formalizing those? There is one more significant criterion that **PC** must be measured against.

5. Is it a reasonably accurate model or our pre-formal notions? Does it require us to give up a great deal of our intuitions about what's true, or what follows from what?

2. The paradoxes of PC

If the moon is made of green cheese, then $2 + 2 = 4$.

PC tells us this is true. That clashes with almost everyone's (naive) intuitions, assuming that we're not using the sentence as hyperbole.

A classical logician might argue that this is a confusion on my part. **PC** doesn't tell us that this proposition is true, but only that its formal version using '$\rightarrow$' is. And '$\rightarrow$' only captures those aspects of 'if ... then ...' which depend solely on the truth-values of propositions. So in this logic the formalization of any 'if ... then ...' proposition with false antecedent or true consequent is true. There's nothing paradoxical about that.

To lessen the shock of having to accept as true the formal version of propositions such as the one above, many people read '$\rightarrow$' in **PC** as 'materially implies' and call the conditional formed using it a 'material implication'. At least they do so when they're being careful: most of the time they forget and read it as 'if ... then ...' because that is what they set out to model. There's just a bad match here which doesn't get better by saying that we weren't really trying to formalize our notion of 'if ... then ...', 'implies', or 'follows from'. These observations were made by C. I. Lewis in *1912* even before the last volume of *Principia Mathematica* was published. Due to his critique the formulas $A \rightarrow (B \rightarrow A)$ and $\neg A \rightarrow (A \rightarrow B)$ have come to be known as the *paradoxes of classical logic.*

PC is only partially successful at capturing our informal notions, our intuitions about what follows from what, about which 'if ... then ...' propositions are true. That's because in almost all reasoning some other property of propositions and some relationship between propositions other than that between their truth-values matters. It seems to me that **PC** is applicable and reasonable only in those areas of propositional reasoning where it can be argued that the truth-values and forms of propositions are all that matter, such as in logic itself or perhaps mathematics and science. But even in mathematics there are serious arguments given by the intuitionists against using **PC** (see Chapter VII).

Still, there are many people who disagree and claim that **PC** is the right logic. There are several arguments to that effect, and I want to look at two here.

First, consider the proposition, 'If triangles had four sides then bananas would be purple'. It seems unreasonable to us to accept this as true. Bennett, *1969,* in his sustained defense of **PC** says that's because any proposition which begins with 'If triangles had four sides ...' will, in his words, 'strike one as implausible, weird, unsatisfactory' because it's antecedent is false.

I don't think that's what makes the conditional weird; rather, it's the total lack of connection between the antecedent and consequent. Nor is the subjunctive mood indicating a counterfactual the point. I'd venture to say that almost everyone would accept as true and not weird or implausible, 'If triangles had four sides, then rectangles would have five sides'.

Bennett's reply to this is that we shouldn't or can't incorporate any notion of connection of meanings or such like into a logical analysis of 'if ... then ...'. Hughes and Cresswell make the same point.

> ... to insist on it [some connection of "content" or "meaning"] seems to introduce into an otherwise clear and workable account of deducibility a gratuitously vague element which will make it impossible to determine whether a formal system is a correct logic of entailment or not.
>
> *Hughes and Cresswell, 1968,* pp. 336–337

Here they are defending classical modal logic (see Chapter VI) which they see as an extension of classical logic. In that logic any proposition with impossible antecedent or necessarily true consequent is adjudged true. Since '2+2 = 4' is considered to be a necessary truth, the initial proposition of this section would be true under their analysis. Comparable to the paradoxes of **PC** we have the *paradoxes of strict implication*, $A \rightarrow (B \rightarrow B)$ and $(A \wedge \neg A) \rightarrow B$.

The argument seems to me to be that classical logic (or its extension, classical modal logic) is the "right" logic because only the truth or falsity (and the possibility or impossibility) of an atomic proposition should matter to logic. These are clear and precise notions. Classical logic is right and where it clashes with our intuitions about reasoning, our intuitions are defective. Logic is primarily prescriptive.

But outside Plato's heaven propositions are full of imprecision and ambiguity. If we are to use logic, and I want to use logic in arguments ethical, mathematical, and political, we must face the fact that in almost all contexts some content of a proposition other than its truth-value matters to reasoning. I will argue in §VI.C.2 that even in classical modal logic an imprecise notion of content is crucial.

We need some general methods for modeling such contents. Before I present those in Chapter IV I want to go to the other extreme from classical logic and show how to incorporate into a formal model of reasoning a very imprecise notion of content: subject matter.

III Relatedness Logic: The Subject Matter of a Proposition – S and R –

In this chapter I motivate and formalize a logic which takes account of the subject matter of propositions. The point of the chapter is to present a particularly simple example of a general method for incorporating an aspect of propositions into the semantics; nothing in later chapters depends on your accepting this analysis of subject matters. In the next chapter I introduce the general approach which, if you prefer, you can read first and then return to this as an example.

D. Walton, R. Goldblatt and I originally developed this logic, which I first presented in *Epstein, 1979* . Walton, *1982* and *1985,* has given applications of it, while Iseminger, *1986,* has discussed whether it is appropriate to view this logic as a formalization of entailment.

A. An Aspect of Propositions: Subject Matter

'If the moon is made of green cheese, then 2+2 = 4' just can't be true. It's not that it's odd: what possible connection does 'the moon is made of green cheese' have with '2+2=4' ? They're about completely different things, they have nothing in common, so how can the latter follow from the former?

Oh, I know it's just my confusion that I think that 'if ... then ...' shouldn't be read truth-functionally in logic. If I'd only remember that. Though I seem to recall that Dostoevsky said in *Letters from the Underworld,*

> Man is so prone to systems and to abstract deductions that he is forever ready to mutilate the truth, be blind to what he sees, and deaf to what he hears, so long only as he can succeed in vindicating his logic.

It's not that I'm confused. I simply believe that in any normal discourse, in any normal reasoning and use of logic, something other than the truth-values of propositions matters in establishing the truth-value of an 'if ... then ...' proposition. And I don't believe other aspects of propositions are too ambiguous or that it would be hopelessly complicated to take account of them.

Even if I can't precisely specify and define, say, the subject matter of a proposition, I can nonetheless tell you in many examples if two propositions have some common subject matter, in what fashion subject matter affects the truth of an 'if ... then ...' proposition, and how the subject matter of a proposition is related to its parts. I mayn't be able to tell you what the subject matter of '2+2=4' is, or of 'the moon is made of green cheese'. But I can tell you that in any reasonable discussion we'll say that they are unrelated. And '2+2=4 and the moon is made of green cheese' does have some subject matter in common with both. If I can formulate rules which you'll agree are faithful to some commonly held notion of subject matter, then we can investigate together whether it's a coherent notion and whether it's suitable to be taken account of in logical argumentation.

The approach I'm going to take is that the subject matter of a proposition isn't so much a property of it as a relationship it has to other propositions. We may agree that 'Ralph is a dog' is, say, true, and that the truth-value is a property of that proposition. But the subject matter of 'Ralph is a dog' is an aspect of it related to other propositions. For example

'Ralph is a dog' is related to {
'George is duck'
'Dogs are faithful'
'Cats are nasty'
⋮

'Ralph is a dog' is not related to {
'2+2=4'
'The action of an internal combustion engine can be described in physics'
⋮

Suppose you stop me here and say, 'Look, no one's going to use both '2+2 = 4' and 'Ralph is a dog' in the same argument, in some logical deduction, so why worry? When we actually make up some particular model of **PC**, some assignment for the propositional variables, we won't include both of these propositions, so **PC** will do fine for us.'

> Only those conditionals are worth affirming which follow from some manner of relevance between antecedent and consequent—some law, perhaps, connecting the matters which these two components describe. But such connection underlies the useful application of the conditional without needing to participate in its meaning.
>
> Quine, *1950*, p.24

Then **PC** isn't universal? It's only applicable if all the propositions in the model have the same subject matter? That is exactly my point.

If I'm going to convince you of something I must assume some shared knowledge—say if I want to convince you of the Pythagorean Theorem I'll need to assume some mathematical propositions are true and accepted so by you, for example, the area of a square with sides of length s is s^2. Similarly, if I want to convince you of the truth of some proposition in a ordinary language setting, say a law court, then I'll have to assume a common stock of true propositions, but I will also have to assume that we share a common notion of subject matter or related propositions. Otherwise you're likely to dismiss my reasoning on the ground that I've introduced unrelated propositions.

Usually the common stock of propositions we accept as true is given implicitly, with only an occasional explicit truth-value stated, sometimes as a working hypothesis. Similarly, it's only with an occasional "odd" proposition that we may find disagreement about whether it's related to certain others. In that case we need a way to survey the consequences of whether it is or is not related to those others.

Suppose we want to take some unproved mathematical conjecture such as Fermat's last theorem as an atomic proposition in a model. We can't agree whether it's true or false, yet we want to see what its logical consequences are. We can do this because we know how to survey the assignments in which it's assumed true, and those in which it's assumed false. Similarly, I don't know whether 'Fermat's last theorem holds' is related in subject matter to 'Light bulbs have copper filaments', but I can work out their consequences if I can survey assignments in which they are assumed to be related and those in which they are not.

The notion of the subject matter of an atomic proposition is apparently both more dependent on context and more holistic than truth. In a particular model we may decide that 'Ralph is a dog' and 'George is a duck' are related because one of the topics under discussion is living creatures. But were our topics restricted to mammals and birds we would, perhaps, take these propositions to be unrelated. The holism of this approach is in part a consequence of the fact that the content of a sentence can only be given by considering all the other atomic sentences in the model to decide which it's related to and which not, for in that survey we may find that certain topics are under discussion. However, in practice we can deal with the notion molecularly by either postulating the topics to begin with or by considering only a small number of propositions in any one disputation. Hence we can give a determinate content to each proposition and build up the contents in a molecular pattern thereafter.

To further explain this notion of common subject matter I need to be able to talk about the internal structure of propositions. I understand two propositions as being related if they share, either explicitly or implicitly, some common predication or both refer to some common object. Thus 'Ralph is a dog' and 'Ralph barks' share a common reference; 'Don is a bachelor' and 'Reginald is a man' share, implicitly, a common predication. Agreements should be established in terms of predicates and names, and that is how I develop the notion with Stanisław Krajewski in Volume 2 using the tools of predicate logic (Krajewski has presented the formal part of that joint work in *Krajewski, 1986*).

Since we're not using the tools of predicate logic here, I have to hope that you find the notion of subject matter plausible and have gained some insight into it from this discussion. Based on that assumption I will motivate structural properties of this aspect of propositions.

Relatedness Assumption A proposition can be viewed as having subject matter.

Here, and until §D, I am using the phrase 'has subject matter' as shorthand for 'is related to certain propositions and unrelated to others'.

The more aspects or the more significant the aspect of a proposition that we model, the better we'll be able to satisfy the criterion for a logic that it concur with our intuitions about what's true and what follows from what until we finally reach the complexity of natural language itself. But the more we model the less simple our logic will be and the harder it will be to apply it. So let's look at only this one new aspect of propositions which we've discussed.

The Relatedness Abstraction The only properties of a proposition which matter to (this) logic are its form, its truth-value, and its subject matter.

But which forms?

B. The Formal Language

Note that I've been discussing the same English language connectives that I did with **PC**. Should we choose some new formal symbols to use in modeling them? That would have the advantage of reminding us that in one case relatedness of propositions matters, and in another it doesn't. But it hides a basic point: both in **PC** and here we're formalizing the same English connectives. A comparison of how different aspects of propositions matter in using these connectives in logic will be obscured if we choose different symbols. Therefore, the formal language I'll use is $L(p_0, p_1, \ldots \neg, \rightarrow, \wedge)$.

I don't include $\vee$ because here, even more than for **PC**, there are many good choices for how to formalize 'or', and arguments about the suitability of one over another are secondary to the main ideas of this chapter. We'll see shortly that all of the choices can be defined from the other primitives.

One question that may strike you is: why don't we take 'relatedness' as a primitive in the language? The answer is because the binary relation of subject matter relatedness, call it R, is not a connective. For example,

R(George is a duck, Ralph is a dog)

is read as

'George is a duck' is related to 'Ralph is a dog'

The original propositions are not used in this, only their names. It's an unnecessary confusion of objects and their names to formalize 'is related to' as a sentence connective. It makes no sense to iterate it, for how could we read: R (R (George is a duck, Ralph is a dog), Dusty is a horse)?

C. Properties of the Primitive: Relatedness Relations

Did you realize that by making the Relatedness Abstraction I'm continuing to assume that every proposition can be viewed as being either true or false? If 'Ralph is a dog then $2+2=4$' is not true, then it's false. Not odd, or unacceptable, or some third truth-value. Just plain false. There are two truth-values: one for propositions we accept and use in deducing "what is the case", and one for those we reject. And part of "what is the case" is how we understand 'if ... then ...'. Between affirming and denying there is no third choice: *tertium non datur*. Throughout this book any oddness, uncertainty or other aspect of a proposition will be taken account of in terms of its content—here, subject matter.

In that case what's the relation between the subject matter of a proposition and its truth? When I suggested that 'Ralph is a dog' is unrelated to '$2+2=4$' I'll wager that you didn't ask whether 'Ralph is a dog' is true. The use or assertion of a proposition carries with it the subject matter of the proposition. But it cannot also carry the truth of it. We may assert a sentence to be true, but that's not to say that it *is* true.

The subject matter of a proposition is independent of its truth-value.

As I see it, a virtual consequence of this assumption is that the logical connectives are neutral with respect to what a proposition is related to: they are syncategorematic, without any content. Roughly speaking, 'Ralph is a dog and $2+2=4$' is related to the same propositions as 'If Ralph is a dog then $2+2=4$' because the predications implicit in and things referred to by the former are the same as for the latter. There are other ways to view the logical connectives, two of which I discuss in § V.A.2, though I think they are inappropriate here. Therefore, I will list this as an additional assumption.

> The logical connectives are syncategorematic: they are neutral with respect to subject matter.

These assumptions rule out the view that a tautology has no subject matter, a view propounded by Prior, *1960*, p. 89. Whatever the notion of subject matter behind that, it surely can't be based on how or to what propositions refer or the predications involved in them.

The assumption that the logical connectives are syncategorematic tells us that A and $\neg$A are related to the same propositions and that $A \rightarrow B$ and $A \wedge B$ have the same subject matter. Using 'R(A, B)' to stand for 'A is related to B' we can symbolize these observations as

1. $\mathsf{R}(A,B)$ iff $\mathsf{R}(\neg A,B)$

2. $\mathsf{R}(A,B\wedge C)$ iff $\mathsf{R}(A,B\rightarrow C)$

I should also list $\mathsf{R}(B,A)$ iff $\mathsf{R}(B,\neg A)$ and similarly for the second. But that's not necessary for I believe you'll agree that subject matter relatedness is symmetric. It doesn't matter in which order we consider 'Ralph is a dog' and 'George is a duck' in determining whether they have some common subject matter. That is,

3. $\mathsf{R}(A,B)$ iff $\mathsf{R}(B,A)$

It seems in keeping with our understanding of this notion to further agree that every proposition is related to itself, that is the relationship R is reflexive.

4. $\mathsf{R}(A,A)$

So the only question left is: what is $A \wedge B$, and hence $A \rightarrow B$, related to?

To begin with, is 'Ralph is a dog' related to 'Ralph is a dog $\wedge$ $2+2=4$' ? Surely yes, as the former is a part of the latter. Roughly speaking the former proposition refers to some of the same things and involves some of the same predications as the latter. Don't confuse truth and subject matter here: the argument isn't that because the latter proposition is a conjunction then if it's true so is the former.

If A is related to $A \wedge B$ shouldn't also A be related to $C \wedge D$ if it's related to

C or to D? And how else can A have something in common with $C \wedge D$ unless it's related to C or to D? Thus

5. $\mathsf{R}(A, B \wedge C)$ iff $\mathsf{R}(A, B)$ or $\mathsf{R}(A, C)$
   $\mathsf{R}(A, B \rightarrow C)$ iff $\mathsf{R}(A, B)$ or $\mathsf{R}(A, C)$

the second equivalence following from the first by (2).

You may have answered my last question differently, claiming that something transcendent happens when we conjoin propositions, so that $C \wedge D$ can be related to propositions other than those that C or D are related to. I don't think that anything unusual or "transcendent" occurs when two propositions are conjoined, but you can view (5) as a simplification if you like, along the lines of a Fregean assumption (§II.A).

To summarize, we have

R1. $\mathsf{R}(A, B)$ iff $\mathsf{R}(\neg A, B)$
R2. $\mathsf{R}(A, B \wedge C)$ iff $\mathsf{R}(A, B \rightarrow C)$
R3. $\mathsf{R}(A, B)$ iff $\mathsf{R}(B, A)$
R4. $\mathsf{R}(A, A)$
R5. $\mathsf{R}(A, B \wedge C)$ iff $\mathsf{R}(A, B)$ or $\mathsf{R}(A, C)$

Let's give the name *subject matter relatedness relation* to any binary relation on propositions of the semi-formal language which we take as establishing subject matter relatedness and which satisfies R1–R5.

R1–R5 allow us to define inductively a unique relatedness relation from a *reflexive* and *symmetric* subject matter relation on atomic propositions. The following lemma establishes that there is a very simple functional relation between the subject matter of a complex proposition and the subject matter of its parts. Thus, as with truth-values, we may take subject matter relatedness as primitive on just the atomic propositions.

Lemma 1 $\mathsf{R}(A,B)$ iff there is some atomic proposition p which appears in A and some atomic proposition q which appears in B such that $\mathsf{R}(p, q)$.

Proof: I include the proof of this lemma as an example of double induction on the length of propositions.

The lemma is immediate if both A and B have length 1. Suppose it's true for all A of length $\leq n$ and for all B of length 1, i.e., B is q. I will show that it holds for all A of length $\leq n+1$ and all B of length 1.

Case 1: A is $C \wedge D$. Then $\mathsf{R}(C \wedge D, q)$ iff $\mathsf{R}(C, q)$ or $\mathsf{R}(D, q)$ by R5 and R3, which by induction is iff some p appears in C and $\mathsf{R}(p, q)$ or some p appears in D and $\mathsf{R}(p, q)$, which is iff some p appears in $C \wedge D$ and $\mathsf{R}(p,q)$.

Case 2: A is $C \rightarrow D$. Then proceed as in Case 1 using R2.

Case 3: A is $\neg C$. Then proceed as in Case 1 using R1.

Now suppose the lemma is true for all A when B has length $\leq n$. I will show it holds for all A, when B has length $\leq n+1$.

Case 1: B is $C \wedge D$. Then $R(A, C \wedge D)$ iff $R(A, C)$ or $R(A, D)$ which by induction is iff for some p in A, and some q in C or q in D, $R(p, q)$, which is iff for some p in A, some q in $C \wedge D$, $R(p, q)$.

The other cases follow similarly. ■

Note: At the same time that I was developing these ideas, B.J. Copeland, *1978* and *1982*, working from somewhat different motivations independently arrived at a notion of relevance that obeys virtually the same laws as R1–R5.

D. Subject Matters As Set-Assignments

We've taken the notion of two propositions being related as primitive. Can we derive a definite content for each proposition from that, one which we could call 'the subject matter of the proposition'? David Lewis has shown how. Though I'll use mathematical notation, no infinitistic assumptions are necessary.

Given a subject matter relatedness relation, R, define *the subject matter of a proposition* A to be

$$s(A) = \{ \{A, B\} : R(A, B) \}$$

We call s the *subject matter set-assignment associated with* R.

Lemma 2 Given a relatedness relation R and s the subject matter set-assignment associated with R, then $R(A,B)$ iff $s(A) \cap s(B) \neq \emptyset$.

I leave the proof to you.

Alternatively, we could take the notion of a definite content for a proposition, a subject matter, as primitive and then define two propositions as being related if they have some subject matter in common. To do that we postulate a set of "topics" which we assign to the propositions under consideration. That is, we have

a set of topics S

and

an assignment $s(p) \subseteq S$ for every atomic proposition p satisfying $s(p) \neq \emptyset$

The last clause is needed so that each proposition will have some subject matter in common with itself.

We extend s to complex propositions by taking

$$s(A) = \bigcup \{ s(p) : p \text{ appears in } A \}$$

The subject matter of a proposition is the union of the subject matter of its parts.

Any such s and S we'll call a *subject matter assignment.*

Define the *relatedness relation associated with* s as

$$\mathsf{R}(A, B) \text{ iff } \mathsf{s}(A) \cap \mathsf{s}(B) \neq \varnothing .$$

Example: For a thoroughly developed example we would need to consider the internal structure of propositions, and subject matters in terms of the subject matter of predicates and names. Even so, a simple example may help you grasp the meaning of these abstractions.

Let the set of topics of our discussion be S = { mammals, birds, dogs, cats, ducks, eagles, wolves, humans, eats, sleeps, runs, flies, swims, barks, quacks, iron, rust, car, drives, squashes, plastic, rubber, tires, engines }. Then we might take s(Ralph is a dog) = { mammals, dogs, eats, sleeps, runs, barks} and s(George is a duck) = { birds, ducks, flies, swims, quacks}, so that R(George is a duck, Ralph is a dog) fails. I'll leave to you to determine s(Reginald drives a car $\wedge$ cars have tires) and whether R(Reginald drives a car $\wedge$ cars have tires, George is a duck).

In this example and throughout I understand the elements of S to be bits of language, inscriptions, much as in the tradition of Buridan (see the Introduction to *T. K. Scott, 1966*, especially pp. 46–47). Someone working in the tradition of Frege, *1892,* might choose to interpret S to be a collection of nonsensible objects such as meanings or senses.

Lemma 3 The relatedness relation associated with a subject matter assignment s satisfies R1–R5.

Proof: First, because for every p, $\mathsf{s}(p) \neq \varnothing$, we have $\mathsf{s}(A) \neq \varnothing$ for every A, so $\mathsf{R}(A, A)$ holds. And R is symmetric. We also have $\mathsf{s}(A) = \mathsf{s}(\neg A)$, so $\mathsf{R}(A,B)$ iff $\mathsf{R}(\neg A, B)$. And $\mathsf{s}(A \wedge B) = \mathsf{s}(A \rightarrow B) = \mathsf{s}(A) \cup \mathsf{s}(B)$ so $\mathsf{R}(A, B \rightarrow C)$ iff $\mathsf{R}(A, B \wedge C)$. Finally,

$$\begin{aligned} \mathsf{R}(A, B \wedge C) \text{ holds } & \text{iff } \mathsf{s}(A) \cap [\mathsf{s}(B) \cup \mathsf{s}(C)] \neq \varnothing \\ & \text{iff } \mathsf{s}(A) \cap \mathsf{s}(B) \neq \varnothing \text{ or } \mathsf{s}(A) \cap \mathsf{s}(C) \neq \varnothing \\ & \text{iff } \mathsf{R}(A, B) \text{ or } \mathsf{R}(A,C) \end{aligned}$$

■

Can we use these two approaches interchangeably? The next lemma shows that for every R there is a unique s, and for every s a unique R such that we can pass back and forth between them without losing any significant information.

Lemma 4

a. Given a relatedness relation R let s be the subject matter assignment associated with it. Let R_{s} be the relatedness relation associated with s. Then $\mathsf{R} = \mathsf{R}_{\mathsf{s}}$.

b. Given a subject matter assignment s let R be the relatedness relation associated with it. Let s_{R} be the subject matter assignment associated with R. Then $\mathsf{s}(A) \cap \mathsf{s}(B) \neq \varnothing$ iff $\mathsf{s}_{\mathsf{R}}(A) \cap \mathsf{s}_{\mathsf{R}}(B) \neq \varnothing$.

Proof: a. $\mathsf{R}_{\mathsf{s}}(A, B)$ iff $\mathsf{s}(A) \cap \mathsf{s}(B) \neq \varnothing$
iff there are C, D such that $\{A, C\} \in \mathsf{s}(A)$ and $\{B, D\} \in \mathsf{s}(B)$ and $\{A, C\} = \{B, D\}$ and $\mathsf{R}(A, C)$ and $\mathsf{R}(B, D)$
iff $A = B$ so $\mathsf{R}(A, B)$, or $A \neq B$, so $C = B$ and $\mathsf{R}(A, B)$.

b. This follows from part (a) by passing to the relatedness relation associated with s_{R}. ■

We don't claim in Lemma 4 that $\mathsf{s} = \mathsf{s}_{\mathsf{R}}$. All we can be sure of is that s_{R} preserves the two-way overlap from s. The following two subject matter assignments result in the same s_{R} in Lemma 4.

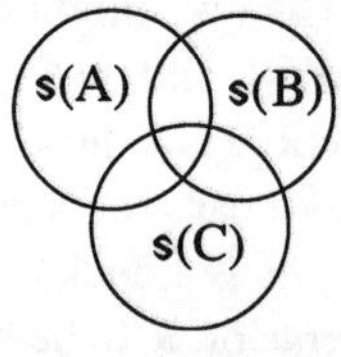

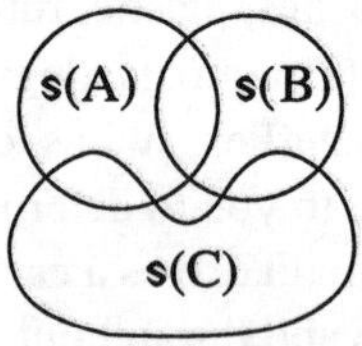

In the left one there is three-way overlap between $\mathsf{s}(A)$, $\mathsf{s}(B)$, and $\mathsf{s}(C)$, whereas in the right hand one there isn't. Because we are investigating only unary and binary connectives at this stage, the difference between these two will be inessential.

E. Truth-Tables

How are we to determine the truth-value of a complex proposition? For the same reasons as for classical logic (§II.A) I am going to invoke the Fregean assumption:

> The truth-value of a complex proposition is determined by its form and the properties of its constituents.

And for the same reasons as before (§II.B) we should invoke the Extensionality Consideration.

> If two propositions have the same semantic properties then they are indistinguishable in any semantic analysis, regardless of their forms. So if A is part of C, then the truth-value of C depends only on the semantic properties and not the form of A, except insofar as the form of A determines the semantic properties of A.

We have already decided with the Relatedness Abstraction which semantic properties of the constituents are to be taken into consideration. So we may conclude that each connective is interpreted as a function of the truth-values of and relatedness between its parts. That is, the connectives will be interpreted as *truth-and-relatedness functions*.

What is the function appropriate for negation? The arguments I gave for the independence of subject matter from truth-value should convince you that the table should be the same as for classical negation.

A	$\neg$A
T	F
F	T

It also seems to me that we should be allowed to conjoin any two propositions without regard to their subject matters: using $\wedge$ is like making a list, for example, 'Ralph is a dog $\wedge$ $2+2=4$ $\wedge$ light bulbs have copper filaments'. So the table I will take for conjunction is the classical one.

A	B	A $\wedge$ B
T	T	T
T	F	F
F	T	F
F	F	F

I believe that these tables are the best choices, but you can view them as an application of a simplicity constraint: the simplest way in which we can take account of subject matter relatedness is to let it affect only the one connective that originally worried us, the conditional.

When is A$\rightarrow$B true? If A and B are related then the usual **PC** truth-table should apply, as suggested in §A: **PC** models are ones in which all propositions are taken to be related. If A and B are unrelated, then A$\rightarrow$B is false: when we wanted to reject 'If Ralph is a dog then $2+2=4$' the question of whether 'Ralph is a dog' is true didn't arise. The lack of subject matter overlap in itself disqualifies A$\rightarrow$B from being true, regardless of the truth-values of A and B. Thus the table for the conditional is

A	B	R(A,B)	A $\rightarrow$ B
any values		fails	F
T	T	holds	T
T	F		F
F	T		T
F	F		T

I call this the *truth-table for the* (*subject matter*) *related conditional*.

Should we take account of relatedness in formalizing 'or'? If not, then we can represent it as $\neg(\neg A \wedge \neg B)$, the truth-functional inclusive 'or'. If we want a related 'or' we can represent it as $\neg A \rightarrow B$ which is true iff A is related to B and at least one of A, B is true. Because these two choices are definable, there seems to me no reason to introduce a controversy here by making a definitive choice for $A \vee B$.

F. The Formal Semantics

1. Models based on relatedness relations

We've made enough agreements now to collect them and present a logic.

A *subject matter relatedness model* is

I

$$L(p_0, p_1, \ldots \neg, \rightarrow, \wedge)$$

$\downarrow$ } realization

$$\{p_0, p_1, \ldots, \text{complex propositions formed from these using } \neg, \rightarrow, \wedge\}$$

$\downarrow$ } v, R, and truth-tables

$$\{\mathsf{T}, \mathsf{F}\}$$

Here $p_0, p_1, \ldots$ are propositions in English which we take to be atomic; v is a way of assigning truth-values to these; and R is a relatedness relation on them. We extend R to all complex propositions inductively by R1–R5, or equivalently by: $\mathsf{R}(A,B)$ iff for some p_i in A, some p_j in B, $\mathsf{R}(p_i, p_j)$. Then v is extended inductively to all propositions by using the classical tables for $\neg$ and $\wedge$ and the table for the related conditional for $\rightarrow$ from the previous section. A proposition A of the semi-formal language is *true* if $\mathsf{v}(A) = \mathsf{T}$, *false* if $\mathsf{v}(A) = \mathsf{F}$.

Since every atomic proposition is abstracted in our model to only its truth-value and subject matter (in terms of its place in a relatedness relation) we can simplify a model of type **I** in the same way we did for classical logic.

II

$$L(p_0, p_1, \ldots \neg, \rightarrow, \wedge)$$

$\downarrow$ v, R, and truth-tables

$$\{\mathsf{T}, \mathsf{F}\}$$

Here v is a way to assign truth-values to the propositional variables; R is a symmetric and reflexive relation on the variables which is extended to all wffs by R1–R5, or equivalently by: $\mathsf{R}(A, B)$ iff for some p_i in A, some p_j in B, $\mathsf{R}(p_i, p_j)$; and v is extended to all wffs by the tables for classical $\neg$ and $\wedge$ and the related conditional for $\rightarrow$. Thus any difference between two models of type **I** is obliterated if they result in the same model of type **II**.

I will usually refer to a relatedness model as $\langle \mathsf{v}, \mathsf{R} \rangle$, where R is either specified on all wffs or only on the propositional variables. Then a *wff* A *is true in*

$<\upsilon,R>$ if $\upsilon(A)=T$, false if $\upsilon(A)=F$. We often write $<\upsilon,R>\models A$, read '$<\upsilon,R>$ validates A', for $\upsilon(A)=T$, and $<\upsilon,R>\not\models A$ for $\upsilon(A)=F$.

Up to this point we have made no infinitistic assumptions, either about the formal language or the semantics. To have full generality and to freely use mathematics in the study of this logic, and for all the reasons and with all the provisos for mathematizing classical logic (§II.F), we will view the collection of propositional variables, PV, and the formal language as completed wholes. And we will further abstract our models.

The Fully General Relatedness Abstraction Any function $\upsilon: PV \to \{T,F\}$ and any symmetric, reflexive relation $R \subseteq PV \times PV$ together form a model $<\upsilon,R>$ when R is extended to all wffs by R1–R5 and υ is extended to all wffs by the truth-tables.

Thus we not only obliterate any difference between models of type **I**, we no longer care where a model of type **II** comes from.

The notions of validity, semantic consequence, and semantic equivalence can now be defined with respect to these models for both propositions and wffs in the same manner as they were for the classical case (§II.F). We say that a wff is a *subject matter relatedness tautology,* or *relatedness tautology* for short, if it is valid: for every $<\upsilon,R>$, $<\upsilon,R>\models A$. (*Subject Matter*) *Relatedness Logic,* denoted **S**, is defined as the collection of all relatedness tautologies. When we want to remind ourselves that these notions are defined with respect to relatedness models (as opposed to, say, classical ones) we'll preface them with '**S**-' as in '**S**-validity', or '**S**-tautology' and notate the semantic consequence relation as '$\models_{\mathbf{S}}$'.

2. Models based on subject matter assignments

Using the observations of §D we can describe **S** in terms of subject matters, too. I will skip directly to the fully general abstraction.

Given any countable set $S \neq \varnothing$ together with any function $s: PV \to$ subsets of S such that $s(p) \neq \varnothing$ for all variables p, and any function $\upsilon: PV \to \{T,F\}$, we will call $<\upsilon,s>$ a *set-assignment model for* **S** if s is extended to all wffs via $s(A) = \bigcup\{s(p)\colon p \text{ appears in } A\}$ and υ is extended to all wffs by the classical tables for $\neg$ and $\wedge$ and the relatedness table for $\to$ reading '$s(A) \cap s(B) \neq \varnothing$' for '$R(A,B)$'. We define $<\upsilon,s>\models A$ iff $\upsilon(A)=T$. Then from Lemma 2, given any model $<\upsilon,R>$, if s is the subject matter set-assignment associated with R we have $<\upsilon,s>\models A$ iff $<\upsilon,R>\models A$. And similarly, by Lemma 3, given $<\upsilon,s>$, if R is the relatedness relation associated with s then $<\upsilon,R>\models A$ iff $<\upsilon,s>\models A$. Hence if we define the notions of validity, semantic consequence, etc. with respect to set assignment models for **S** they will coincide with those we define for relatedness models. Therefore, I will use both kinds of models as the occasion arises, calling them collectively *models for* **S**.

3. The logic R

In *Epstein, 1979,* I called **S** 'Symmetric Relatedness Logic'. That distinguished it from **R** which is the collection of tautologies of models <v, R> defined as above except that symmetry is no longer required of R. That is, the class of relations on Wffs which can be used in models of **R** is characterized by conditions R1, R2, R4, R5 (p. 67), and

R6. R(B, A) iff R(B, ¬A)
R7. R(B∧C, A) iff R(B→C, A)
R8. R(B∧C, A) iff R(B, A) or R(C, A)

Alternatively, we take any reflexive $R \subseteq PV \times PV$ plus any $v: PV \to \{T, F\}$ to establish a model, extending R to all wffs by R1, R2, and R4–8. At that time I called **R** 'Relatedness Logic', but I now think it more apt to call it *Nonsymmetric Relatedness Logic*.

Nonsymmetric Relatedness Logic was devised by D. Walton, R. Goldblatt, and myself to use in action theory, see *Walton, 1979* . Walton, *1985*, has also applied it to analyses of *ad hominem* fallacies. I will use the term *relatedness logic* (without capitals) to refer to both systems and the semantic analyses of this chapter. Only when both systems **R** and **S** are under consideration at the same time will I use the full names 'subject matter relatedness logic', 'subject matter relation', etc.

G. Relatedness Logic Compared to Classical Logic

1. $\mathbf{S} \subset \mathbf{PC}$

I said earlier that the classical logician, in justifying **PC**, views every pair of propositions as logically related (cf. *Bennett, 1969*). Or at least he will argue that in any useful model of **PC** every two propositions are related (see the quote from Quine in §A, p.63). The semantics for subject matters reflect this. Let U denote the *universal relation* on Wffs, that is, for every A and B we have U(A, B). This is a relatedness relation. In any model <v, U> the relation contributes nothing to the evaluation of the truth-value of a wff: $\langle v, U\rangle \models_{\mathbf{S}} A$ iff $v \models_{\mathbf{PC}} A$. Hence the collection of wffs true in all models <v, U> is exactly **PC**. So $\mathbf{S} \subseteq \mathbf{PC}$.

This is as we want, for we have put further conditions for the formalization of an 'if ... then ...' proposition to satisfy for it to be true in the semantics for **S**. Many **PC**-tautologies fail to be **S**-tautologies, in particular the paradoxes of classical logic as we shall see momentarily. So $\mathbf{S} \subsetneq \mathbf{PC}$.

However, if we think of **PC** as formalized in the language of ¬ and ∧ then **S** can be viewed as an extension of **PC** in the language of ¬, ∧, and →. If we had chosen a different symbol for the relatedness formalization of the conditional then our language would have suggested this point of view. I believe it is the wrong way to think of **S**, for both **S** and **PC** are formalizing the same English propositions, interpreting 'if ... then ...' differently, **S** more stringently than **PC**, so it is proper to think of $\mathbf{S} \subset \mathbf{PC}$.

It's also possible to see **S** as definable within **PC** via a translation given in §X.A.4.

2. Some classical tautologies which aren't relatedness tautologies

Before I exhibit some classical tautologies which fail to be relatedness tautologies, I will show how we can decide for any wff whether it is a relatedness tautology or not. The method is a variation on the one for classical logic (§II.G.5). Given a wff A in $L(p_0, p_1, \ldots \neg, \to, \wedge)$ it is mechanically computable (though for long wffs unfeasible) to list each of the finite number of ways to assign truth-values to the variables appearing in A and to symmetrically and reflexively relate those variables, and then to evaluate the wff for each assignment by the tables for $\neg, \wedge, \to$. If there is no assignment for which the wff comes out false, then A is a relatedness tautology. If there is one, then A is not a relatedness tautology, for any such valuation and binary symmetric reflexive relation can be extended to all variables, for example, assign F to all other variables, and, except for relating each variable to itself, relate no others. In that model A is false.

Let's now turn to some examples of schemas of wffs which are **PC**-tautologies to see if they are **S**-tautologies. In falsifying some of these I will give general schematic conditions on a model which ensure that the schema is false.

1. $B \to (A \to B)$
This is not a relatedness tautology for it is false in those models where $v(B) = T$ and $R(A, B)$ fails.

2. $\neg A \to (A \to B)$
This is false if $v(A) = F$ and $R(A, B)$ fails.

These "paradoxes" of **PC** are resolved by Relatedness Logic. In any reasonable model the formalization of 'If Socrates was not French, then if Socrates was French then the sum of any two odd integers is even' will be false because 'Socrates was French' and 'the sum of any two odd integers is even' are unrelated.

The "paradoxes" of strict implication also fail in **S**.

3. $(A \wedge \neg A) \to B$
4. $A \to (B \to B)$
Each is true in a model $\langle v, R \rangle$ iff $R(A, B)$ holds in that model.

Thus the formalization of 'If Socrates was Greek and Socrates was not Greek, then 2+2=4' is true in a model just in case 'Socrates was Greek' has some subject matter in common with '2+2=4'.

In Relatedness Logic a model is sensitive to the placement of premisses.

5. *Exportation* $((A \wedge B) \rightarrow C) \rightarrow (A \rightarrow (B \rightarrow C))$
This is false if A, B, C are all true, R(A,C) holds, and R(B,C) fails.

6. *Importation* $(A \rightarrow (B \rightarrow C)) \rightarrow ((A \wedge B) \rightarrow C)$
This is false if A is false, R(A,B) holds, and both R(A,C) and R(B,C) fail.

7. $(A \rightarrow (B \wedge C)) \rightarrow ((A \rightarrow B) \wedge (A \rightarrow C))$
This is false if A is false and R(A,B) holds, but R(A,C) fails.

On the other hand, here are some **PC**-tautologies from §II.G.6 which are relatedness tautologies.

8. $(A \wedge (A \rightarrow B)) \rightarrow B$
9. $\neg B \rightarrow ((A \rightarrow B) \rightarrow \neg A)$
10. $(A \rightarrow B) \rightarrow ((A \rightarrow B) \rightarrow \neg A)$

And if we take $A \vee B$ to be interpreted as either the classical or related version of 'or' of §E, then the Disjunctive Syllogism is a relatedness tautology too.

11. $((A \vee B) \wedge \neg A) \rightarrow B$

These examples may suggest to you general observations about the form of **PC**-tautologies in which $\rightarrow$ is the main connective which also are **S**-tautologies. In § V.E I'll give some syntactic criteria for that when I compare **S** to similar logics.

3. On the transitivity of $\rightarrow$

A noteworthy aspect of this logic is that transitivity of the conditional fails. Neither

$$((A \rightarrow B) \wedge (B \rightarrow C)) \rightarrow (A \rightarrow C)$$

nor

$$(A \rightarrow B) \rightarrow ((B \rightarrow C) \rightarrow (A \rightarrow C))$$

are relatedness tautologies. In any model $<v,R>$ each is true for all A, B, C iff R is transitive. Indeed,

$$\vDash_S (p_1 \wedge \neg p_1) \rightarrow (p_1 \wedge (p_2 \wedge \neg p_2)) \text{ and } \vDash_S (p_1 \wedge (p_2 \wedge \neg p_2)) \rightarrow (p_2 \wedge p_3)$$

yet

$$\nvDash_S (p_1 \wedge \neg p_1) \rightarrow (p_2 \wedge p_3)$$

So why not require every R to be transitive? The only transitive relatedness relation is the universal relation on Wffs: for every A, B we have $R(A, A \rightarrow B)$ and $R(A \rightarrow B, B)$, so transitivity will yield R(A,B). Note that we may have that $R \subseteq PV \times PV$ is transitive, yet that still does not guarantee that the extension of R to all wffs is transitive as the example above illustrates.

The argument from transitivity of relatedness to every proposition being related is the one that is often given by classical logicians to establish that there is no nontrivial notion of subject matter relatedness. For instance, 'John loves Mary' is related to 'Mary has 2 apples' which in turn is related to '$2+2=4$'. By the transitivity of relatedness, 'John loves Mary' would be related to '$2+2=4$'. Not surprisingly, if you think that these two propositions are related then you'll probably believe that any two propositions are related. It may be comforting to mathematicians to think that love and mathematics have something in common, but for most of us that sounds like sophistry.

It has been argued that one cannot do logic if '$\rightarrow$' is not transitive, and I think that this is what lies at the bottom of the insistence that relatedness is transitive. Yet here we have a logic based on some (reasonable) intuitions about truth and language in which '$\rightarrow$' is not transitive. T. J. Smiley, *1959,* has also developed a logic in which '$\rightarrow$' is not transitive and in that paper he reviews and argues against the views that this is unacceptable. He says,

> The need for an unrestrictedly transitive entailment-relation for serious logical work is no reason at all against accepting a relation which is not unrestrictedly transitive as being a satisfactory reconstruction of an intuitive idea of entailment. But the need itself is undeniable: the whole point of logic as an instrument, and the way it brings us new knowledge, lies in the contrast between the transitivity of 'entails' and the non-transitivity of 'obviously entails,' and all this is lost if transitivity cannot be relied on. Of course if there is an effective way of predicting when transitivity will hold then most of the objection vanishes; ...
>
> Smiley, *1959,* p. 242

In relatedness logic we do have that the consequence relation (§II.D.2, pp. 19–20) is transitive, for it takes into account only truth-values: if $A \models B$ and $B \models C$, then $A \models C$.

E. P. Specker, *1960,* has also discussed the problem of a nontransitive '$\rightarrow$' in relation to a logic of quantum mechanics, suggesting that a further condition needed to infer $A \rightarrow C$ from $A \rightarrow B$ and $B \rightarrow C$ is that A and C be simultaneously decidable.

H. Functional Completeness of the Connectives and The Normal Form Theorem for S

Though we didn't take relatedness as a connective, examples (3) and (4) above show how we may define a formula that corresponds to it. I will use (4) as it involves only the conditional.

$$R(A, B) \equiv_{Def} A \rightarrow (B \rightarrow B)$$

Then $<\upsilon,R>\vDash R(A,B)$ iff $\mathsf{R}(A,B)$.

This abbreviation is useful for axiomatizing **S** in the next section. But I do not view 'R' as a connective, only as an abbreviation useful in metalogical investigations.

If one were to take 'R' as a primitive along with $\neg$ and $\wedge$ then one could define $A\rightarrow B$ as $\neg(A\wedge\neg B)\wedge R(A,B)$. Axiomatizing would then be easy: take the axiomatization of **PC** in the language of $\neg$ and $\wedge$ of §II.K.6 (p.57) and add axioms corresponding to conditions R1–R5 which relatedness relations must satisfy. Looked at in this way **S** can be seen as an extension of **PC**, as I discussed in §G.1. But to justify that viewpoint I think one would be obligated to give arguments for why this approach is not based on a use-mention confusion, as I argued in §B.

We can reduce the number of primitives of **S** by defining $\wedge$ in terms of $\neg$ and $\rightarrow$. In **S** we have that

$$A\wedge B \text{ is semantically equivalent to } \neg(A\rightarrow(B\rightarrow \neg((A\rightarrow B)\rightarrow(A\rightarrow B))))$$

Indeed, it is possible to define every truth-and-relatedness connective from $\neg$ and $\rightarrow$, as I'll now demonstrate.

Informally, a truth-and-relatedness functional connective is some connective $\gamma(A_1, \ldots, A_n)$ which is evaluated semantically by a truth table which assigns T or F to the entire formula for each of the various ways T or F can be assigned to the A_i's and the various A_i's related, subject to the condition that they are related reflexively and symmetrically. More formally, γ is an n-ary *truth-and-relatedness functional connective* if it is interpreted semantically as a function from $\{T,F\}^n\times\{T,F\}^m$ to $\{T,F\}$ where $m = n(n-1)/2$, and the ith entry of the sequence for $i\leq n$ indicates whether $\upsilon(A_i)=T$ or $\upsilon(A_i)=F$, while the succeeding entries stipulate whether $\mathsf{R}(A_1,A_2)$, $\mathsf{R}(A_1,A_3), \ldots, \mathsf{R}(A_1,A_n)$, $\mathsf{R}(A_2,A_3), \ldots, \mathsf{R}(A_2,A_n), \ldots, \mathsf{R}(A_{n-1},A_n)$ hold (notated T), or fail (notated F) (this covers all cases since R is always reflexive and symmetric).

Theorem 5 $\{\neg,\rightarrow\}$ is truth-and-relatedness functionally complete for **S**.

That is, in **S** every truth-and-relatedness functional connective is semantically equivalent to a schema in which only $\neg$ and $\rightarrow$ are used.

Proof: We've already shown how to define $\wedge$ and R in terms of $\neg$ and $\rightarrow$. In this proof I'll take $(A\vee B)$ to be defined as $\neg(\neg A\wedge\neg B)$, the truth-functional inclusive 'or'.

If $\gamma(A_1, \ldots, A_n)$ always takes the value F, then it is semantically equivalent to $(A_1\wedge\neg A_1)\vee(A_2\wedge\neg A_2)\vee \cdots \vee (A_n\wedge\neg A_n)$, associating to the right. Otherwise let those sequences which are assigned T by the function for γ be

$$\delta_1 = (\alpha_{11}, \ldots, \alpha_{1n} ; \beta_{11}, \ldots, \beta_{1m}), \quad \ldots \quad, \delta_k = (\alpha_{k1}, \ldots, \alpha_{kn} ; \beta_{k1}, \ldots, \beta_{km})$$

Define for $i = 1, \dots, k$,

$$\text{for } j = 1, \dots, n, \quad B_{ij} = \begin{cases} A_{ij} & \text{if } \alpha_{ij} = \mathsf{T} \\ \neg A_{ij} & \text{if } \alpha_{ij} = \mathsf{F} \end{cases}$$

for $j = 1, \dots, m$,

$$C_{ij} = \begin{cases} R(A_s, A_t) & \text{if } \beta_{ij} = \mathsf{T} \text{ and the } j^{\text{th}} \text{ entry refers to whether } \mathsf{R}(A_s, A_t) \text{ holds} \\ \neg R(A_s, A_t) & \text{if } \beta_{ij} = \mathsf{F} \end{cases}$$

Lastly define $D_i = (B_{i1} \wedge B_{i2} \wedge \cdots \wedge B_{in}) \wedge (C_{i1} \wedge C_{i2} \wedge \cdots \wedge C_{im})$, associating the conjuncts to the right. Then $\gamma(A_1, \dots, A_n)$ is semantically equivalent to $D_1 \vee \cdots \vee D_k$, associating to the right. ■

Example

A_1	A_2	$\mathsf{R}(A_1, A_2)$	$A_1 * A_2$
T	T	T	F
T	T	F	F
T	F	T	T
T	F	F	F
F	T	T	T
F	T	F	T
F	F	T	F
F	F	F	F

I write $A_1 * A_2$ for $\gamma(A_1, A_2)$. Then $\delta_1 = (\mathsf{T}, \mathsf{F}; \mathsf{T})$, $\delta_2 = (\mathsf{F}, \mathsf{T}; \mathsf{T})$, and $\delta_3 = (\mathsf{F}, \mathsf{T}; \mathsf{F})$. And $A_1 * A_2$ is semantically equivalent in **S** to:

$$((A_1 \wedge \neg A_2) \wedge R(A_1, A_2)) \vee ((\neg A_1 \wedge A_2) \wedge R(A_1, A_2)) \vee ((\neg A_1 \wedge A_2) \wedge \neg R(A_1, A_2))$$

The proof of Theorem 5 yields a normal form for every wff with respect to the semantics for **S**.

Corollary 6 **(*A Normal Form Theorem for* S)**

Given any wff A there is a wff B which contains exactly the same propositional variables and is a truth-functional disjunction of truth-functional conjunctions of either propositional variables or their negations or wffs of the form $R(p, q)$ or $\neg R(p, q)$, such that $\langle v, \mathsf{R} \rangle \vDash A$ iff $\langle v, \mathsf{R} \rangle \vDash B$. Moreover, for every C, $\mathsf{R}(A, C)$ iff $\mathsf{R}(B, C)$.

It's possible to reduce the primitives still further by defining them all from just one, a relatedness version of the Sheffer stroke: $A \mid B$ is true in $\langle v, \mathsf{R} \rangle$ iff at least one of A, B is false and $\mathsf{R}(A, B)$ holds. Then $\neg A$ is equivalent to $A \mid A$, and $A \rightarrow B$ is equivalent to $A \mid \neg B$.

Relative to the subject matter assignment models there is a broader notion of a truth-and-subject matter assignment connective which is sensitive to n-ary overlap between subject matters for $n > 2$. For that notion $\{\neg, \rightarrow\}$ is not functionally complete, as you can show using the diagram of §D.

J. An Axiom System for S

Here is an axiom system for **S**. It is a correction to the one I gave in *Epstein, 1979.* Carnielli, *1987*, has also produced an elegant and natural syntactic formulation of **S** in terms of tableaux.

S, Subject Matter Relatedness Logic
in $L(\neg, \rightarrow, \wedge)$

$R(A,B) \equiv_{Def} A \rightarrow (B \rightarrow B)$

$A \leftrightarrow B \equiv_{Def} (A \rightarrow B) \wedge (B \rightarrow A)$

In axiom 4 the abbreviation $C \vee D$ is used for $\neg(\neg C \wedge \neg D)$; that can be replaced by $\neg C \rightarrow D$ without affecting any of the subsequent results.

axiom schemas

1. $R(A,A)$
2. $R(B,A) \rightarrow R(A,B)$
3. $R(A,\neg B) \leftrightarrow R(A,B)$
4. $R(A, B \rightarrow C) \leftrightarrow (R(A,B) \vee R(A,C))$
5. $R(A, B \wedge C) \leftrightarrow R(A, B \rightarrow C)$
6. $(A \wedge B) \rightarrow A$
7. $A \rightarrow (B \rightarrow (A \wedge B))$
8. $(A \wedge B) \rightarrow (B \wedge A)$
9. $A \leftrightarrow \neg\neg A$
10. $(A \rightarrow B) \leftrightarrow (\neg(A \wedge \neg B) \wedge R(A,B))$
11. $A \rightarrow (\neg(B \wedge A) \rightarrow \neg B)$
12. $\neg(A \wedge B) \rightarrow (\neg(C \wedge \neg B) \rightarrow \neg(A \wedge C))$
13. $\neg((C \rightarrow D) \wedge (C \wedge \neg D))$

rule $\dfrac{A,\ A \rightarrow B}{B}$

The syntactic consequence relation is defined in the usual manner (see §II.H.2, p.41). Rather than subscript it as '$\vdash_S$' I will use '$\vdash$' throughout this chapter to refer to this consequence relation only.

In Chapter V (§V.A.7) I give a nonconstructive proof of the strong completeness of a logic similar to **S**. That proof can be modified in a straightforward way to apply here, and the axioms have been chosen with that in mind. It would be a worthwhile project to give a constructive proof of completeness for **S**, perhaps using the Normal Form Theorem above.

Theorem 7 **(*Strong Completeness for* S)** $\Gamma \vdash A$ iff $\Gamma \vDash A$

For the logic **R** (§F.3) we can obtain a strongly complete axiomatization by modifying the axiom system of **S** in the following way.

R

in $L(\neg, \rightarrow, \wedge)$

Delete axiom schema 2 from the axiom system for **S** and add:

$R(B, A) \leftrightarrow R(\neg B, A)$

$R(B \rightarrow C, A) \leftrightarrow (R(B, A) \vee R(C, A))$

$R(B \wedge C, A) \leftrightarrow R(B \rightarrow C, A)$

It's easy to show that $\mathbf{S} = \mathrm{Th}(\mathbf{R} \cup \{\text{all instances of } R(A,B) \rightarrow R(B,A)\})$.
In **S** the rule of substitution

$$\frac{\vdash A(p)}{\vdash A(B)} \quad \text{where } B \text{ is substituted uniformly for } p$$

is a derived rule: a proof can be given along the lines of the one for Theorem II.2.m, using the Completeness Theorem for **S**.

Recall that in **PC** we also had the derived rule of substitution of logical equivalents

$$\frac{A \leftrightarrow B}{C(A) \leftrightarrow C(B)}$$

(Corollary II.17). This fails in **S**. If we have a model in which $p_1 \leftrightarrow p_2$ is true and $R(p_1, p_3)$ holds but $R(p_2, p_3)$ fails, then $(p_1 \rightarrow p_3) \leftrightarrow (p_2 \rightarrow p_3)$ is false, so that the rule leads from a true wff to a false one. Even the rule of substitution of provable logical equivalents

$$\frac{\vdash A \leftrightarrow B}{\vdash C(A) \leftrightarrow C(B)}$$

fails in **S**. For example, take A to be $\neg(p_1 \wedge (p_2 \wedge \neg p_2))$ and B to be $\neg(p_2 \wedge \neg p_2)$ and C to be $p_1 \rightarrow p_3$ where the substitution is for p_3. To ensure that a substitution of B for A in C leads from a true wff to a true wff in a model we need that A and B have the same truth-value and that they are related to precisely the same subformulas of C, which is more than logical equivalence can guarantee.

K. The Deduction Theorem

For **S**, unlike **PC**, the metalogical formalization of 'follows from' as semantic consequence does not coincide with the validity of the corresponding conditional. For every A, B we have that if $\models A \to B$ then $A \models B$. But the converse fails, for example, $p_1 \models p_2 \to p_2$, but $\not\models p_1 \to (p_2 \to p_2)$. The semantic consequence relation is sensitive only to the relative truth-values of the wffs, while we have formalized an additional property of 'if ... then ...' with '$\to$'. Because we designed the relation of syntactic consequence for **S** to be the same as the relation of semantic consequence, the same break occurs between syntactic consequence and the theoremhood of the corresponding conditional. To summarize

Theorem 8 *a.* If $\models A \to B$, then $A \models B$.
b. If $\vdash A \to B$, then $A \vdash B$.
c. The converses of (a) and (b) fail.

The connective whose validity corresponds to semantic consequence is the material conditional.

***Theorem 9* (*The Material Implication Form of the Deduction Theorem*)**

$\Sigma, A \models B$ iff $\Sigma \models \neg(A \wedge \neg B)$

$\Sigma, A \vdash B$ iff $\Sigma \vdash \neg(A \wedge \neg B)$

We can, however, prove a restricted form of the Deduction Theorem for '$\to$', which I will leave to you.

Theorem 10 If A and B share a propositional variable, then

$\Sigma, A \models B$ iff $\Sigma \models A \to B$

$\Sigma, A \vdash B$ iff $\Sigma \vdash A \to B$

Theorems 8 and 9 might suggest that there is a serious break between our logic and metalogic. But Theorem 10 may be all we need for our metalogic. We could also argue that in our metalogic, being the study of logic, we can rightly assume that all the propositions under consideration share the common subject matter of logic.

Or we could argue (though I won't) that the deducibility of B from A is properly modeled by $\vdash A \to B$. Then the semantic and syntactic consequence relations are only curiosities, a viewpoint taken by modal logicians (§VI.D).

We do not have $\vdash (A \wedge \neg A) \to B$ but we do have $(A \wedge \neg A) \vdash B$, and it is worth noting a particular derivation of that. I'll use the Completeness Theorem in two places, taking $\vee$ to be defined as in the axiomatization.

1. $A \wedge \neg A$ — hypothesis
2. A — *modus ponens* on (1) and axiom 6

3. $\neg A \wedge A$	*modus ponens* on (1) and axiom 8
4. $\neg A$	*modus ponens* on (3) and axiom 6
5. $\vdash A \rightarrow (A \vee B)$	by the Completeness Theorem
6. $\vdash ((A \vee B) \wedge \neg A) \rightarrow B$	by the Completeness Theorem
7. $A \vee B$	*modus ponens* on (2) and (5)
8. $(A \vee B) \wedge \neg A$	*modus ponens* twice on (7), (4), and axiom 6
9. B	*modus ponens* on (6) and (8)

If $A \vdash B$ then there is a chain of related implications that leads from A to B, for there is a proof each step of which is of the form C, $C \rightarrow D$, D where C and D are related in any model in which A is true. Classical logicians argue that such a chain establishes the relatedness of A to B and hence the appropriateness of $\vdash A \rightarrow B$ (see for example *Bennett, 1969*). However, the previous deduction shows that this is wrong, at least relative to the notion of subject matter relatedness which we have formalized. In relatedness logic the deduction is not denied, but exhibiting the chain matters. For any wffs which involve only $\neg$ and $\wedge$ we have $A \vdash_{PC} B$ iff $A \vdash_S B$, for both kinds of models evaluate $\neg$ and $\wedge$ classically.

Is there perhaps some way we could modify the formal notions of consequence to make them coincide with the validity or provability of our formalization of the corresponding 'if ... then ...' proposition? Here are some possibilities.

1. We could change the definition of truth in a model to:

$$\langle v, s \rangle \models A \text{ iff } v(A) = T \text{ and } s(A) = S$$

Then $\models A \rightarrow B$ iff $A \models B$. This is similar to the approach of both modal logic (Chapter VI) and intuitionistic logic (Chapter VII), but I can't see how to reconcile it with the notion of subject matter here.

2. We could change the definition of 'proof', requiring further relatedness conditions to enter in. We already have that R(A,B) is deducible if we have deduced $A \rightarrow B$. If we want to detach B we might also require that R(C,B) be deduced for every C preceding B in the proof. That will guarantee a deduction theorem. But I don't see how to provide a complete axiomatization using this notion of proof. Moreover, strengthenings of the definition of 'proof' along this line have the flaw that we can't automatically use a theorem we've proved in the proof of another theorem.

3. Newton da Costa suggested the following idea. Keep the definition of $\vdash A$ and $\models A$ as before. Then define

$\vdash A$ iff $\vdash A$

$\Sigma \vdash A$ iff there are $A_1, \ldots, A_n$ in Σ such that $\vdash (A_1 \wedge \cdots \wedge A_n) \rightarrow A$

So we have $A \vdash B$ iff $\vdash A \rightarrow B$. The appropriate notion of semantic consequence to correlate to $\vdash$ is

$A \models B$ iff in every model A is related to B and if A is true then B is true

Then $\Sigma \models A$ should mean that in every model there is some C in Σ such that C is related to A and if all the propositions in Σ are true then A is true. In that case we have $\Sigma \models A$ iff $\Sigma \vdash A$.

But the full form of the Deduction Theorem still fails: because Importation fails (example 6 of §F, p.76) we may have $\Sigma \vdash A \rightarrow B$ but $\Sigma, A \nvdash B$, for example, $\{p \wedge \neg p\} \vdash p \rightarrow q$, but if p and q are distinct $\{p \wedge \neg p, p\} \nvdash q$. Similarly, because Exportation fails (example 5 of §F) we may have $\Sigma, A \vdash B$ but $\Sigma \nvdash A \rightarrow B$.

L. A Criticism of Relatedness Logic

A criticism of relatedness logic which I have heard several times is the following. Suppose our formalization of 'If Ralph is a dog then $2+2=4$' is taken as false in a model due to the antecedent and consequent being unrelated. Nonetheless, the formalization of 'If Ralph is a dog and (if 1=1 then 1=1), then $2+2=4$' will be true, since we'll certainly agree that '1=1' and '$2+2=4$' are related.

'Therefore, ... what?' I always want to say. Invariably the offending proposition is pointed to as if it were self-evident that it is counterintuitive. I believe that's based on the view that a tautology has no subject matter and shouldn't contribute to establishing relatedness of propositions. That view is incompatible with the notion of subject matter which has been formalized here, as I pointed out in §C.

What if you wished to formalize a different notion, say one in which **PC**-tautologies have no subject matter, or in which the content of a wff is its consequences in **PC**? How could you go about it? Is there some general form of semantics, some framework we can use as a guide? In the next chapter I propose one.

IV A General Framework for Semantics for Propositional Logics

The general form of semantics which I present here is based on a view of propositional logics as a spectrum, all of the same general form, each based on some aspect of propositions in addition to form and truth-value, with the exception of classical logic which ignores all other aspects. As we vary the aspect we vary the logic.

Much of this has been motivated in the previous chapters, though it is not necessary to have read those. I will repeat some of the arguments and definitions which I have made previously in more specific contexts. I give a completely formal mathematical treatment of the framework in Appendix 2 to this chapter.

A. Aspects of Sentences

1. Propositions

In many kinds of reasoning some aspect of a proposition matters beyond just its truth-value and form. For instance,

its subject matter
the possible ways in which it could be true
its constructive mathematical content
the situations in which the event described by it can be measured
its consequences in some particular logic
its likelihood
its connotation or sense

If we wish to take account of some other aspect of propositions in our formal models of logic how are we to incorporate it as a new primitive?

We should not change our notion of a proposition. Perhaps this other aspect seems significant only for some restricted class of propositions: after all, who would be concerned with the constructive mathematical content of 'Ralph is a dog' ? We might wish to specify that restricted class when we set out a logic, but just as often we discover that class by using the logic. We can do that only if we retain the same meaning for 'a proposition': a written or uttered declarative sentence which we agree to view as being either true or false, but not both (Chapter I) .

You may balk here, thinking of some of these aspects I've mentioned, and say, 'Why should a complex proposition, or for that matter an atomic proposition, be true or false? why not simply nonsensical, or meaningless, or unacceptable?' Why? Because there are only two classes of propositions: those which are true, which correspond to the case, part of which is how we understand the connectives; and those which are not, which are false. Third truth-values, undefined truth-values, dual truth-values, levels of plausibility, all these will be taken into account as the content of a proposition. There are only two mutually exclusive truth-values, and every logician ascribes to something like this in that, in the end, he parcels out propositions into those which are acceptable to proceed on as the basis of reasoning in determining what is the case, and those which are not. Between affirming and denying there is no third choice.

Given a background of agreements which may be adopted for metaphysical, psychological, physical, or pragmatic reasons (it may not be possible to know which of these reasons prevails), then I agree with C. I. Lewis and C. H. Langford, if by 'fact' we understand a true proposition:

> We make an inference upon observation of a certain relation between facts. Whether the facts have that relation or not we do not determine. But whether we shall *be observant of* just this particular relation of facts, and whether we shall *make that relation the basis of our inference* are things which we do determine.
>
> *Lewis and Langford, 1932*, p.258

If you wish you could view this dichotomy of propositions as a simplification, a simplicity constraint. But if it is then I think it's one that is embedded in the very way we perceive the world.

In Chapter I.B and Chapter XI I discuss more fully the notion of a proposition, how we can view a proposition as a type, and why truth-values should be viewed hypothetically. In § I.B and throughout I have explained how the technical work which I present might be interpreted in platonist terms, viewing a proposition as an abstract object which sometimes can be correlated to sentences.

2. The logical connectives

We should also retain the same formal language when incorporating another aspect of propositions into our logic. Almost all reasoning involves 'and', 'not', and to a lesser extent 'or'. And if our logic is to have any notion of 'if ... then ...', or 'implies', any notion whose converse is 'follows from' then we should use the symbol '$\rightarrow$' for it. There are two reasons for this.

First, it will allow us to compare how these various aspects affect our logics. And second, the comparisons are valid because these aren't different notions of 'follows from' but the same notion taking into account different aspects of propositions.

> Now all the Dialecticians agree in asserting that a conditional holds whenever its consequent follows from its antecedent; but as to when and how it follows, they disagree with one another and set forth conflicting criteria for this "following."
>
> Sextus Empiricus, *Against the Mathematicians*, VIII, 112
> translated in *Mates, 1953*, p.97

We aren't talking different languages; we are paying attention to different things. In that sense a translation of one logic into another either is, or ought to be accompanied by a reduction of one aspect of propositions, for example constructive mathematical content, to another, say the possible ways in which a proposition could be true.

So each formal logic owes us an explanation of these four connectives, though by defining some in terms of the others fewer may be taken as primitive. Other connectives may become important relative to a particular aspect of propositions which we are studying, and the general semantics I present in the next section can easily be extended to accommodate them.

Therefore, I am going to assume that we will be working with the same formal language $L(p_0, p_1, \ldots \neg, \rightarrow, \wedge, \vee)$. In Chapter II I've given a formal definition of this language and discussed the relation of it to reasoning in ordinary language.

3. Two approaches to semantics

There are two ways in which we can view how an aspect of propositions can affect the truth-values of compound propositions. If what is of primary interest to us is how the aspect affects the truth-value of '$\rightarrow$' propositions, then we can view it as a primitive relationship between propositions, as when we say that one proposition is relevant to another. That's how Relatedness Logic was first developed in Chapter III.

Alternatively, we can view the primitive as a property of propositions. Each proposition has some content. That's how subject matters were treated in §III.D and §III.F.2

This latter approach seems to me a bit more intuitive to describe in the general case. I'll present it first, then explain the binary relation approach and the connection between the two.

B. Set-Assignment Semantics

1. Models

To give a *model*, or interpretation, of the formal language we first realize $p_0, p_1, \ldots$ as particular sentences in English, $p_0, p_1, \ldots$ which we take to be atomic propositions. That is, we agree to view each as either true (T) or false (F) but not both, and the internal structure of each will not be taken into account in this model. We will symbolize this assumption for a proposition p by writing $\upsilon(p) = \mathsf{T}$ or $\upsilon(p) = \mathsf{F}$. The assignment υ is called a *valuation*.

From these we can form compound *semi-formal propositions* using the formal connectives $\neg, \wedge, \vee, \rightarrow$ on the pattern of the formal language. Thus, 'Ralph is a dog' might be assigned to p_0 (and hence is p_0) and '$2+2=4$' could be assigned to p_1, (and hence is p_1). Then the interpretation of $p_0 \wedge p_1$ is 'Ralph is a dog $\wedge$ $2+2=4$' (cf. §II.C). The purpose of the model is to assign truth-values to all of these compound propositions.

To do that we first choose for this model a nonempty collection S which represents the least bits of *content* in terms of the aspect which is being formalized. Then every proposition, both complex and atomic, is assigned a collection from within S, the method of assignment being denoted s. For example, the modal logics of Chapter VI take S to be a collection of possible worlds and the content of a proposition to be those worlds in which it is true. Often we specify s(A) for every A and then the collection S is given implicitly as the union of those. Depending on what aspect is being formalized there may be only one S for all models (cf. the logic **DPC**, §V.F), or S may be any of a range of specified sets. In §D I discuss the relation between the content of atomic propositions and the content of compound ones formed from them and to what extent that should be a regular functional relationship.

Note that we do not necessarily assume that s is independent of v. In some logics it is, for example, Relatedness and Dependence Logics (Chapters III and V), in some logics not, for example, Heyting's intuitionistic logic (Chapter VII).

In the model we are giving here, the only properties of a proposition which matter are its form, its truth-value, and its content relative to the one aspect being considered. So surely the truth-value of a complex proposition must be a function of these properties of its parts. If not, what else would determine it? If something transcendent occurs when a connective joins two propositions then how are we to reason? Reasoning requires some regularity. We name this assumption.

The Fregean Assumption The truth-value of a compound proposition is determined by its form and the truth-values and contents of its parts.

Here 'parts' is to be understood as 'proper parts' and not the whole.

Clearly the truth-value of a compound proposition must depend on its form, but all the semantic properties of the constituents should be accounted for by their truth-values and contents. Certainly that's true when the constituents are atomic, but to ensure that that is also the case when the constituents are compound we have to invoke the following.

The Extensionality Consideration If two propositions have the same semantic properties then they are indistinguishable in any semantic analysis, regardless of their forms. So if A is part of C, then the truth-value of C depends only on the semantic properties and not the form of A, except insofar as the form of A determines the semantic properties of A.

This rules out, for example, the evaluation of a negation, $\neg A$, by first taking into account the truth-value and content of A, and then counting how many negation signs A begins with, using one analysis for less than three negations and another for more than three. For example, we might argue that any excessive use of 'no' or 'not' in English serves only to emphasize the initial use, as in 'No, no, no, cat's aren't nice'. So the truth-value of $\neg\neg\neg\neg p$ should simply be T if p is F, and F if p is T. In that case we are assigning semantic content to the appearance of the negation signs, and therefore their presence or absence should be noted as part of the content of A.

So we can conclude from the Fregean Assumption and the Extensionality Consideration that the connectives will operate semantically as truth-and-content functions in all circumstances, regardless of whether their constituents are compound or atomic, for the only semantic properties of propositions we are concerned with are truth-value and content. It only remains to decide which truth-and-content functions correspond to the connectives. The conditional is the heart of the matter.

Where there is the correct "connection of meaning" between antecedent and consequent then only the truth-values matter; in that case the classical table is appropriate. Where there fails to be the connection then the conditional is false, for we may not infer truths from this.

What is this connection? It will vary depending on the aspect being modeled. For example, in Relatedness Logic (Chapter III) we require that the antecedent, A, and consequent, B, have some subject matter in common, and hence the relation is $\mathsf{s}(A) \cap \mathsf{s}(B) \neq \varnothing$. For example, we could have a model in which 'Ralph is a dog $\rightarrow$ $2+2=4$' is false because the antecedent and consequent are viewed as having no common subject matter. For most modal logics (Chapter VI) the connection is that the ways in which the antecedent could be true are all ways that the consequent could be true, and so in structural terms it is $\mathsf{s}(A) \subseteq \mathsf{s}(B)$. For a logic of equality of contents designed as a logic of sense and reference (§ V.D) it is $\mathsf{s}(A) = \mathsf{s}(B)$. In each case the connection is taken to be some fixed relationship, B, on contents of propositions. We will write $\mathsf{B}(A,B)$, though it will be clear from context that we mean $\mathsf{B}(\mathsf{s}(A), \mathsf{s}(B))$. Thus the truth-table for the conditional is:

(1)

A	B	B(A,B)	A→B
any values		fails	F
T	T		T
T	F	holds	F
F	T		T
F	F		T

Often content is not considered significant in evaluating negations so that the classical table for $\neg$ applies. However, in some logics the content of a proposition must have some particular property, symbolized here by N, or else its negation cannot be taken as true. For example, in Heyting's intuitionistic logic (Chapter VII) the property is $s(A) = \varnothing$. Thus the general table for negation is:

(2)

A	N(A)	¬A
any value	fails	F
T	holds	F
F		T

Here, too, we write N(A) when N(s(A)) is meant.

I have not encountered a logic in which conjunction can not be interpreted classically, but I see no reason why the contents of the conjuncts should not be taken into consideration. For example, we might want to distinguish between 'He took off his clothes and went to bed' and 'He went to bed and took off his clothes'. Thus the general table for conjunction is:

(3)

A	B	C(A,B)	A∧B
any values		fails	F
T	T	holds	T
T	F		F
F	T		F
F	F		F

Here C is a fixed relation between the contents of propositions, where we write C(A,B) for C(s(A), s(B)). Unless specified otherwise, however, C will be assumed to be the universal relation holding between every two propositional contents.

Also unless specified otherwise ∨ will be a defined connective. When it isn't, its truth-table will be governed by a relation A between contents of propositions: when the relation holds the table for classical inclusive ∨ is used; when the relation fails the alternation is false.

(4)

A	B	A(A,B)	A∨B
any values		fails	F
T	T	holds	T
T	F		T
F	T		T
F	F		F

I've chosen the letters 'N', 'C', 'A' as mnemonics for 'the relations governing the table for negation, conjunction, alternation,' and 'B' to stress the binary aspect of the relation governing the table for the conditional. I will, however, vary the symbols used for these relations to reflect the aspect being considered in particular logics. These are collectively referred to as the *relations governing the truth-tables.*

Using these tables the truth-value of each proposition of the semi-formal language of the model may be uniquely evaluated. We write $\upsilon(A) = \mathsf{T}$ or $\upsilon(A) = \mathsf{F}$ according to whether A is true or false in the model.

Schematically then, a model for our formal language is

I

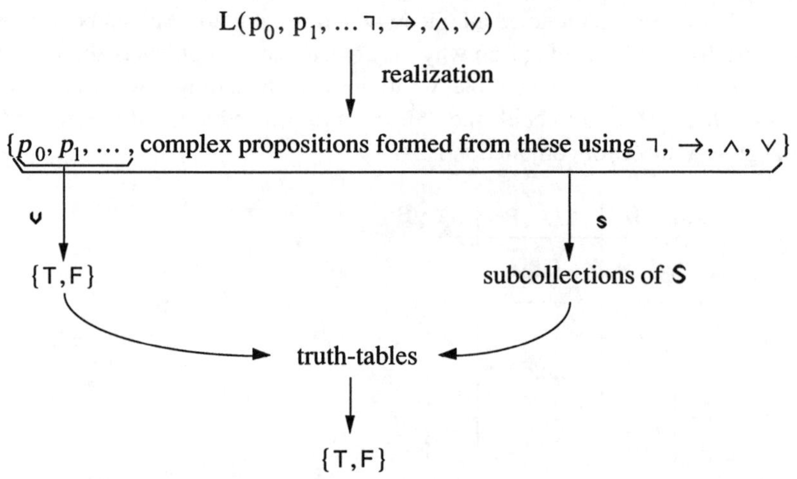

Though we've allowed for each of the four basic connectives to depend only on the truth-values of the constituent propositions by taking the appropriate relation to be universal, we haven't made any provision for them to depend only on content. For the most part, that is outside the scope of this project: such connectives in English seem to me to be part of a project of giving semantics to more of language than we are concerned with in logic. However, there is one such interpretation of the conditional which arises in modal logics (Chapter VI) which roughly corresponds to 'if ... then ...' in the subjunctive mood, usually used with false antecedent (the counterfactual conditional). The subjunctive conditional formed from 'Socrates is alive' and 'abortion is legal' is 'If Socrates were alive, then abortion would be legal'. This connective differs markedly from others used in logic in that the truth-value of the compound does not depend on the truth-values of the "constituent" propositions, as can be seen in the example. Hence, for this notion of 'if ... then ...' *modus ponens* can fail: we cannot conclude B from A and $A \rightarrow B$. This one connective depending only on contents is nonetheless a very weak form of implication. So I'll extend our list of possible tables to include the *weak table for the conditional*:

(5)

A	B	$s(A) \subseteq s(B)$	$A \to B$
any values		fails	F
any values		holds	T

Explicit mention needs to be made when this table is substituted for table (1) above.

2. Abstract models

In a model all that matters about an atomic proposition is its truth-value and the content assigned to it. So we can simplify our models to

II $$L(p_0, p_1, \ldots \neg, \to, \wedge, \vee)$$

$$\downarrow \quad v,\ s, \text{ and truth-tables governed by } \mathsf{B}, \mathsf{N}, \mathsf{C}, \text{ and } \mathsf{A}$$

$$\{T, F\}$$

A wff is true or false in this model according to whether $v(A) = T$ or $v(A) = F$. We have explicitly recognized that all other aspects of the propositions in our model are to be ignored.

What is the simplest, most general way to conduct investigations of these models? Up to this point our models have been built from propositions which are actual English sentences, and content sets are motivated and argued for with respect to the aspect of propositions under consideration, as are the relations B, N, C, and A. Even in diagram **II** the assumption is that the model arose from one as in diagram **I**.

We are not obliged to carry out our logical investigations with any greater abstraction than this. But to achieve full generality, to simplify our proofs and to isolate the structural nature of these models, it is useful and now customary to make a number of assumptions relative to models of type **II**. These assumptions allow us to apply mathematics in our studies.

The first assumption we make is to view the set of propositional variables as a completed whole, PV. We also view the collection of well-formed-formulas as a completed totality, Wffs, treating these mathematically as sets. Then we make the following abstraction, writing 'Sub S' for 'subsets of S'.

The Fully General Abstraction

1. Any function $v : \mathrm{PV} \to \{T, F\}$ is suitable to use in a model as the truth-value assignment to the variables.
2. Any set S and any function $s : \mathrm{Wffs} \to \mathrm{Sub}\, S$ which (together with v) satisfy the structural rules for modeling the aspect under consideration are

suitable to use as the content assignment in a model.

3. Given v and a pair s and S as in (1) and (2), then any relations B, N, C, and A on the sets assigned to the variables which satisfy the structural rules for modeling the relations governing the truth-tables are suitable to use in a model.

To use parts (2) and (3) we need to have proceeded beyond examples or the ability to produce models of type **I** to a structural analysis of contents of propositions.

The Fully General Abstraction obliterates the differences between aspects of propositions which satisfy the same formal rules. For example, the same structural analysis, $s(A) = \bigcup\{ s(p) : p \text{ appears in } A\}$, is used for the referential content of a proposition (§ V.A) and the content of a proposition taken to be its consequences in classical logic (§ V.F).

We could make the Fully General Abstraction and use mathematical tools on our models without assuming anything about infinite completed totalities such as PV, Wffs, or S. That is fairly unusual and in that case I would say that we make a *Finitistic Fully General Abstraction.* More usual is to restrict the sets S to be *countable*, which is the assumption I make in this book unless noted otherwise.

If we make these abstractions it's because we believe that the results we obtain using them when applied to propositions in fully motivated models of type **I** won't contradict our original assumptions and intuitions, that is, our semantics. In §II.F I have discussed more fully the role of mathematical abstraction in logic as well as its relation to a platonist view of propositions.

3. Semantics and logics

What the Fully General Abstraction does is reduce our semantics to a translation from one language, $L(p_0, p_1, \dots \neg, \rightarrow, \wedge, \vee)$, to another, the (informal) language of mathematics. That by itself is no semantics at all, it is just a translation. The fully general abstract semantics are rules governing the form of meanings, comparable to an axiomatization of the logic in the more formal language $L(p_0, p_1, \dots \neg, \rightarrow, \wedge, \vee)$. Another syntactic characterization, from a different point of view, does not give meaning.

Semantics are only given when arguments are made, agreements based on common understandings are reached, our intuitions sharpened so that we can give an analysis in terms of models of type **I**. Questions of meaning, truth, reference, and our relation to the world must be dealt with. Otherwise we have no semantics but only an empty formalism.

To summarize, what I have presented here is the framework for semantics for a propositional logic. It is fleshed out into a logic when:

1. We choose an aspect of propositions to formalize or choose to formalize none except truth-value.
2. We explain that aspect in simple terms, giving examples, developing an intuition, justifying the assumption that *a proposition can be viewed as having this aspect*. In doing so we may come to restrict ourselves to only a certain class of propositions to which this assumption applies, either ordinary language propositions or ones from some technical scientific language. We may also choose to add further connectives to the formal language.
3. We argue that in the context of the deductions we are formalizing it is reasonable to ignore all other aspects of propositions, justifying the abstraction that *the only properties of a proposition which matter to* (*this*) *logic are its form, its truth-value, and this one additional aspect*.
4. We stipulate a class of content-assignments which are appropriate to model this aspect, usually through rules which they must obey; that is, a structural analysis of content is given. In particular, we decide whether content assignments are independent of the assignment of truth-values to the atomic propositions.
5. We explain the connections of meaning which govern the truth-tables and stipulate a class of relations B, C, A, and property N which are appropriate to model them. These classes, too, are usually presented on the basis of a structural analysis.
6. We choose whether to make the Fully General Abstraction of models.
7. We return to our examples and show that the formalization of them and the resulting semantic analysis are reasonable in terms of the original motivation of (2). When our pre-formal intuitions clash with the formal analysis we may explain why our semantics are appropriate to act prescriptively.

The *logic* then is this analysis: the syntax and the formal semantics developed through (1)–(7), and I will often use that term in this manner.

For the rest of this chapter I am going to discuss structural questions of propositional logics under the assumption that in each case we do make the Fully General Abstraction. The models I will refer to are of type **II**. Indeed, *throughout this volume I will assume the Fully General Abstraction for every logic unless I specify otherwise.*

Also, throughout this book I'll assume that arguments need to be given for why negation and conjunction should *not* be interpreted classically. Changes from classical logic need to be argued for, not conformity with it.

4. Semantic and syntactic consequence relations

There are two major concerns of formal logic which result in other uses of the term 'a logic'.

As logicians we cannot specify any particular model as being an accurate interpretation of the world. All we can do is say what compound propositions are true given a specific truth- and content-assignment. However, certain propositions will be true regardless of the assignment. They are true (in this logic) due to their (propositional) form only. Their schematic form as wffs will be true in every model. These are the "universal truths" (of this logic). We call the class of wffs true in every model the *tautologies* of this logic and say that they are *valid.* I have discussed more fully the notion of validity in §II.D.1, and that discussion applies generally if references to classical logic are deleted.

When the classification of propositions true due to their form only becomes the main focus of an investigation then the class of tautologies is sometimes referred to as the logic, and on occasion I will use the name of a logic to refer to that class.

Although as logicians we cannot say what *is* true, we can say what follows on the basis of certain propositions. There are two distinct semantic ways we can do that.

First, we may argue that this is precisely what our formalization of 'if ... then ...' was meant to capture: 'B follows from A' just means $A \rightarrow B$ is valid.

Alternatively, we may say that what we are concerned with is whether B follows from the assumption of the truth of A, or from the truth of a collection of propositions Γ. Given a logic **L**, we write $A \vDash_{\mathbf{L}} B$ to mean that in every model of **L** in which A is true so is B. And we write $\Gamma \vDash_{\mathbf{L}} B$ to mean that in every model in which every A in Γ is true, so is B. This is the *semantic consequence relation for* **L**. In this notation $\vDash_{\mathbf{L}} A$ means that A follows from no propositions and hence is true in all models; that is, A is a tautology. I have discussed these notions as well as semantic equivalence and the relation of them to deduction in ordinary language in §II.D.2 in a way which generalizes if references to classical logic are deleted. In §II.G.2 I have given a more formal presentation, and Theorem II.2.(a)–(g) (p.31) applies generally to the semantic consequence relation of any logic in this book.

Which is the more appropriate way to formalize the notion of B following from A? The answer to that depends upon the logic being considered. For modal logics (Chapter VI) it is usually argued that it is the validity of an '$\rightarrow$' wff. For classical logic (Chapter II) it is arguably the semantic consequence relation. For some logics no choice need be made, for the two formalizations are equivalent: $A \vDash_{\mathbf{L}} B$ iff $\vDash_{\mathbf{L}} A \rightarrow B$. This is a version of the Semantic Deduction Theorem, which I discuss in §E below.

The phrase 'A *implies* B' or 'A *entails* B' is often used to indicate that B follows from A. Whether that phrase is an appropriate reading of $A \vDash_{\mathbf{L}} B$ or of $\vDash_{\mathbf{L}} A \rightarrow B$ will depend on the logic in question. There has been much discussion of whether implication is a metalogical notion or a connective and whether formalizing it as a connective involves a use-mention confusion, a topic I treat in §II.D.2, §III.K, §VI.B.2, and §VI.D.

Given a class of models and associated notion of validity there may be other ways to define a semantic consequence relation which is peculiar to the logic in question. In §III.K, for instance, other proposals for a semantic consequence relation for Relatedness Logic are discussed; Smiley, *1976*, discusses three different consequence relations for a particular many-valued logic. Usually the intention is to ensure a semantic deduction theorem. Hence the term 'a logic' is sometimes reserved for the chosen semantic consequence relation. Throughout this book the semantic consequence relation for a logic will always be derived from the notion of a wff being true in a model as defined above. Wójcicki, *1989*, discusses consequence relations more generally.

There is yet another conception of logic which is concerned primarily with form and not meaning. As I discussed in §II.H.1 and §II.K.1, earlier in this century the notion of a logic was usually understood solely in terms of formal systems of axioms and rules which were meant to formalize the notion of proof. In §II.H I present the general form of one such formalization due to Hilbert which I use throughout this book. On this conception the notion of B following from A is understood to mean that we can prove B on the hypothesis of A. A logic is then either the collection of theorems, that is wffs which can be proved from the axioms only, or the syntactic consequence relation. It will be clear from context whenever I use the term in one of these senses.

In §II.H.4 I have summarized these four ways to present a logic.

C. Relation Based Semantics

In a set-assignment model of type **II** the content of a proposition is of significance only in establishing what other propositions it is related to by the relations on contents B, C, A, and whether the content has property N. Hence a model could be given by stipulating these relations and property of propositions, ignoring the contents altogether.

Given any binary relations B, C, A on propositions (wffs), and N a property of propositions (wffs), and assignment v of truth-values to the atomic propositions (propositional variables) we have a model

II $L(p_0, p_1, \ldots \neg, \to, \land)$

$\downarrow$ v, s, and truth-tables governed by B, N, C, and A

$\{T, F\}$

The truth-value assignment is extended to all propositions (wffs) in the model by tables (1)–(4).

If models are stipulated by giving only relations B, C, A and property N, then I will say that we have *relation based semantics*. That is how Relatedness Logic was first presented in Chapter III. The Fregean Assumption and the Extensionality Consideration yield that the truth-value of a compound proposition is determined by its principal form, the truth-values of its parts, and the relations holding between its parts. That is, the connectives are truth-and-relation functions (cf. §III.H). I will leave to you the modification of the summary of how to give semantics ((1)–(7) of §B.3, p.95) and the Fully General Abstraction so as to apply here.

I will say that a set-assignment semantics and a relation based semantics *give rise to the same logic* or simply *are the same logic* if their semantic consequence relations are the same. This ignores the fact that there may be no simple translation between the two kinds of semantics. In §X.B.6 I return to the question of when we are justified in calling two different presentations 'the same logic'.

Every set-assignment semantics gives rise to a relation based semantics. What is the relation between these two?

Q1 If the set-assignment semantics have a simple presentation, do the relation based semantics have a simple presentation, too?

Q2 Given a relation based semantics having a simple presentation, does it arise from some set-assignment semantics? Does it arise from some set-assignment semantics having a simple presentation?

To answer these we need a precise definition of 'simple presentation'. On the other hand, trying to answer these may lead to what definitions we choose. I'd like to put forward tentative definitions based on my experience working with various applications of these semantics.

D. Semantics Having a Simple Presentation

Let's first look at set-assignment semantics in which only the relation governing the table for '$\to$', B, is not universal, and s and B are independent of v. An example is Subject Matter Relatedness Logic of Chapter III.

What should we mean by 'a simple presentation for B'? B should be a relation on sets assigned to formulas which satisfies some equation using the Boolean operations union, intersection, and complementation, as well as possibly some (small) finite number of designated subsets of S, that is, some parameters.

Example

1.	B(A,B) iff $s(A)\cap s(B)\neq\varnothing$	Subject Matter Relatedness, Chapter III
2.	B(A,B) iff $s(A)\supseteq s(B)$	Dependence Logic, Chapter V
3.	B(A,B) iff $s(A)=s(B)$	**Eq**, Chapter V

4. B(A,B) iff $(s(A) \cap T) \subseteq (s(B) \cap T)$ for a parameter $T \subseteq S$
modal logic **T**, Chapter VI

Note that containments, as in (2), can be defined using equations and the Boolean operations.

The rules which pick out the class of set-assignments should similarly be a finite number or possibly a simple recursive set of equations between sets assigned to formulas and their subformulas, using the Boolean operations and a finite number of parameters, or truth-functional conditions involving such equations. For instance, many modal logics in Chapter VI use: if $s(A) = S$ then $s(A \to (A \to A)) = S$. Truth-functional connectives are the appropriate ones here because working in the metalogic it's reasonable to make the mathematician's assumption that all propositions about wffs of the formal language and semantics are "connected in meaning."

For comparable relation based semantics in which only B is not universal and B is independent of v, a simple presentation of the semantics ought to be a finite number or a simple recursive set of formulas which determine the class of relations. Each formula should be built from the symbol B for the relation, variables for wffs and their subformulas, truth-functional propositional connectives, and possibly a finite (or recursive) set of parameter wffs. As an example, one of the formulas establishing the class of relations B for Relatedness Logic (Chapter III) is B(A, B→C) iff B(A,B) or B(A,C).

If the semantics allow v to enter into the determination of s or of B then '$v(A) = T$' and '$v(A) = F$' should also be allowed into the conditions, as it is for Heyting's intuitionistic logic (Chapter VII) which uses 'if $v(A) = T$ then $s(A) = S$'. In that case it may be difficult to present simple relation based semantics since this condition may not be expressible in terms of the relations governing the truth-tables.

I'll let you generalize my remarks to semantics in which C, N, or A is not universal.

Should we require a version of the Fregean Assumption for contents of wffs? For example,

(A) The content of a complex proposition is determined by its form and the contents of its proper parts, as well as possibly the truth-values of the parts.

My first inclination is to say, 'yes'. But what if the answer to Q2 is 'no'? Then we might have a simple relation based semantics, perhaps satisfying the following version of a Fregean Assumption and yet have no simple set-assignment semantics which satisfy (A).

(B) Whether two propositions are related is determined by their forms and the relations between their proper parts, as well as possibly the truth-values of the parts.

Nonsymmetric Relatedness Logic (reflexivity only) of Chapter III satisfies (B) yet I don't know any set-assignment semantics for it, much less ones which satisfy (A). Dependence Logic (§V.A) has very simple set-assignment semantics satisfying (A), yet the corresponding relation based semantics do not satisfy (B).

Moreover, in coming to grips with some difficult notion, some aspect of propositions which is vague or hard to formalize, we may wish to begin our investigations using a logic in which there is no, or at least no simple functional relation between the set assigned to a complex proposition and the sets assigned to its parts. Here I'm thinking of some preliminary investigations I made of the referential content of negated propositions where I could at best say that $\mathsf{s}(\neg A) \subseteq \overline{\mathsf{s}(A)}$.

And sometimes, though the semantics do not satisfy (A) or (B), the notion of content may nonetheless be reducible to some other primitive which does satisfy a functional relation between the parts and the whole. For example, modal logics (Chapter VI) where $\mathsf{s}(A)$ is understood as the possible worlds in which A is true and a simple translation to Kripke models is given, or the logic of §V.F in which the content of a proposition is its consequences in **PC**. I would like to think that where Fregean assumptions like (A) and (B) both fail there is (or should be) some underlying, more primitive notion to which contents can be reduced and which satisfies a functional relation between the part and the whole.

It's worth remarking that I haven't allowed quantification in the formulas or equations of semantics having a simple presentation, other than implicit universal quantification over all wffs. This has to do with the possibility of representing the relation syntactically and axiomatizing the logic, as discussed in questions Q5 and Q6 below.

E. Some Questions

In the Introduction I have discussed the general plan of this book: to present examples of logics, to explain why the general framework gives good readings of these, and to account for the diversity and objectivity of logics.

Here I'd like to present a technical guide for the following chapters by posing some structural questions about the general form of semantics I've presented. Some of them are admittedly vague for the same reason as the previous two were: the best way of making the question precise may come from trying to answer it.

Q3 The Deduction Theorem

In some logics the two ways of semantically formalizing that B follows from A turn out to be equivalent: $A \vDash B$ iff $\vDash A \rightarrow B$. This is called *The Semantic Deduction Theorem.*

Relatedness Logic (Chapter III) does not satisfy that, but does satisfy: $A \vDash_S B$ iff $\vDash_S \neg(A \wedge \neg B)$. In general it is useful and important to know when there is a connective, whether primitive or defined, whose validity corresponds to semantic consequence. Can conditions be given for when relation based or set-assignment semantics determine a logic which satisfies a

Semantic Deduction Theorem:

There is a schema $\theta(A,B)$ such that $A \vDash B$ iff $\vDash \theta(A,B)$.

If we have a Semantic Deduction Theorem then we can represent any finite semantic consequence in terms of validity. By induction we can show that for every $n \geq 1$ there is a schema θ_n such that for all $A_1, \ldots, A_n, B$,

$$\{A_1, \ldots, A_n\} \vDash B \text{ iff } \vDash \theta_n(A_1, \ldots, A_n, B)$$

For θ_1 can be θ, and $\theta_n(A_1, \ldots, A_{n+1}, B)$ can be $\theta(A_1, \theta_n(A_2, \ldots, A_{n+1}, B))$. We call such a reduction of finite consequences to validity *a Semantic Deduction Theorem for Finite Consequences*.

Only if the *semantics are compact*, that is, for all Γ and A, $\Gamma \vDash A$ iff there is some finite $\Delta \subseteq \Gamma$ such that $\Delta \vDash A$, can we use a Semantic Deduction Theorem to reduce the entire semantic consequence relation to the set of tautologies. The standard way to show that semantics are compact is to prove a strong completeness theorem for a particular axiomatization, since our notion of syntactic consequence is compact (§II.H.2). The proof of strong completeness, as opposed to completeness or finite strong completeness (§II.H.4), usually requires nonconstructive infinitistic assumptions (cf. §II.J.3 and §II.K.1).

In demonstrating that a particular axiomatization is complete it may be useful to show that the syntactic consequence relation satisfies a

Syntactic Deduction Theorem

There is a schema $\theta(A,B)$ such that $A \vdash B$ iff $\vdash \theta(A,B)$.

from which follows a Syntactic Deduction Theorem for Finite Consequences. I reserve the term '*The* Syntactic Deduction Theorem' for the case where $\theta(A,B)$ is $A \rightarrow B$. In general, whether a logic satisfies a Syntactic Deduction Theorem is not a separate question from the semantic one.

Blok and Pigozzi, *1982* and *198?*, have studied these questions from an algebraic viewpoint (see Q8 below), and Porte, *1982*, has an extensive bibliography in his survey of deduction theorems.

Q4 Functional completeness of the connectives

We began by formalizing 'not', 'and', 'or', 'if ... then ...' as $\neg, \wedge, \vee, \rightarrow$. In classical logic we showed that every truth-functional connective could be defined in terms of these (§II.G.3). For Relatedness Logic these four interpreted connectives also sufficed to define every truth-and-relatedness connective.

Given a semantics determined by a class of quadruples of relations B, C, A, and N governing the truth-tables, we'll say that θ *is a truth-and-relation functional connective* for these semantics if there is a table which determines the truth-value of $\theta(A_1, \ldots, A_m)$ in a model solely on the basis of the truth-values of the A_i's and whether or not the various A_i's are related by B, C, A, or have property N. When the relations arise from set-assignments we say that θ is a *truth-and-content functional connective*.

Given a logic (a language plus semantics) we'll say that a collection of connectives of the language, $\{\theta_1, \ldots, \theta_n\}$, is *functionally complete* if every truth-and-relation (truth-and-content) functional connective can be defined in terms of these. That is, given a truth-and-relation functional connective θ (if necessary extending the language to include it) there is a schema S using only $\theta_1, \ldots, \theta_n$ such that $\theta(A_1, \ldots, A_m)$ and $S(A_1, \ldots, A_m)$ have the same truth-value in every model. That is, $\theta(A_1, \ldots, A_m)$ is semantically equivalent to $S(A_1, \ldots, A_m)$.

The primitive connectives for Łukasiewicz's three-valued logic (§VIII.C.1) are not functionally complete. What structural conditions on the semantics of a logic ensure that the connectives chosen as primitive are functionally complete?

There are other notions of functional completeness specific to the original semantics of other logics studied in this volume for which a stronger notion of functional completeness is needed in the general framework. For example, in the paraconsistent logic $\mathbf{J_3}$ the connective which always takes value 1/2 cannot be defined in terms of the original semantics (§IX.B.3); nonetheless, we can find a schema which is semantically equivalent to it. That is because the notion of definability above takes into account only the truth-values of wffs. In §X.B.7 I discuss a stronger notion of definability of connectives in terms of semantically faithful translations.

Q5 Representing the relations governing the truth-tables within the formal language
Certain truth-and-relation functional connectives are especially important to represent in the language. To demonstrate the functional completeness of the connectives and to axiomatize Relatedness Logic (Chapter III), we first showed that we could find a formula R(A, B) in the language such that R(A, B) is true in a model iff R(A, B) holds, where R is the relation governing the truth-table for '$\rightarrow$'. Given a logic, that is, a language with relation or set-assignment semantics, when is there a formula θ(A, B) which is true in a model iff B(A, B) holds, and similarly for the other relations?

Q6 Characterizing the relations governing the truth-tables in terms of valid schemas
Even if we can represent the relations governing the truth-tables as defined connectives, a further problem often remains before we can axiomatize the logic: can the stipulations which determine the class of relations be given in propositional form? Sometimes it's not hard: in Relatedness Logic (Chapter III) conditions such

as R(A,¬B) iff R(A,B) can be represented via the defined connective R as R(A,¬B) ↔ R(A,B). Sometimes it's surprisingly complicated even when the set-assignment semantics are simple, as in the logic of equality of contents of §V.D. Can some general criteria be given for when there's a positive answer? Better yet, is there some general way to translate from rules governing the relations into propositional formulas?

It seems likely to me that if the equations or formulas defining the relations of the semantics necessarily involve existential quantifiers over propositions then it will not be possible to define the class of relations using propositional schemas.

Q7 Translating other semantics into the general framework

Some of the logics considered in this volume are given set-assignment semantics by a translation from Kripke-style possible-world semantics or many-valued matrices. Are there general translations of these kinds of semantics into set-assignment or relation based semantics? And conversely, when can set-assignment or relation-based semantics be translated into one of these?

Q8 Providing algebraic semantics

Relation based semantics seem to be more general than algebraic semantics. For instance, a Lindenbaum–Tarski algebra cannot be formed for **S**, Subject Matter Relatedness Logic (Chapter III), for neither semantic equivalence, $A \models_S B$ iff $B \models_S A$, nor provable equivalence, $\vdash_S A \leftrightarrow B$, is a congruence relation on the algebra of formulas. Can conditions be given for a logic presented by relation based semantics to have algebraic semantics?

In *Epstein, 1987*, I give algebraic semantics for Dependence Logic (§V.A), but I also show that that logic is not algebraizable. That is, not every consequence relation on the base algebra corresponds to a theory in the logic. So a separate question is: what conditions guarantee that the logic is algebraizable? Blok and Pigozzi, *1982* and *1989*, discuss this question.

Q9 Decidability

We say that a logic is *decidable* if there is a computable procedure to determine for any wff whether it is valid, or, if the logic is presented syntactically, whether it is a theorem.

If the logic has a Deduction Theorem then the decidability of validity implies that the finite consequence relation, $\{A_1, \ldots, A_n\} \models B$, is decidable, too. There is no corresponding notion of decidability of the full consequence relation, $\Gamma \models A$, for in general it is not even decidable whether a given wff belongs to an infinite collection Γ.

Decidability is a desirable property for a propositional logic. We say that a relation based semantics is *compactly decidable* if given any wff we can determine if it is valid by surveying the finite number of ways to assign truth-values to the variables in it and ways to relate its parts by the relations governing the truth-tables: the wff is valid if and only if the wff comes out true for each assignment which can be evaluated using the truth-tables. An analogous definition can be given for set-assignment semantics where the finite number of ways to assign sets to its parts in terms of Venn diagrams are surveyed instead. The logics of Chapters III and V have compactly decidable set-assignment semantics.

For some logics it may be that an assignment which makes a wff false cannot be extended to an assignment for all wffs, and hence there is no model corresponding to it. In that case it seems that the assumptions about infinite totalities have entered more strongly into the semantics. Can structural conditions on relation based or set-assignment semantics be given which guarantee that the resulting logic is decidable? is compactly decidable?

Q10 Extensionally equivalent propositions and the rule of substitution

Two propositions are said to be extensionally equivalent relative to a particular logic if they have the same semantic properties in a model, regardless of their form. They mean the same.

In classical logic the only semantic property under consideration is truth-value, so if $A \leftrightarrow B$ is true (in a model) then A and B are extensionally equivalent. From this we have the valid rule of substitution: for any C,

$$\frac{A \leftrightarrow B}{C(A) \leftrightarrow C(B)}$$

where B is substituted for some but not necessarily all occurrences of the subformula A of C.

For Relatedness Logic I remarked (p.81) that there is no schema S(A,B) whose truth guarantees that A is extensionally to B. In particular, $A \leftrightarrow B$ does not. Nor is there any schema for which

$$\frac{S(A,B)}{C(A) \leftrightarrow C(B)}$$

is a valid rule.

For Łukasiewicz's three-valued logic $\mathbf{L}_3$ (§VIII.C.1) there is a schema whose truth guarantees extensional equivalence, though $A \leftrightarrow B$ does not. Letting the formula IA be defined as $A \leftrightarrow \neg A$, then $(A \leftrightarrow B) \wedge (IA \leftrightarrow IB)$ is true in a model of $\mathbf{L}_3$ iff A and B have the same semantic properties (p.240). The appropriate rule of substitution for $\mathbf{L}_3$ is then

$$\frac{(A \leftrightarrow B) \wedge (IA \leftrightarrow IB)}{(C(A) \leftrightarrow C(B)) \wedge (IC(A) \leftrightarrow IC(B))}$$

which is valid, while the weaker rule

$$\frac{(A \leftrightarrow B) \wedge (IA \leftrightarrow IB)}{C(A) \leftrightarrow C(B)}$$

is valid, too.

Generally for any logic it is an important question whether there is a schema S(A, B) such that in every model S(A, B) is true iff A and B are extensionally equivalent. If there is such a schema, then the Extensionality Consideration should guarantee that

$$\frac{S(A,B)}{S(C(A), C(B))}$$

is a valid rule, a correct rule of substitution.

The solution of that question is closely related, I believe, to whether there is a congruence relation on the algebra of formulas and whether algebraic semantics can be given for the logic.

Appendix 1 On the Unity and Division of Logics

A. Quine on Deviant Logical Connectives

The views of Quine on logical connectives (*Quine, 1970*, Chapter 6) are in marked contrast to those of the previous sections and I shall comment on them because of their influence.

Suppose we were to say that subject matters affect conjunction (I wouldn't, but I suppose you could). And suppose further that we denied that 'Ralph is a dog $\wedge$ $2+2=4$' is true, while assenting to each conjunct. Quine would, I believe, argue that the use of the symbol '$\wedge$' is inappropriate.

> Here, evidently, is the deviant logician's predicament: when he tries to deny the doctrine he only changes the subject.
>
> *Quine, 1970*, p.81

I certainly agree with him that we would no longer be talking of classical conjunction. But he has stressed that we were, first of all, discussing natural language 'and'. Speaking of translations he says,

> If a native is prepared to assent to some compound but not to a constituent, this is a reason not to construe the construction as conjunction.
>
> *Quine, 1970*, p.82

I assume Quine would say similarly that if "the native" were to assent to both constituents but not the compound we should not construe the construction as conjunction.

But no native speaks using $\wedge$, $\neg$, $\rightarrow$ or words like 'conjoin', unless he is a Western logician. He speaks an ordinary language, using words such as 'and' or 'et'. Moreover, no one, I would venture to say, invariably uses 'and' truth-functionally. Yet it would only be on the basis of everyone doing so that Quine could claim that logical truths are obvious.

> Naturally, the native's unreadiness to assent to a certain sentence gives us reason not to construe the sentence as saying something which would be obvious to the native at the time. ... I must stress that I am using the word 'obvious' in an ordinary behavioral sense, with no epistemological overtones. When I call '1+1 = 2' obvious to a community I mean only that everyone, nearly enough, will unhesitatingly assent to it, for whatever reason; ... Logic is peculiar; every logical truth is obvious, actually or potentially.
>
> *Quine, 1970*, p.82

Simple logical "truths" involving the formal '$\wedge$' which we explicitly agree to interpret truth-functionally are certainly obvious, but not in the sense of Quine for no one speaks a language with that "word" in it. Logical truths, whatever they may be, are not obvious in his sense if it means only that, nearly enough, people will unhesitatingly assent to them. For 'nearly enough' conceals an immense problem of explication; for example, we know that 'or' is not used nearly always in the truth-functional inclusive sense. Moreover, subjects of a tyrant may unhesitatingly assent to 'The tyrant is good'; logic is not so peculiar. The reason for assent matters.

There is evidence in our language and reasoning to support many different models of the same connectives. And, you may recall, we chose the connectives to investigate and built our formal language first, then decided to use the classical interpretation in Chapter II. We are speaking the same language; we are paying attention to different aspects or uses of it. There is a change of subject in that sense, but we may still have real disagreements about which model represents the way people do reason (if you're listening to the natives) or should reason. Quine's sense of obviousness is accommodated and better understood as agreement between, archetypally, two participants in a logical discussion. That explanation and how agreements are founded upon a background understanding of the world I have developed in Chapter XI.

B. Classical vs. Nonclassical Logics

Susan Haack, *1974*, has developed what I take to be Quine's viewpoint and has divided propositional logics according to whether they are rivals or supplements to classical logic, based on syntactic comparisons. In this book I hope to have shown that the same viewpoint underlies both kinds of logics. Classical logic stands out because it has the simplest semantics, resulting from the greatest abstraction from actual propositions.

Within the framework here, a division of the sort Haack suggests might be justified on the basis of some semantic division of aspects of propositions into those which extend classical logic and those which are deviant. Here are some possibilities.

1. If the aspect of propositions of significance to the logic can be defined with reference to, or is reducible to the semantics of classical logic one might classify the logic as nondeviant. For example, the modal logics of Chapter VI would not be deviant, for the content of a proposition is taken to be the ways in which it could be true and that is explained in terms of classical models. However, the originator of modern modal logics, C. I. Lewis, *1912* and *1932*, formulated them as rivals to classical logic.

Another logic which would be nondeviant by this criteria is the one discussed in § V.F in which the content of a proposition is taken to be its consequences in classical logic.

2. One might argue that a logic is compatible with classical logic if it has relation based semantics in which any valuation v can be paired with the universal relations for B, C, A, and the universal property for N to yield a model. In that case the collection of tautologies of the logic will be contained in those of classical logic.

3. If $\neg$ and $\wedge$ are interpreted classically according to some relation based semantics for the logic, then the logic could be viewed as extending classical logic. Or one could be more generous and take the logic as a classical extension if $\neg$ and at least one of $\wedge$, $\vee$, $\rightarrow$ is interpreted classically.

By both criteria (3) and (4) Subject Matter Relatedness Logic (Chapter III) would be nondeviant, yet the aspect of propositions on which it is based doesn't seem reducible to the semantics of classical logic.

4. Another possibility is to call a logic classical if i. the syntactic notions of a consistent theory and a complete theory are the same as for classical logic, and ii. a set of propositions is a consistent and complete theory if and only if it has a model. In this case Relatedness Logic (Chapter III) is classical, while Łukasiewicz's three-valued logic (§ VIII.C.1) is not.

5. One might also try to divide logics by using the notion of a translation discussed in Chapter X. Unfortunately, the obvious divisions yield counterintuitive results: if we say that a logic is nondeviant if there is a semantically faithful translation of classical logic into it, then Łukasiewicz's three-valued logic (§ VIII.C.1) is not deviant. If we say a logic is compatible with classical logic if it can be grammatically translated into classical logic, then I know of no example of a nondeviant logic other than classical logic itself.

Appendix 2 A Mathematical Presentation of the General Framework

with the assistance of **Walter Carnielli**

A. Languages

Denote by **L** the class of languages $L(p_0, p_1, \ldots ; \gamma_0, \gamma_1, \ldots, \gamma_n)$ where p_i is a propositional variable and γ_j is an i_j-ary connective for a natural number $i_j \geq 0$. The well-formed-formulas (*wffs*) of the language are defined inductively:

1. (p_i) is a wff for each $i = 0, 1, 2, \ldots$.
2. If $A_1, \ldots, A_r$ are wffs and $i_j = r$, then $\gamma_j(A_1, \ldots, A_r)$ is a wff
3. No other concatenations of symbols are wffs.

Throughout, when I write '$\gamma_j(A_1, \ldots, A_r)$' I am assuming that $i_j = r$. Note that we allow γ_j to be a 0-ary connective, that is, a *constant*.

For a proof of the unique readability of wffs and further definitions concerning formal languages see §II.G.1. The set of propositional variables is denoted as 'PV', and the set of wffs of the language as 'Wffs'.

I restrict attention to languages with a finite number of connectives. This excludes from consideration certain modal logics of belief, knowledge, action, and so on, in which there is a distinct modal operator for each of an infinite number of individuals, for example, $K_x A$ read as 'x knows that A'. The definitions below can be extended to such languages, but it seems to me that then we are doing predicate logic, for example, $\forall x\, ((K_x A) \rightarrow A)$.

B. Formal Set-Assignment Semantics

I stress the word 'formal' here to distinguish these technical mathematical structures from semantics for logics as described in §B.3 above.

Let $L \in$ **L** be a fixed language throughout the following.

Given an r-ary connective γ_j, a *truth-function* for it is a function

$$f_j : \{T, F\}^{i_j} \rightarrow \{T, F\}$$

We reserve the symbols $\neg$, $\rightarrow$, $\wedge$, and $\vee$ for connectives which are assigned the classical truth-functions as illustrated in tables (1)–(4) of §II.B (p.15). We call these, respectively, *negation*, *the conditional*, *conjunction*, and *disjunction*. For example if γ_j is $\wedge$, then $i_j = 2$ and

$$f_j : \{T, F\} \times \{T, F\} \rightarrow \{T, F\}$$

is given by

$$f_j(T,T) = T, \quad f_j(T,F) = F, \quad f_j(F,T) = F, \text{ and } f_j(F,F) = F$$

We reserve the symbol $\perp$ for 0-ary connectives which are assigned the constant truth-function F.

A *formal set-assignment model* for L with respect to truth-functions $f_0, \dots, f_n$ for the connectives $\gamma_0, \dots, \gamma_n$ is

$$M = \langle v, s, S, (R_j : 0 \le j \le n) \rangle$$

where

v is a *valuation*, $v: \mathrm{PV} \to \{T, F\}$
S is a set
$s: \mathrm{Wffs} \to \mathrm{Sub}(S)$ is a *set-assignment*, where, unless otherwise noted, $S \neq \varnothing$
$R_j \subseteq \mathrm{Sub}(S)^{i_j}$ is the *relation governing the truth-table for* γ_j

The valuation v is extended inductively to all wffs by:

$$v(\gamma_j(A_1, \dots, A_{i_j})) = T \text{ iff } \begin{cases} R_j(s(A_1), \dots, s(A_{i_j})) \\ \text{and} \\ f_j(v(A_1), \dots, v(A_{i_j})) = T \end{cases}$$

This is the *truth-table* for γ_j.

If in every model the relation R_j is the *universal* relation, that is, $R_j(s(A_1), \dots, s(A_{i_j}))$ holds for every sequence of wffs $A_1, \dots, A_{i_j}$, then we say that γ_j is *truth-functional*. If one of $\neg, \to, \wedge, \vee$ is truth-functional, then we say it is a *classical connective* .

We write $M \models A$ if $v(A) = T$, and $M \models \Gamma$ if for every $B \in \Gamma$, $v(B) = T$.

A *formal set-assignment semantic structure* for a language L is a choice of truth-functions $f_0, \dots, f_n$ for $\gamma_0, \dots, \gamma_n$ and a collection of models for L with respect to them. A formal semantic structure will be denoted as $\mathcal{M}$.

Given a formal semantic structure $\mathcal{M}$ for L, the class of *tautologies with respect to* $\mathcal{M}$ is $\{A : M \models A \text{ for every } M \in \mathcal{M}\}$. The *consequence relation with respect to* $\mathcal{M}$ is the relation $\models_{\mathcal{M}}$ on $\mathrm{Sub}(\mathrm{Wffs}) \times \mathrm{Wffs}$ given by: $\Gamma \models_{\mathcal{M}} A$ iff for every $M \in \mathcal{M}$, if $M \models \Gamma$ then $M \models A$. Here 'if ... then ...' is understood classically.

I denote by $\mathcal{SA}_L$ the class of all formal set-assignment semantic structures for the language L, and by $\mathcal{SA}$ the class of all formal set-assignment structures for all languages in $\mathbf{L}$.

C. Formal Relation Based Semantics

A *formal relation based model* for $L \in \mathbf{L}$ with respect to truth-functions $f_0, \dots, f_n$ for $\gamma_0, \dots, \gamma_n$ is

$$M = \langle v, (R_j : 0 \le j \le n) \rangle$$

where

υ is a *valuation*, $\upsilon: PV \rightarrow \{T, F\}$

$R_j \subseteq \mathrm{Sub(Wffs)}^{i_j}$ is the *relation governing the truth-table for* γ_j

and υ is extended inductively to all wffs by:

$$\upsilon(\gamma_j(A_1, \dots, A_{i_j})) = T \quad \text{iff} \quad \begin{cases} R_j(A_1, \dots, A_{i_j}) \\ \text{and} \\ f_j(\upsilon(A_1), \dots, \upsilon(A_{i_j})) = T \end{cases}$$

A *formal relation based semantic structure* for L, the class of tautologies, and the consequence relation for such a semantic structure are defined as in the previous section only with respect to relation based models.

I denote by RB_L the class of all formal relation based semantic structures for the language L, and by *RB* the class of all formal relation based structures for all languages in **L**.

Denote by $C(\mathit{SA})$ and $C(\mathit{RB})$ the class of consequence relations of all set-assignment semantics structures, and of all relation based semantic structures respectively. We have $C(\mathit{SA}) \subseteq C(\mathit{RB})$, for any model given by relations on sets assigned to wffs can be viewed as given by those same relations on the wffs themselves. I do not know if $C(\mathit{SA}) = C(\mathit{RB})$.

D. Specifying Semantic Structures

To specify a semantic structure

1. We first specify the truth-functions associated with the connectives.

For example, for the conditional we adopt the classical truth-function (table (4) of §II.B). That is how we know that in each of the models we are talking about the same connective, the conditional.

Specifying the truth-functions first accords with our hypothesis that every proposition is to be viewed as having a truth-value and that an additional aspect of propositions can be factored into the semantics in terms of content or relations between propositions. Thereafter,

2. We specify the class of models comprising the semantic structure by specifying the class of relations governing the truth-tables (and the class of set-assignments for structures of *SA*).

In this section I want to examine how to do (2).

1. Set-assignments and relations for *SA*

We say that a *connective* γ_j *is uniformly interpreted in* $\mathcal{M}$, where $i_j = r$, if there is a single set-theoretic condition $C(x_1, \dots, x_r)$ such that for each formal model $M \in \mathcal{M}$,

$\mathsf{R}_j(\mathsf{s}(A_1), \ldots, \mathsf{s}(A_{i_j}))$ iff $C(\mathsf{s}(A_1), \ldots, \mathsf{s}(A_r))$

For example, for most of the modal logics of Chapter VI the relation governing the table for '$\rightarrow$' is $x_1 \subseteq x_2$; for Relatedness Logic (Chapter III) the relation governing the table for '$\rightarrow$' is $x_1 \cap x_2 \neq \varnothing$.

Here 'a set-theoretic condition' is meant to be any formula of first-order set theory with exactly i_j free variables. The complexity, variety, and pathological possibilities for such set-theoretic conditions makes it important to restrict our attention to some class of "simple" uniform interpretations, as suggested in §D of this chapter.

The *class of set-assignments of* $\mathcal{M}$ *is uniformly presented* if there is a finite number or recursive set of set-theoretic conditions on contents of wffs which determine the class. The class of assignments is *equational* if each condition can be expressed as a universally quantified equation. For example, the conditions on contents of wffs specifying the class of union set-assignments used by the logics of Chapter V are: $\mathsf{s}(\neg A) = \mathsf{s}(A)$, $\mathsf{s}(A \wedge B) = \mathsf{s}(A) \cup \mathsf{s}(B)$, $\mathsf{s}(A \rightarrow B) = \mathsf{s}(A) \cup \mathsf{s}(B)$.

I will not try to make explicit the language in which such conditions are to be expressed, but it is clear that it should contain function symbols, variables ranging over Wffs and variables ranging over PV, a symbol for the relation 'is a subformula of', as well as the language of set theory with equality.

The class of set-assignments of $\mathcal{M}$ is *presented uniformly depending on truth-values* if the conditions on set-assignments are allowed to include expressions of the form '$\upsilon(A) = \mathsf{T}$' and '$\upsilon(A) = \mathsf{F}$'. For example, one of the conditions for the modal logic **S5** (§VI.G.1) is: if $\upsilon(A) = \mathsf{T}$, then $\mathsf{s}(\Diamond A) \neq \varnothing$.

It seems to me that only uniformly presented relations and set-assignments are of importance to logic: an infinite collection of *ad hoc* cases does not amount to a logic, for reasoning requires uniformity.

2. Relations for *RB*

A *connective* γ_j *is uniformly interpreted* in a relation based semantic structure if the class of relations governing its truth-table is uniformly presented. That is, there is a finite number or recursive collection of conditions on relations which determine the class. For example, the conditions defining the class of relations governing the table for '$\rightarrow$' for Dependence Logic (§V.A.4) are: 1. $\mathsf{D}(A,A)$, 2. $\mathsf{D}(\neg A,A)$, 3. $\mathsf{D}(A,\neg A)$, 4. if $\mathsf{D}(A,B)$ and $\mathsf{D}(B,C)$, then $\mathsf{D}(A,C)$, 5. $\mathsf{D}(A,B\wedge C)$ iff $\mathsf{D}(A,B)$ and $\mathsf{D}(A,C)$, and 6. $\mathsf{D}(A,B\rightarrow C)$ iff $\mathsf{D}(A,B)$ and $\mathsf{D}(A,C)$.

Again, without trying to make explicit the language in which such conditions are to be expressed, I'll simply note that it should have variables ranging over Wffs and variables ranging over PV, a relation to be interpreted as 'is a subformula of', and relation symbols.

The class of relations governing the truth-table is *uniformly presented depending on truth-values* if the conditions on set-assignments are allowed to include expressions of the form '$v(A) = T$' and '$v(A) = F$'.

E. Intensional Connectives

The semantic structures above allow us to consider truth-functional interpretations of connectives, but do not provide for an evaluation of a connective solely in terms of content, for example, $v(\Box A) = T$ iff $s(A) = S$. As I said in §B.1, such connectives seem to me to be part of a project of giving semantics to more of English than we are usually concerned with in logic. They are entirely intensional.

However, there are certain intensional uses of 'if ... then ...' in modal logics (the subjunctive conditional) and of 'not' in paraconsistent logics which seem best explained in terms of such evaluations (cf. §VI.J.1, §VI.K.2, and §IX.F). Therefore I will extend the definitions of the previous sections to accommodate these.

A *formal model with intensional connectives* is defined as above except that a connective γ_j need not have a truth-function assigned to it. In that case γ_j is called a *wholly intensional connective* in the model and is evaluated by the table:

$$v(\gamma_j(A_1, \ldots, A_{i_j})) = T \text{ iff } R_j(A_1, \ldots, A_{i_j})$$

where in the case of a set-assignment model R_j is understood to be a relation on the sets assigned to $A_1, \ldots, A_{i_j}$.

In a model with intensional connectives we may use the symbols $\neg, \rightarrow, \wedge, \vee$ for wholly intensional connectives, but only if there is no other connective which is assigned the classical truth-function for that symbol.

A *formal semantic structure* for $L(p_0, p_1, \ldots ; \gamma_0, \gamma_1, \ldots, \gamma_n)$ *with intensional connectives* is defined as a choice of truth-functions $f_{i_1}, \ldots, f_{i_m}$ for $\gamma_{i_1}, \ldots, \gamma_{i_m}$ and a collection of models with respect to those. That is, if γ_j is interpreted intensionally in one model, it is interpreted intensionally in all.

Any intensional connective can be modeled in the semantics of the previous sections by taking the associated truth-function for it to have constant output T, though in most cases this will be counterintuitive. Thus, except for relaxing the conditions on interpretations of $\neg, \rightarrow, \wedge$, and $\vee$, semantic structures with intensional connectives give rise to no new consequence relations. Their role should be seen as providing for more intuitively acceptable presentations.

F. Truth-Default Semantic Structures

The semantic structures outlined above could be termed *falsity-default* structures: in a model, $\gamma_j(A_1, \ldots, A_{i_j})$ is evaluated to be *false* if $R_j(A_1, \ldots, A_{i_j})$ does not hold; otherwise it is evaluated by the truth-function for γ_j. I have argued for this two-valued falsity-weighted analysis of semantics for logics throughout this volume, partly by presenting a wide variety of logics in that form. I have also developed that theme in *Epstein, 198?*.

One could, however, devise a mirror image of these semantic structures by taking truth to be the default value. That is, in a model $\gamma_j(A_1, \ldots, A_{i_j})$ is evaluated to be *true* if $R_j(A_1, \ldots, A_{i_j})$ does not hold; otherwise it is evaluated by the truth-function for γ_j. Thus $\gamma_j(A_1, \ldots, A_{i_j})$ is true unless it passes two tests to be false, instead of false unless it passes two tests to be true.

$$\upsilon(\gamma_j(A_1, \ldots, A_{i_j})) = \mathsf{F} \text{ iff } \begin{cases} R_j(A_1, \ldots, A_{i_j}) \\ \text{and} \\ f_j(\upsilon(A_1), \ldots, \upsilon(A_{i_j})) = \mathsf{F} \end{cases}$$

This is what I call a *truth-default truth-table for* γ_j. See pp. 285–286 for the truth-default tables for $\neg, \rightarrow, \wedge, \vee$ (reading $\neg$ for $\sim$).

We may define *formal truth-default semantic models*, structures, consequence relations, and so on, as in the previous sections except using these tables.

Given any such table there is another table which uses the negation of that relation. So the same class of models, etc., can be defined by using the following, which I also call *truth-default tables*.

$$\upsilon(\gamma_j(A_1, \ldots, A_{i_j})) = \mathsf{T} \text{ iff } \begin{cases} R_j(A_1, \ldots, A_{i_j}) \\ \text{or} \\ f_j(\upsilon(A_1), \ldots, \upsilon(A_{i_j})) = \mathsf{T} \end{cases}$$

Here 'or' is to be understood as classical inclusive 'or'.

From this viewpoint truth-default semantics take the relation governing the table for, e.g., '$\rightarrow$' as an additional way for the conditional to be true even if the antecedent is true and consequent false.

It is my opinion that such an interpretation of $\neg, \rightarrow, \wedge$, or $\vee$ is not of a piece with our usual understandings of 'not', 'if ... then ...', 'and', 'or'. I discuss an example in §X.G, the paraconsistent logic $\mathbf{J_3}$, where I suggest that '$\neg$' interpreted by a truth-default table is better understood as an intensional connective.

However, within a falsity-default semantic structure it may be possible to define truth-default connectives, and vice versa (certainly this is so if the logic has a functionally complete set of connectives). An important example is $\Diamond A$ in various modal logics (table (14) of §VI.C.1 using $s(A) = \varnothing$ in place of $s(A) \neq \varnothing$). Further research may suggest a greater reliance on truth-default semantics or a mixing of truth-default with falsity-default connectives in the definition of a logic.

G. Universal Tautologies and Anti-Tautologies

I call a wff of the language $L(\neg, \rightarrow, \wedge, \vee)$ a *falsity-default tautology* if it is true in every falsity-default semantic structure, a *falsity-default anti-tautology* if it is always false. Similarly we can define *truth-default tautologies* and *truth-default anti-tautologies.*

There are no falsity-default tautologies: for a wff to be true it must pass two tests, one of which refers to its content (or place in a relation).

There are, however, falsity-default anti-tautologies:

$A \wedge \neg A$ is a falsity-default anti-tautology.

If A, B are falsity-default anti-tautologies, then $A \vee B$ is a falsity-default tautology, and for any wff C, so are $A \wedge C$, $C \wedge A$.

I suspect that these conditions completely characterize the class of falsity-default anti-tautologies. In particular, note that for any A, $\neg A$ cannot be an anti-tautology, since that would require that A be a tautology; and $A \rightarrow B$ cannot be an anti-tautology, since that would require that A be a tautology

There are no truth-default anti-tautologies. There are, however, truth-default tautologies:

$A \vee \neg A$ is a truth-default tautology.

If A, B are truth-default tautologies, then $A \wedge B$ is a truth-default tautology, and for any wff C, so are $A \vee C$, $C \vee A$, $C \rightarrow A$.

I suspect that these conditions completely characterize the class of truth-default tautologies. In particular, note that $\neg A$ cannot be a truth-default tautology for any A since that would require that A be an anti-tautology.

V Dependence Logics – D, Dual D, Eq, and DPC –

The first system studied in this chapter is Dependence Logic. It is a case study of how to use the general framework of the last chapter to develop a logic based on a particular aspect of propositions. Thereafter I develop several closely related logics on the same themes.

The aspect of propositions of concern here is the referential content of a proposition, a notion closely allied to the idea of analyticity in logic. It is similar to the notion of the subject matter of a proposition (Chapter III) and differs perhaps only in applications: subject matters divide predicates and propositions roughly according to the traditional categories of philosophy with the concern being whether the subject matter of one proposition overlaps that of another; analyticity divides propositions hierarchically according to what is implicit in them, and containment of analytic content is the significant relationship between propositions.

I will give Hilbert-style axiomatizations of the logics here. Walter Carnielli, *1987*, has produced natural and elegant syntactic formulations of Dependence and Dual Dependence Logic in terms of analytic tableaux.

I am grateful to Roger Maddux for conversations and comments that helped me to develop the technical work of §A.

A. Dependence Logic

1. The consequent is contained in the antecedent

One of the most persistent ideas concerning the conditional is that $A \rightarrow B$ can hold only if the consequent is contained in the antecedent.

> And those who judge by "suggestion" [that is, 'the power of signifying more than is expressed', i.e. what is implicit] declare that a conditional is true if its consequent is in effect included in its antecedent. According to these, 'If it is day, then it is day,' and every repeated conditional will probably be false, for it is impossible for a thing to be included in itself.
>
> Sextus Empiricus, *Outlines of Pyrrhonism*, II, 111
> translated in *Mates*, *1953*, p.109

There have been many explanations of what it is that's contained and the manner of containment. Sometimes it's said that the 'information conveyed by B is included in A' (*Hanson*, *1980*, p.659); Leibniz spoke of the terms of the consequent being implicitly contained in those of the antecedent. There seems to be general agreement now that, in contradistinction to the Sextus Empiricus quote, containment need not be strict.

Two examples will show that classical logic does not adequately model this conception of the conditional. The first is from *Smullyan, 1980.*

1. There must be a person, call him a, such that if a drinks then everyone drinks. That is, 'if a drinks then everyone drinks' is true. For if everyone drinks, then anyone will do for a. And if there's someone who doesn't drink, then choose one, call him a, and the proposition is true as the antecedent is false.

2. One of the following must be true:
   If $2+2=4$, then Fermat's Last Theorem holds.
   If $2+2=4$, then Fermat's Last Theorem fails.

There is nothing paradoxical about these examples. They seem odd because in

PC we identify an atomic proposition with its truth-value. That the consequent of each example "conveys more information" than the antecedent is irrelevant to the classical analysis of truth. We therefore turn to another analysis.

Consider how Prior explains the nature of containment.

> One proposition entails another ... means rather that the fact expressed by the consequent is identical with that expressed by the antecedent, or at least is "not a new fact in addition to" the first. ... The proposition that John owns a mare entails the proposition that John owns a horse, while the *fact* that John owns a horse does not entail but contains the fact that John owns a horse.
>
> Prior, *1948*, p.62

Here facts are understood by Prior not as true propositions nor as abstract objects, but as arrangements or configurations of the world. And ways in which the world is contain other ways in which the world is. But if that is how we understand 'facts' then we have to account for negative facts and possible facts, which lead to abstraction piled upon abstraction.

It seems to me that we have propositions with which we describe the world, but beyond those there is only the world itself. There is no further object called 'a fact'. To postulate a further entity, a kind that can contain or be contained in, is to point to what can't be said. For 'Ralph is a dog' to be true the world must be some way, but the way the world must be is: Ralph is a dog. There is not some further fact over and above the proposition being true. "The way the world is" is what our propositions are meant to describe.

Therefore, I would rather say that the *content* of the proposition 'John owns a horse' is contained in the content of 'John owns a mare.' I explain that further by saying that anything referred to or any predicate implicit in the one proposition is referred to or implicit in the other. This is the notion of *the referential content of a proposition* or its *conceptual framework* which I will develop in this chapter.

Compare Prior's comparable explanation.

> The fact that John does not own a mare "contains" the sex of the animal that John does not own (not in the sense in which a fact contains other facts, but in the sense in which a fact contains the objects it is about) ...
>
> Prior, *1948,* p.66

But what objects are contained in possible facts or negative facts: is the sex of an animal a thing ?

I prefer to say that certain predicates, for example, 'is a mare', have implicit in them certain others, in this case 'is a horse', 'is female', 'is an animal', ... Whether we wish to include on our list 'has lungs' or 'has over 5,000,000 red blood cells' will depend on the context established by our discussion. There is no fixed eternal body of predicates which constitute the content of a proposition unless predicates and propositions are taken to be eternal, timeless abstract objects, for ordinary language is fluid and therein lie our predicates. The belief that the conceptual framework of a proposition is fixed for all time seems a striving after certainty in an uncertain world.

This aspect of propositions is apparently more holistic than truth, depending on the range of propositions under discussion, but in practice it is molecular since normally only a finite number of predicates are under discussion (cf. the discussion of subject matters in §III.A). A fuller explanation of the referential content of propositions requires an analysis of predicates, predication, and the entire apparatus of predicate logic. This I attempt in Volume 2. I hope here to have done enough to convince you that the following assumption is reasonable.

Referential Content Assumption A proposition can be viewed as having referential content.

But suppose you were to criticize this notion by arguing that the only way we could discern if the content of one proposition contains that of another is by recognizing whether the corresponding entailment holds. Even granting that, we may be able to set out how referential content functions in logic by making certain observations or agreements concerning how some containments are derivative from others. Thus we might decide that if the content of B is contained in that of A then it is contained in that of $A \wedge C$ for any C. And in the process of reckoning the logical structure of referential content we will sharpen that notion, for as it stands now there are a number of different interpretations we could make.

2. The structure of referential content

We want to examine the view that B is *dependent* on A if, and only if, B does not go beyond the conceptual framework established by A.

This is a weaker view than that A must analytically imply B. Consider: 'John is a bachelor' analytically implies 'John is a man' if 'bachelor' and 'man' have their usual meanings. But 'John is a bachelor' does not analytically imply 'John is not a man', though the former does establish the conceptual framework for the latter. This, of course, only holds if we take the view that the logical connectives are neutral in establishing what is referred to by a proposition:

The logical connectives are syncategorematic: they have no referential content.

Thus, informally, 'Ralph is a dog' and 'Ralph is not a dog' have the same content; and 'Ralph is a dog and George is a duck' has the same content as 'If Ralph is a dog then George is a duck.'

Under this view the truth of a proposition is not part of its content. The predicates implicit in a proposition and the truth-value of the proposition are distinct, for the referential content remains unchanged regardless of whether the proposition is true or false. The assertion or use of a proposition carries with it the contents of the proposition, but it cannot also carry the truth of it. We may say that when we assert a sentence then we assert that it is true, but that is not to say it *is* true. Yet what we take as its contextual frame of reference does stay constant. Our assumption, then, is:

The referential content of a proposition is independent of its truth-value.

Based on these assumptions the *simplest* and, I believe, the most reasonable agreement we can make about the content of a compound is that what the whole refers to and predicates is exactly what its parts refer to and predicate. This is in accord with the Fregean Assumption (§IV.D, p.99) that the content of the whole is a function of the content of its parts.

The Structure of Referential Content Assumption The referential content of a compound proposition is the union of the contents of its parts.

Thus the content of 'John is a bachelor' does contain that of 'John is not a man'. But in any reasonable model 'If John is a bachelor, then John is not a man' fails because the truth-value of the parts will matter in establishing the truth-value of the compound sentence. Hence we begin to approximate a notion of analytic implication when we evaluate 'if ... then ...' propositions using both the referential content and truth-values of their parts. And these will be the only aspects of propositions we'll consider.

Dependence Logic Abstraction The only properties of a proposition which matter to (this) logic are its form, its truth-value, and its referential content.

There are two other views which could perhaps be put forward for what the referential content of a compound sentence is. First, one could argue that, say, the connective 'not' does not change the content of 'Ralph is a dog' but rather augments it. That is, 'Ralph is not a dog' speaks of the content of 'Ralph is a dog' and of 'negation'. Similarly, the content of an 'if ... then ...' sentence might be said to include 'hypothetical assertion.' If, on this view, the logical connectives have distinct contents then we would never have any conditional evaluated as true if the consequent had a connective not appearing in the antecedent, that is unless we had a hierarchy of contents of logical connectives. Alternatively, if the content of each of the connectives is the same, namely 'logical connective', then it's clear that nothing new would be added to our logic except that 'if A, then B' would fail if A is atomic and B isn't. For an argument that relies on the logical connectives having content see *Prior, 1964*, pp. 161–162, where Prior says, 'I cannot see how the sense of a sentence can ever be identical with a logical complication of itself.' K. Fine, *1979*, discusses that view, too.

The second view is that negating a sentence distorts the reference of it. Asserting 'Ralph is not a dog' includes a wider reference than that of 'Ralph is a dog.' This view is difficult to accommodate into the motivation I've given as it seems to confuse the truth of a sentence with its content.

There is, moreover, a prevalent view that tautologies have no content: see *Prior, 1960,* for an application of that. Whatever the motivation of such a view it cannot be that the predications expressed by the parts of the proposition contribute uniformly to the content of the whole. Searle, *1970*, p.124, discusses this point.

3. Set-assignment semantics

We can now present a logic which incorporates the assumptions we've made by filling out the general form of semantics for propositional logics.

Our assumptions dictate the class of set-assignments appropriate to model the notion of the referential content of a proposition. They are those satisfying

(†) $$s(A) = \bigcup \{ s(p) : p \text{ is atomic and } p \text{ appears in } A \}$$

which we call the *union set-assignments.* Perhaps it would be in keeping with our motivation to require that $s(A) \neq \varnothing$ for all A, but as that won't affect the resulting logic I don't make that additional restriction.

For a sketch of such an assignment, compare the example for subject matters in §III.D, p.69, reading 'predicates' for 'topics'.

What remains to specify are the connections which govern the truth-tables. The relation governing the conditional is containment, as we first stated.

Dependent Implication

A	B	$s(A) \supseteq s(B)$	$A \to B$
any values		fails	F
T	T		T
T	F	holds	F
F	T		T
F	F		T

That negation is classical seems to follow from the fact that the referential content of a proposition is independent of its truth. And barring any good reason why content should affect conjunction, I'll assume that conjunction, too, is classical:

Negation and conjunction are classical.

If you wish you can view this as a simplicity constraint: we allow the contextual framework of a proposition to affect only the table for the conditional as that is the main concern of the motivating discussion. However, I think that these are the natural interpretations of the connectives in terms of the notions of referential content and analyticity.

There seem to me to be three reasonable ways to interpret $\vee$. We could take it to be classical and hence definable as $\neg(\neg A \wedge \neg B)$. Or we could interpret $A \vee B$ to be true iff at least one of A, B is true and the content of A contains that of B and that of B contains the content of A, what I call *strongly dependent disjunction*, definable as $(\neg A \rightarrow B) \wedge (\neg B \rightarrow A)$. Or we could interpret $A \vee B$ as being true iff either A or B is true and either the content of B is contained in A or vice versa, what I call *weakly dependent disjunction*, definable as $\neg(\neg(\neg A \rightarrow B) \wedge \neg(\neg B \rightarrow A))$. Arguments for each of these could be made but they would be inessential to the main analysis of this chapter. I will leave to you the choice of one of these or some other interpretation of $\vee$ for it won't substantially affect what follows.

Therefore, the formal language we will use is $L(p_0, p_1, \ldots \neg, \rightarrow, \wedge)$. In terms of this language we can give an alternate specification of the union set-assignments as being those that satisfy the following for all wffs.

U1. $\mathsf{s}(\neg A) = \mathsf{s}(A)$
U2. $\mathsf{s}(A \wedge B) = \mathsf{s}(A) \cup \mathsf{s}(B)$
U3. $\mathsf{s}(A \rightarrow B) = \mathsf{s}(A) \cup \mathsf{s}(B)$

Now we can specify the models for our logic, recalling that we have agreed that truth-values and contents are independent.

A *dependence model* is

I

$$L(p_0, p_1, \ldots \neg, \rightarrow, \wedge)$$

$\downarrow$ } realization

$$\{p_0, p_1, \ldots, \text{complex propositions formed from these using } \neg, \rightarrow, \wedge\}$$

$\downarrow$ } v, s, and truth-tables

$$\{T, F\}$$

Here v is a valuation for the atomic propositions; s is a union set-assignment on the atomic propositions which is extended to all propositions of the semi-formal language by (†) or inductively by U1–U3. And v is extended to all propositions of the semi-formal language by using the classical tables for $\neg$ and $\wedge$, and the table for dependent implication for $\rightarrow$.

By virtue of the Dependence Logic Abstraction, nothing matters about an atomic proposition except its truth-value and content. So we can simplify a model.

II

$$L(p_0, p_1, \ldots \neg, \rightarrow, \wedge)$$
$$\Big\downarrow \quad v,\ s, \text{ and truth-tables}$$
$$\{T, F\}$$

And we make a Fully General Abstraction that any function $v: PV \rightarrow \{T, F\}$ together with any *countable* set S and map $s: PV \rightarrow \text{Sub } S$ can comprise a model $<v,s>$. Equivalently we can take $s: \text{Wffs} \rightarrow \text{Sub} S$ satisfying U1–U3. Unless previously specified otherwise, whenever I write '$<v, s>$' the set S is to be taken to be $\bigcup\{s(A): A \text{ is a wff}\}$.

As usual, $<v,s> \models A$ means that $v(A) = T$, and $<v,s> \models \Gamma$ means that $v(A) = T$ for every A in Γ. The semantic consequence relation is the usual one: $\Gamma \models A$ iff if $<v,s> \models \Gamma$ then $<v,s> \models A$. A proposition is a *dependence tautology* if it is true in every dependence model (under the Fully General Abstraction). The collection of these I'll notate as **D** and refer to it or to the semantic consequence relation as *Dependence Logic* as the context will make clear. When other logics are under consideration I will write $\models_{\mathbf{D}}$ for $\models$.

4. Relation based semantics

These set-assignment semantics give rise to a relation based semantics in the usual way. Call a binary relation D on wffs a *dependence relation* if there is a union set-assignment s such that $D(A,B)$ iff $s(A) \supseteq s(B)$. Dependence logic can be viewed as the class of all tautologies for all models $<v,D>$ where D is a dependence relation, $\neg$ and $\wedge$ are interpreted classically, and the table for $\rightarrow$ is governed by D. We can characterize the class of dependence relations as follows.

Lemma 1 D is a dependence relation iff the following all hold:

1. D is reflexive.
2. D is transitive.
3. $D(A, B \wedge C)$ iff $D(A,B)$ and $D(A,C)$
4. $D(A, B \rightarrow C)$ iff $D(A,B)$ and $D(A,C)$
5. $D(\neg A, A)$
6. $D(A, \neg A)$

Proof: If D is a dependence relation then (1)–(6) all hold. If (1)–(6) hold, set $t(A) = \{B: D(B,A)\}$. Then by (1) and (2), $D(A,B)$ iff $t(A) \subseteq t(B)$. By (3) and (4), $t(A \rightarrow B) = t(A \wedge B) = t(A) \cap t(B)$, and by (5) and (6), $t(\neg A) = t(A)$. Set $s(A) = \text{Wffs} - t(A)$, that is, $s(A) = \{B: B \text{ is a wff and } B \notin t(A)\}$. Then s is a union set-assignment and $D(A,B)$ iff $s(A) \supseteq s(B)$. ■

Despite the simplicity of the conditions in Lemma 1, dependence relations do not satisfy a version of the Fregean Assumption (see (B) of §IV.D, p.99). It is possible to extend a binary relation on the propositional variables to a dependence relation on the collection of all wffs in two distinct ways. For instance, we may have that $\mathsf{D}(p,q)$ and $\mathsf{D}(p,r)$ both fail, yet we can have that either $\mathsf{D}(p \wedge r,q)$ holds, or $\mathsf{D}(p \wedge r,q)$ fails, as the diagrams below illustrate.

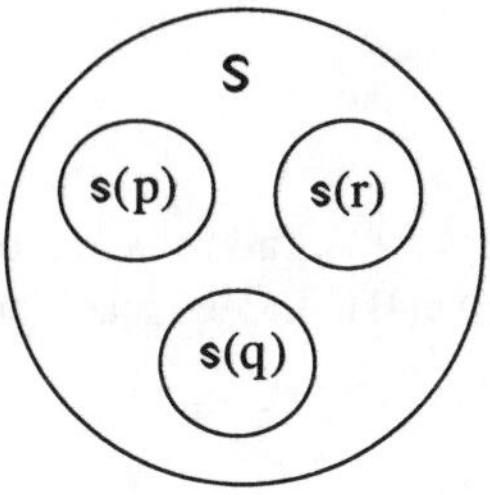

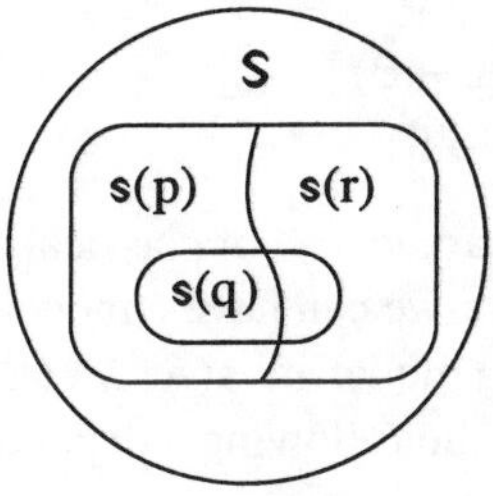

A dependence relation cannot be specified locally; it is a global notion. However, we have the following, which you can prove using the previous lemma.

Lemma 2 D is a dependence relation iff the following all hold:

a. D is reflexive.

b. D is transitive.

c. $\mathsf{D}(A,B)$ iff $\mathsf{D}(A,p)$ for every p in B.

Originally I started with Lemma 2, using it to establish Lemma 1 by building a set-assignment s corresponding to a given D. Each $s(p_i)$ was built up as bits corresponding to whether for each A, $\mathsf{D}(A,p_i)$ holds and $\mathsf{D}(p_i,A)$ holds. I outlined that remarkably complicated construction in *Epstein and Maddux, 1981*. It is of significance partly for giving a set-assignment that has a good reading in terms of the given D, which the s produced in the proof of Lemma 1 does not (that proof was supplied by an anonymous referee). More importantly, such a construction is an attempt to reconcile the holism of the notion of referential content with some regular inductive procedure for producing contents in order to show that our infinitistic assumptions have not led to counterintuitive results.

5. A comparison of dependence and classical tautologies

We have that $\mathbf{D} \subseteq \mathbf{PC}$ because the universal relation is a dependence relation. So let's consider some **PC** tautologies which fail to be dependence ones. First we need to note that **D** is compactly decidable: given any wff A, if every assignment of truth-values and Venn diagram assigning sets to the variables appearing in A validates A, then A is a dependence tautology. Otherwise, pick an assignment which falsifies A and extend it to all other variables, say by assigning T and the empty set to each of them. In that model A fails, and hence A is not a dependence tautology. So we can show the following schemas are not dependence tautologies.

1. $(A \wedge \neg A) \to B$
2. $A \to (B \to B)$
3. $\neg A \to (A \to B)$
4. $B \to (A \to B)$

The "paradoxes" of classical and modal logic, (1)–(4), fail to be dependence tautologies, for we can have a model in which $\mathfrak{s}(A) \not\supseteq \mathfrak{s}(B)$. Indeed, each of the first three hold in a model iff $\mathfrak{s}(A) \supseteq \mathfrak{s}(B)$.

Neither is the following a dependence tautology.

5. $\neg(A \wedge \neg B) \to (A \to B)$

We cannot conclude that there is someone, *a*, such that 'If *a* drinks then everyone drinks' is true. From $\neg(A \wedge \neg B)$ we cannot conclude $A \to B$ because the former does not guarantee that the content of A contains that of B. For a dependent implication to hold it is necessary but not sufficient that the material implication holds.

Comparisons with relatedness logic are also illuminating.

6. $(A \to (B \wedge C)) \to ((A \to B) \wedge (A \to C))$

is a dependence tautology but fails to be a relatedness tautology (see (7) p.76). Relatedness Logic isn't suitable to model the idea that for an implication to hold the antecedent must contain the consequent. But the first two examples of this chapter already show that, for in any plausible model the subject matter of '*a* drinks' overlaps that of 'everyone drinks'.

A more important difference from relatedness logic is that '$\to$' is transitive:

7. $((A \to B) \wedge (B \to C)) \to (A \to C)$

is a dependence tautology.

If we take '$\vee$' to be either classical, weakly dependent, or strongly dependent disjunction (see §3 above), then the following two schemas fail to be dependence tautologies

8. $A \to (A \vee B)$
9. $(A \to B) \to (A \to (B \vee C))$

though both of the following are dependence tautologies:

10. $((A \vee B) \wedge \neg A) \to B$ a propositional form of the *Disjunctive Syllogism*

11. $(A \to (B \vee C)) \to ((A \to B) \vee (A \to C))$

The following are not dependence tautologies:

12. $(A \to B) \to (\neg B \to \neg A)$ *Transposition*

13. $((A \wedge B) \to C) \to (A \to (B \to C))$ *Exportation*

14. $(A \to (B \to C)) \to (B \to (A \to C))$ *Interchange of Premisses*

Transposition also fails in Nonsymmetric Relatedness Logic, but is a Relatedness tautology. On the other hand,

15. $(A \to (B \to C)) \to ((A \wedge B) \to C))$ *Importation*

is a dependence tautology, but not a relatedness one.

Detachment and *modus tollens* fail in these propositional forms:

16. $A \to ((A \to B) \to B)$

17. $\neg B \to ((A \to B) \to \neg A)$

But these conjunctive forms are dependence tautologies:

18. $(A \wedge (A \to B)) \to B$

19. $(\neg B \wedge (A \to B)) \to \neg A$

In §E I'll discuss further the form of dependence tautologies in which '$\to$' is the principal connective.

6. Functional completeness of the connectives

We can define

$$D(A,B) \equiv_{Def} A \to (B \to B)$$

Then $D(A,B)$ is true in a model $<\mathsf{v}, \mathsf{s}>$ iff $\mathsf{s}(A) \supseteq \mathsf{s}(B)$.

This is the same formula which we called 'R(A,B)' for Relatedness Logic. It's right to give it a new name here because we're not defining "the same connective" of English. 'D' is not a connective at all, but an abbreviation useful for the metalogic: 'D(A,B)' is read as 'the proposition named 'B' is dependent on the proposition named 'A'.' It makes no sense to iterate 'D'. See the comments on 'R' in §III.B.4.

We also have that

$A \wedge B$ is semantically equivalent to $\neg((((A \to B) \to (A \to B)) \to A) \to \neg B)$

Using this it is possible to define every truth-and-content functional connective from just $\neg$ and $\to$; the proof is exactly the same as for Relatedness Logic (pp.78–79) reading 'D' for 'R'. A normal form theorem then follows as for Relatedness Logic (Corollary III.6, p.79).

7. Axioms and a completeness proof for D

D, Dependence Logic
in $L(\neg, \rightarrow, \wedge)$

$D(A,B) \equiv_{Def} A \rightarrow (B \rightarrow B)$
$A \leftrightarrow B \equiv_{Def} (A \rightarrow B) \wedge (B \rightarrow A)$

axiom schemas

1. $(D(A,B) \wedge D(B,C)) \rightarrow D(A,C)$
2. a. $D(A, \neg A)$
   b. $D(\neg A, A)$
3. $D(A, B \rightarrow C) \leftrightarrow D(A,B) \wedge D(A,C)$
4. $D(A, B \wedge C) \leftrightarrow D(A,B) \wedge D(A,C)$
5. a. $(A \wedge B) \rightarrow A$
   b. $(A \wedge B) \rightarrow (B \wedge A)$
6. $(D(A,B) \wedge \neg(A \wedge \neg B)) \leftrightarrow (A \rightarrow B)$
7. $\neg\neg A \leftrightarrow A$
8. $(\neg(A \wedge B) \wedge B) \rightarrow \neg A$
9. $(\neg(A \wedge \neg B) \wedge \neg(B \wedge C)) \rightarrow \neg(A \wedge C)$
10. $(A \leftrightarrow B) \leftrightarrow (\neg A \leftrightarrow \neg B)$

rules *modus ponens* adjunction

$$\frac{A,\ A \rightarrow B}{B} \qquad \frac{A,\ B}{A \wedge B}$$

The definition of proof is the usual one from §II.H. Throughout §A I will use $\vdash$ for the syntactic consequence relation of this system. I suspect that these axioms are not independent.

We could take $\wedge$ as a defined symbol. In that case axiom 4 is superfluous. But conjunction plays an important role in the completeness proof and for that we need the rule of adjunction, for $A \rightarrow (B \rightarrow (A \wedge B))$ is not a dependence tautology. So it seems best to take $\wedge$ as primitive. It's definability in terms of $\neg$ and $\rightarrow$ in this axiom system will follow from the completeness theorem, which it is now our goal to prove.

We begin by choosing appropriate notions of consistency and completeness. First of all, Γ is a *dependence theory* if Γ is closed under deduction. That is, if $\Gamma \vdash A$ then $A \in \Gamma$. In particular note that every theorem of this system is in every dependence theory.

Γ is a *complete dependence theory* if Γ is a dependence theory and for every A, one of $A, \neg A$ is in Γ.

Γ is *consistent* if there is no A such that $\Gamma \vdash A$ and $\Gamma \vdash \neg A$.

A *model* of Γ is a dependence model in which every wff in Γ is true.

Note that these are the same definitions as for classical logic.

We have to establish some derivations before we can embark on the proof of the completeness theorem. I'll rarely give explicit syntactic derivations, but rather sketch them listing the "crucial" axiom(s) or ones being used for the first time.

Lemma 3

a. If $\Gamma \vdash A \to B$ and $\Gamma \vdash B \to C$, then $\Gamma \vdash A \to C$.

b. $\vdash A \to A$

c. $\vdash D(A, A)$

d. $\vdash \neg(A \wedge \neg A)$

e. $\vdash \neg((A \to B) \wedge \neg\neg(A \wedge \neg B))$

f. $\Gamma \vdash (A \wedge B) \wedge C$ iff $\Gamma \vdash A \wedge (B \wedge C)$ iff $\Gamma \vdash A \wedge (C \wedge B)$

Proof: a. If $\Gamma \vdash A \to B$ and $\Gamma \vdash B \to C$, then by axioms 5 and 6 the following are all consequences of Γ: $D(A,B)$, $D(B,C)$, $\neg(A \wedge \neg B)$, $\neg(B \wedge \neg C)$. By adjunction, $\Gamma \vdash D(A,B) \wedge D(B,C)$ as well as $\Gamma \vdash \neg(A \wedge \neg B) \wedge \neg(B \wedge \neg C)$, so by axioms 1 and 9 $\Gamma \vdash D(A,C)$ and $\Gamma \vdash \neg(A \wedge \neg C)$. Using adjunction and axiom 6 we get $\Gamma \vdash A \to C$.

b. By axioms 7 and 5, $\vdash A \to \neg\neg A$ and $\vdash \neg\neg A \to A$, so by (a), $\vdash A \to A$.

c. This follows from (b) using axioms 5 and 6.

d. As for (c).

e. By axioms 6 and 5 and (a), $\vdash (A \to B) \to \neg(A \wedge \neg B)$. So by those same axioms again, $\vdash \neg((A \to B) \wedge \neg\neg(A \wedge \neg B))$.

f. By axiom 5 and the rules. ■

Lemma 4

Γ is a complete consistent dependence theory iff there is a dependence model of Γ.

Proof: The implication from right to left is easy.

Suppose now that Γ is a complete and consistent dependence theory. Define

$$\upsilon(p_i) = \begin{cases} \mathsf{T} & \text{if } p_i \in \Gamma \\ \mathsf{F} & \text{if } p_i \notin \Gamma \end{cases}$$

and

$\mathsf{D}(A,B)$ iff $D(A,B) \in \Gamma$

Then D is a dependence relation by the characterization of dependence relations in Lemma 1 and the fact that all instances of axiom schemas 1–4 are in Γ. I will show by induction on the length of wffs that $\langle \upsilon, \mathsf{D} \rangle \models A$ iff $A \in \Gamma$.

It is true if $A \in PV$. Now suppose the equivalence holds for all wffs shorter than A. If A is $\neg B$, then

$\neg B \in \Gamma$ iff $B \notin \Gamma$ by consistency

iff $\upsilon(B) = \mathsf{F}$ by induction
iff $\upsilon(\neg B) = \mathsf{T}$

If A is $B \wedge C$, then

$(B \wedge C) \in \Gamma$ iff $B, C \in \Gamma$ by axiom 5 and the rules
iff $\upsilon(B) = \upsilon(C) = \mathsf{T}$ by induction
iff $\upsilon(B \wedge C) = \mathsf{T}$

Now suppose that A is $B \rightarrow C$. If $\upsilon(B \rightarrow C) = \mathsf{T}$ then $\mathsf{D}(B,C)$, and $\upsilon(B) = \mathsf{F}$ or $\upsilon(C) = \mathsf{T}$. Thus $D(B,C) \in \Gamma$. I'll show that $\neg(B \wedge \neg C) \in \Gamma$, which by axiom 6 gives $B \rightarrow C \in \Gamma$. Suppose to the contrary that $\neg(B \wedge \neg C) \notin \Gamma$. Then by completeness $(B \wedge \neg C) \in \Gamma$. So by axiom 5, $B \in \Gamma$ and $\neg C \in \Gamma$, hence by induction $\upsilon(B) = \mathsf{T}$ and $\upsilon(\neg C) = \mathsf{T}$, so $\upsilon(C) = \mathsf{F}$, which is a contradiction. Thus $(B \rightarrow C) \in \Gamma$.

Lastly, suppose that $(B \rightarrow C) \in \Gamma$. Then by axiom 6, $D(B,C) \wedge \neg(B \wedge \neg C) \in \Gamma$, so $D(B,C) \in \Gamma$ and $\neg(B \wedge \neg C) \in \Gamma$. Hence $\mathsf{D}(B,C)$. Also $(B \wedge \neg C) \notin \Gamma$. So one of $B, \neg C \notin \Gamma$, because if both were then we'd have a contradiction using the rule of adjunction. If $B \notin \Gamma$ then $\upsilon(B) = \mathsf{F}$; if $\neg C \notin \Gamma$ then $C \in \Gamma$, so $\upsilon(C) = \mathsf{T}$. In either case $\upsilon(B \rightarrow C) = \mathsf{T}$. ■

The Syntactic Deduction Theorem played an important role at this point in the proof of the completeness theorem for **PC**. But for **D** we may have $A \models B$ yet $\not\models A \rightarrow B$ (I will discuss this further in the next section). Hence we will have to rely on a different proof-theoretic device. I introduce an auxiliary notion of consistency drawn from *Segerberg, 1968* .

Define Γ to be $\wedge$*-inconsistent* if there are $A_1, \ldots, A_n \in \Gamma$ and some way of placing parentheses to associate them, which I'll ambiguously notate as $A_1 \wedge \cdots \wedge A_n$, such that $\vdash \neg(A_1 \wedge \cdots \wedge A_n)$. Γ is $\wedge$*-consistent* if it's not $\wedge$-inconsistent. Note that by axioms 5, 10, and Lemma 3.f, the A_i's may also be permuted, which is why I included axiom 10.

Lemma 5 **a.** If $\nvdash A$, then there is a complete consistent dependence theory Σ such that $\neg A \in \Sigma$.

b. If $\Gamma \nvdash A$, then there is a complete and consistent dependence theory Σ such that $\Gamma \cup \{\neg A\} \subseteq \Sigma$.

Proof: a. Let $B_0, B_1, \ldots$ be a listing of all wffs (see §II.G.1, p.30). Define

$$\Sigma_0 = \{\neg A\}$$

$$\Sigma_{j+1} = \begin{cases} \Sigma_j \cup \{B_j\} & \text{if this is } \wedge\text{-consistent} \\ \Sigma_j & \text{otherwise} \end{cases}$$

$$\Sigma = \bigcup_j \Sigma_j$$

I'll show that Σ is the required theory.

First, Σ_0 is $\wedge$-consistent for if not then $\vdash \neg\neg A$, and then by axiom 7, $\vdash A$. So by construction, Σ is $\wedge$-consistent.

Next we show that if $\vdash A$ then $A \in \Sigma$. Suppose not. Then $\vdash B$ and $B \notin \Sigma$. Since for some j, B is B_j, $B \notin \Sigma_{j+1}$. Thus there are $A_1, \dots, A_n \in \Sigma_j$ such that $\vdash \neg(A_1 \wedge \cdots \wedge A_n \wedge B)$. By using adjunction and axiom 8, $\vdash \neg(A_1 \wedge \cdots \wedge A_n)$, which contradicts the $\wedge$-consistency of Σ_j.

Exactly one of B and $\neg B$ must be in Σ. They cannot both be in Σ as we would get an $\wedge$-inconsistency via Lemma 3.d. And one must be in Σ for if not then for some Σ_k, B is B_h, $\neg B$ is B_j, with $h, j \leq k$, and $\Sigma_k \cup \{B\}$ and $\Sigma_k \cup \{\neg B\}$ are both $\wedge$-inconsistent. So there would be wffs $A_1, \dots, A_n, B_1, \dots, B_m \in \Sigma_k$ such that $\vdash \neg(A_1 \wedge \cdots \wedge A_n \wedge \neg B)$ and $\vdash \neg(B \wedge B_1 \wedge \cdots \wedge B_m)$. Thus by axiom 9, $\vdash \neg(A_1 \wedge \cdots \wedge A_n \wedge B_1 \wedge \cdots \wedge B_m)$, contradicting the $\wedge$-consistency of Σ.

Σ is closed under the rule of adjunction, for if $A, B \in \Sigma$ and $A \wedge B \notin \Sigma$ then there are $A_1, \dots, A_n \in \Sigma$ such that $\vdash \neg(A_1 \wedge \cdots \wedge A_n \wedge (A \wedge B))$, a contradiction on the $\wedge$-consistency of Σ.

Lastly, Σ is closed under *modus ponens* for suppose not. Then there are wffs A, B such that $A \rightarrow B \in \Sigma$, $A \in \Sigma$, and $B \notin \Sigma$. By the completeness of Σ, $\neg B \in \Sigma$. So by adjunction, $(A \wedge \neg B) \in \Sigma$. Therefore, by completeness, $\neg\neg(A \wedge \neg B) \in \Sigma$, for otherwise $\neg(A \wedge \neg B) \in \Sigma$, contradicting the $\wedge$-consistency of Σ. But by Lemma 3.e this, too, contradicts the $\wedge$-consistency of Σ. So $B \in \Sigma$.

b. Define $\wedge$-*consistent relative to* Γ by replacing '$\vdash$' by '$\Gamma\vdash$' in the definition of $\wedge$-consistency above. Using this notion, Σ can be defined as in part (a) starting with $\Sigma_0 = \Sigma \cup \{\neg A\}$. ■

Theorem 6 **(*Strong Completeness for* D)** $\Gamma \vdash A$ iff $\Gamma \models A$

Proof: Suppose $\Gamma \vdash A$ and $\langle v, s \rangle \models \Gamma$. Every axiom is true in $\langle v, s \rangle$, as you may check. Every wff in Γ is true in $\langle v, s \rangle$, and the rules of proof preserve truth in a model. Hence, $\langle v, s \rangle \models A$.

Now we prove that if $\Gamma \nvdash A$, then $\Gamma \nvDash A$. Suppose $\Gamma \nvdash A$. By Lemma 5.b, there is a complete and consistent dependence theory Σ such that $\Gamma \cup \{\neg A\} \subseteq \Sigma$. By Lemma 4 there is a model of Σ. ■

This proof of the Strong Completeness Theorem is nonconstructive: given a tautology it tells us that there is a derivation for it, but does not tell us how to find one. It would be a worthwhile project to give a constructive proof of finite strong completeness, perhaps using the normal forms discussed in §6 above (cf. §II.K.1).

8. A Deduction Theorem

The Semantic Deduction Theorem fails: $(A \wedge \neg A) \models B$ for every A and B, but $\nvDash (A \wedge \neg A) \rightarrow B$ if B contains a propositional variable that A does not. By the

Completeness Theorem this is also the case for syntactic deductions. However, we do have:

Lemma 7 ***(The material implication form of the Deduction Theorem)***
$\Sigma, A \vdash B$ iff $\Sigma \vdash \neg(A \wedge \neg B)$

We could axiomatize **D** using material detachment,

$$\frac{A, \neg(A \wedge \neg B)}{B}$$

as the only rule, as I described for Relatedness Logic (§III.H, pp.74–75). I gave reasons there why I think that's the wrong approach. It's also the wrong approach from an algebraic viewpoint: the relation $A \sim B$ iff $\vdash_{\mathbf{D}} A \leftrightarrow B$ establishes a congruence relation on Wffs from which we can define algebraic semantics for **D**, as I show in *Epstein, 1987*. However, the relation $A \sim B$ iff $\vdash_{\mathbf{D}} \neg(A \wedge \neg B) \wedge \neg(B \wedge \neg A)$, that is, A is extensionally equivalent to B, is not a congruence relation on Wffs.

Here, unlike for Relatedness Logic, the rule of substitution of logical equivalents,

$$\frac{A \leftrightarrow B}{C(A) \leftrightarrow C(B)}$$

is a derived rule, for it preserves truth in a model: if $\langle v, s \rangle \models A \leftrightarrow B$, then $v(A) = v(B)$ and $s(A) = s(B)$. That is, A is extensionally equivalent to B.

Aside: Because **D** is closed under this rule, it's easy to prove that the Lewis system **S1** (see *Hughes and Cresswell, 1968*, p.217) is a subsystem of **D**, reading strict implication as $\rightarrow$. This suggests the following question: is there a modal logic into which **D** can be translated grammatically (cf. Chapter X)? More particularly, is there a modal logic **L** and a schema $\theta(A,B)$ in its language such that $\mathbf{L} \vdash \theta(A,B)$ iff $\mathbf{D} \vdash A \rightarrow B$?

I don't know if it's possible to give Kripke-style possible world semantics. It is not possible to give finite many-valued semantics for **D** (Corollary VIII.22, p.254).

Notice that except for the question of whether **D** has Kripke-style semantics I've answered all of the questions of Chapter IV as they apply to Dependence Logic.

9. History

I first presented Dependence Logic in a preprint in 1978. I was stimulated to do so by conversations I'd had with Niels Egmont Christensen the previous summer. We'd been discussing Relatedness Logic and I found that I could model his notion of entailment by changing the relationship governing the truth-table for the conditional from nonempty intersection to containment. If you look at the wffs on p.81 of *Christensen, 1973*, you will see that all those under '+' are dependence tautologies, while all those under '–' are not, with the exception of $p \to (\neg p \to p)$.

I later realized that I'd been influenced by a brief conversation I'd had with Michael Dunn that summer. He had taken a system which Parry, *1933*, devised to axiomatize analytic implication and modified it slightly, presenting semantics for the modification in *Dunn, 1972*. In the last section of that paper Dunn reports that R. K. Meyer had noted that the set-assignment semantics of what I have since called Dependence Logic are complete for Dunn's axiom system. So Dunn's logic and mine are equivalent. Parry, *1971* and *197?*, has detailed the history of this and related syntactical approaches to entailment. Parry's syntactic motivation for his logic provides an interesting comparison with that in §A.1 and §A.2 above.

> Our conception of deducibility may be clarified thus: Q is deducible from P if, in any system in which P is asserted, the assertion of Q is justifiable, assuming a reasonably complete logic. Now, we ask, if a system of Euclidean geometry contains the assertion that two points determine a straight line, are we justified in asserting in this system: 'Either two points determine a straight line or some mice like cheese'? No, this strange disjunction is not a legitimate assertion in a system of Euclidean geometry, for the simple reason that no such system contains the terms 'mice' or 'cheese', nor can one define by geometric concepts a type of cheese any self-respecting mouse would nibble at. We conclude, then, that it is not true that the statement that either two points determine a straight line or some mice like cheese is deducible from the statement that two points determine a straight line. So the statement that either p or q is not always deducible from the statement that p; putting it the other way around, we find that p does not always entail that p or q.
>
> *Parry, 1971*, p.5

K. Fine, *1979*, has also discussed Parry's system and his intuitions seem similar to mine. Kielkopf, *1977*, surveys many syntactic and semantic theories of entailment and compares the work of Parry, Dunn, Fine, and others.

Hanson, *1979*, has a particularly stimulating discussion of implication as containment of logical consequences, a topic that I take up in §F below.

B. Dependence-Style Semantics

There are other ways to interpret 'content' in accord with the idea that for $A \to B$ to be true the content of A must contain that of B. And often the idea of dependent implication is allied to a view of negation and conjunction as classical. This suggests looking at a range of logics which use the truth-tables of Dependence Logic.

I call the following tables the *dependence truth conditions*:

A	$\neg$A
T	F
F	T

A	B	A $\wedge$ B
T	T	T
T	F	F
F	T	F
F	F	F

A	B	$s(A) \supseteq s(B)$	A $\rightarrow$ B
any values		fails	F
T	T	holds	T
T	F		F
F	T		T
F	F		T

I'll say that a logic has or uses *dependence-style semantics* if it is presented by set-assignment semantics which use these tables. If the relation governing the conditional is $\supsetneq$ instead of $\supseteq$ then I'll say that we have *weak* dependence truth-conditions, and *weak* dependence-style semantics. In any such logic we can define A $\vee$ B in any of the three ways I described in §A.3, p.121.

Actually we already have a very weak semantics here. We could analyze the tautologies of all models $<v, s>$ which use dependence-style semantics with any set-assignment s. But that's too wide a class of set-assignments. For one thing, the only tautologies of the form A $\rightarrow$ B would be A $\rightarrow$ A. More importantly, though, there is no clear sense of what idea of content could be modeled by the class of all set-assignments. We have here only the outline of content, the idea of an idea of content.

Suppose we're given dependence-style semantics for a logic where the class of set-assignments is defined by equations which use only the Boolean operations of union, intersection, and complementation. Then we can give dual semantics for that logic by

1. dualizing the equations (that is, interchanging union and intersection and interchanging $\varnothing$ and S throughout)
2. evaluating negation and conjunction classically

and

3. evaluating the conditional by the following table:

A	B	$s(A) \subseteq s(B)$	$A \to B$
any values		fails	F
T	T		T
T	F	holds	F
F	T		T
F	F		T

I call these the *dual dependence truth conditions*, and any set-assignment semantics using them *dual dependence-style semantics*. For instance, **D** can be presented as a logic which uses dual dependence truth-conditions and the *intersection set-assignments*, those satisfying $s(A) = \bigcap \{ s(p) : p \text{ appears in } A \}$, or equivalently: $s(A) = s(\neg A)$ and $s(A \to B) = s(A \wedge B) = s(A) \cap s(B)$.

In the same way we can translate dual dependence-style semantics to dependence-style semantics. So if it is clear how to dualize the semantics I'll say that a logic has dependence-style semantics even though I present it using dual dependence-style semantics. For instance, I will say that many of the modal logics of Chapter VI use dependence-style semantics though they're presented using dual dependence truth-conditions.

C. Dual Dependence Logic, Dual D

A more interesting duality arises from Dependence Logic if we retain the same class of set-assignments, that is, those satisfying $s(A) = \bigcup \{ s(p) : p \text{ appears in } A \}$, and use the dual dependence truth-conditions. I call the logic presented in that way *Dual Dependence Logic,* **Dual D**.

Walton, *1981*, has proposed using Dual Dependence Logic for reasoning about actions where we get progressively more information over time. In that paper he compares **D**, **Dual D**, and **R** (Nonsymmetric Relatedness Logic) as logics of actions.

There is a clear sense in which **D** and **Dual D** are dual. The map

$(p_i)^* = p_i$

$(\neg A)^* = \neg(A^*)$

$(A \wedge B)^* = A^* \wedge B^*$

$(A \to B)^* = \neg B^* \to \neg A^*$

establishes a translation of **Dual D** into **D** satisfying $\Gamma \vDash A$ iff $\Gamma^* \vDash A^*$, where $\Gamma^* = \{B^* : B \in \Gamma\}$. At the same time it is also a translation of **D** into **Dual D** which preserves semantic consequence. I discuss in §X.B.6 whether this should lead us to consider **D** and **Dual D** to be the "same logic".

We have the following axiom system for **Dual D**.

Dual D, Dual Dependence Logic
in $L(\neg, \rightarrow, \wedge)$

$D(A,B) \equiv_{Def} A \rightarrow (B \rightarrow B)$

$A \leftrightarrow B \equiv_{Def} (A \rightarrow B) \wedge (B \rightarrow A)$

axiom schemas

1. $D(A,B) \rightarrow \neg(D(B,C) \wedge \neg D(A,C))$
2. a. $D(A, \neg A)$
   b. $D(\neg A, A)$
3. $D(A \rightarrow B, C) \leftrightarrow (D(A,C) \wedge D(B,C))$
4. $D(A \wedge B, C) \leftrightarrow (D(A,C) \wedge D(B,C))$
5. a. $A \rightarrow (B \rightarrow (A \wedge B))$
   b. $(A \wedge B) \rightarrow (B \wedge A)$
6. a. $D(A,B) \rightarrow (\neg(A \wedge \neg B) \rightarrow (A \rightarrow B))$
   b. $(A \rightarrow B) \rightarrow D(A,B)$
   c. $(A \rightarrow B) \rightarrow \neg(A \wedge \neg B)$
7. $\neg\neg A \leftrightarrow A$
8. $(\neg(A \wedge B) \wedge B) \rightarrow (\neg A \wedge B)$
9. $(\neg(A \wedge \neg B) \wedge \neg(B \wedge C)) \rightarrow (\neg(A \wedge \neg B) \wedge \neg(A \wedge C))$
10. $(A \leftrightarrow B) \leftrightarrow (\neg A \leftrightarrow \neg B)$
11. $\neg A \rightarrow \neg(A \wedge B)$

rules $\dfrac{A, A \rightarrow B}{B}$ and $\dfrac{A \wedge B}{B}$

I'll write $\vdash_{\mathbf{Dual\ D}}$ for the consequence relation of this system and $\models_{\mathbf{Dual\ D}}$ for the semantic consequence relation of the semantics.

I've chosen these axioms to make it as easy as possible for you to prove the following completeness theorem by a straightforward modification of the proof of the completeness theorem for Dependence Logic (§A.7).

***Theorem 8** (**Strong Completeness for** **Dual D**)* $\Gamma \vdash_{\mathbf{Dual\ D}} A$ iff $\Gamma \models_{\mathbf{Dual\ D}} A$

The facts we proved about **D** can be duplicated here with the appropriate changes. For instance, **Dual D** is compactly decidable. And $\wedge$ is definable from $\neg$ and $\rightarrow$ since

$A \wedge B$ is semantically equivalent to $\neg(A \rightarrow (B \rightarrow \neg((A \rightarrow B) \rightarrow (A \rightarrow B))))$ (the same equivalence as for **S**, p.78).

Walter Carnielli has suggested calling **Dual D** an *external dualization* of **D**. That is, it arises by retaining the same class of set-assignments and dualizing the relations governing the nonclassical truth-tables, here changing $\supseteq$ to $\subseteq$. This is in contrast to an *internal dualization* which arises by retaining the truth-tables but dualizing the conditions establishing the class of set-assignments, here changing from union to intersection set-assignments. I have not studied the logic which is the internal dualization of **D**, except to note that it is not **D**, for $A \rightarrow (B \rightarrow (A \wedge B))$ is valid in it, nor is it **Dual D**, for taking $\vee$ to be defined classically from $\neg$ and $\wedge$ we have that $[(A \vee B) \rightarrow C] \rightarrow [(A \rightarrow C) \vee (B \rightarrow C)]$ is valid in it.

D. A Logic of Equality of Contents, Eq

1. Motivation

One criticism of Dependence Logic which I have heard is the following. Suppose we have a model $<v, s>$ in which $s(p) \not\supseteq s(q)$ and yet p is false or q is true, and so $\neg(p \wedge \neg q)$ is true. Then $p \rightarrow q$ fails in the model; but if we take any proposition r, either true or false, which satisfies $s(r) \supseteq s(q)$ then $(p \wedge r) \rightarrow q$ is true. For instance,

'The moon is made of green cheese $\rightarrow 1 = 0$' is false.

and

'(The moon is made of green cheese and $2+2=4$) $\rightarrow 1 = 0$' is true.

I don't see what's wrong with this. The antecedent does provide the contextual framework for the consequent. I think that the latter proposition is unacceptable to some people because they want the antecedent to have the *same* referential content as the consequent.

That's precisely what Hans Sluga, *1986*, suggests as a model of Frege's notion of sense. The logic he appears to be proposing is what I call **Eq**, and differs from Dependence Logic only in this one matter of the relation governing the truth-table for $\rightarrow$.

2. Set-assignment semantics

We take the formal language to be $L(p_0, p_1, \ldots \neg, \rightarrow, \wedge)$. **Eq** is the logic given by the class of models $<v, s>$ which use union set-assignments (that is, $s(A) = \bigcup\{s(p) : p \text{ appears in } A\}$), evaluate $\neg$ and $\wedge$ classically, and use the following table for the conditional:

A	B	$s(A) = s(B)$	$A \rightarrow B$
any values		fails	F
T	T		T
T	F	holds	F
F	T		T
F	F		T

We assume the Fully General Abstraction for these models.

For **Eq** I'll write $\vDash_{\mathbf{Eq}}$ instead of $\vDash$. Note that this logic is compactly decidable by the same methods as for **D** (§V.A.5).

As for **D** and **S** there are various definable choices for $\vee$ which will not affect what follows.

We can view **Eq** as arising from **D** by defining in **D**

$$A \Rightarrow B \equiv_{\mathrm{Def}} (A \rightarrow B) \wedge (\neg B \rightarrow \neg A)$$

Then we have the grammatical translation

$$(p)^* = p$$
$$(\neg A)^* = \neg(A^*)$$
$$(A \wedge B)^* = A^* \wedge B^*$$
$$(A \rightarrow B)^* = A^* \Rightarrow B^*$$

That is, $\Gamma \vdash_{\mathbf{Eq}} A$ iff $\Gamma^* \vdash_{\mathbf{D}} A^*$, where $\Gamma^* = \{A^* : A \in \Gamma\}$.

Unfortunately, this translation doesn't give us much grasp on the form of tautologies of **Eq** or how to derive them. For example, imagine trying to prove a simple tautology of **Eq** such as $((A \rightarrow B) \wedge \neg B) \rightarrow (\neg A \wedge \neg B)$ from the axiomatization of **D** by means of this translation.

We need to axiomatize **Eq** directly. The hard step for that is to characterize the class of binary relations on wffs that arise from union set-assignments where equality of contents is the condition for A to be related to B. Note, to begin with, that if we define

$$E(A, B) \equiv_{\mathrm{Def}} A \rightarrow (B \rightarrow B)$$

then for **Eq**-models, $\langle v, s \rangle \vDash E(A, B)$ iff $s(A) = s(B)$.

3. Characterizing the Eq-relations

I'll say that E is an *Eq-relation on wffs* if there is a union set-assignment s such that $\mathsf{E}(A, B)$ iff $s(A) = s(B)$.

Theorem 9 E is an Eq-relation iff the following all hold:

1. E is an equivalence relation.
2. If E(A,B), then E(A∧C,B∧C).
3. E(A,¬A)
4. E(A∧B,B∧A)
5. E(A→B,A∧B)
6. E(A∧(B∧C),(A∧B)∧C)
7. If E(A, B∧C) and E(B, A∧D), then E(A,B).
8. E(A, A∧A)

Proof: If E is an Eq-relation then (1)–(8) all hold. For the converse I'll need some more properties of E.

Lemma 10 If E satisfies (1)–(8) then

a. If E(A,B) and E(C,D), then E(A∧C, B∧D).
b. If E(A,B) and E(A,C), then E(A, B∧C).
c. E(B∧C, B∧B∧C) for either way of associating.
d. E(A→B,B→A)
e. E((A→B)→C, A→(B→C))
f. E(A,B) iff E(A,B†), where B† = ⩕{p: p appears in B} and the conjuncts appear in the numerical order of their subscripts (⩕ indicates the conjunction over all of the elements in the set, associating to the right).
g. Given any schema $\varphi(A_1, \ldots, A_n, C)$ where $A_1, \ldots, A_n$ are fixed wffs in the language, then: if E(A,B), then E(φ(A), φ(B)), where the substitutions are made uniformly for C.

Proof: a. If E(A,B) then by (2), E(A∧C,B∧C). If E(C,D) then by (2), E(C∧B, D∧B). By (4), E(B∧C, C∧B), so by (1), E(A∧C,D∧B). Finally by (2), E(D∧B, B∧D), so by (1), E(A∧C, B∧D).

Parts (b) and (c) follow from (2), (8), and (1).

d. This follows from (5), (4), and (1).

e. This follows from (4), (5), (6), and (1).

f. I will show that for every A, E(A,A†), from which (f) will follow by (1).

The proof is by induction on the length of A. It is immediate if the length of A is 1. So suppose it's true for all wffs of length $\leq$ n and A has length n+1. By (1), E(A,A). We now have three cases.

Case i. A is ¬B. Then E(A,B) by (3), and E(B,B†) by induction. So E(A,B†) by (1).

Case ii. A is B∧C. Then by induction, E(B,B†) and E(C,C†). So by part (a), E(B∧C, B†∧C†). To finish, we need E(B†∧C†, (B∧C)†) which follows from (4), (6), and (8), using (1).

Case iii. A is $B \rightarrow C$. Then $E(A, B \wedge C)$ by (5), and we can proceed as in Case ii.

g. Since C is acting as a variable let's replace it by X; then $\varphi(X)$ can be viewed as a propositional function, φ: Wffs $\rightarrow$ Wffs. If we treat X as having length 1, then we can assign a length to $\varphi(X)$ in the usual way. To prove (g) we then induct on the length of $\varphi(X)$.

For length 1 it is immediate. Suppose it's true for length n, and $\varphi(X)$ has length n+1. Then $\varphi(X)$ is one of i. $\neg\psi(X)$, ii. $\psi(X) \rightarrow D$, iii. $D \rightarrow \psi(X)$, iv. $\psi(X) \rightarrow \gamma(X)$, v. $\psi(X) \wedge D$, vi. $D \wedge \psi(X)$, or vii. $\psi(X) \wedge \gamma(X)$, where D is a fixed wff and ψ, γ are propositional functions. For case (i) the proof is easy by (3). For (ii), if $E(A,B)$ then by induction, $E(\psi(A), \psi(B))$, so by (2) and (5) we are done. The other parts follow similarly. ∎

We return to the proof of the theorem. Given E satisfying (1)–(8), define

D(A,B) iff there is a C such that E(A,C) and B is a subformula of C

I'll prove that D is a dependence relation. So we can conclude that there is a set-assignment s such that D(A,B) iff $s(A) \supseteq s(B)$. I'll complete the proof by showing that for that s, E(A,B) iff $s(A) = s(B)$.

Lemma 11 D is a dependence relation.

Proof: I'll use the characterization of dependence relations given in Lemma 1. First, D is transitive: suppose D(A,B) and D(B,C). Then there are $\varphi(B)$, $\psi(C)$ such that $E(A, \varphi(B))$ and $E(B, \psi(C))$. From the latter and part (g) of the previous lemma we get $E(\varphi(B), \varphi(\psi(C)))$. So $E(A, \varphi(\psi(C)))$, and so D(A,C).

D is reflexive as E is.

$D(A, \neg A)$ and $D(\neg A, A)$ follow from (3) via (1).

Finally we need to show that $D(A, B \wedge C)$ iff D(A,B) and D(A,C), and then the comparable fact for $B \rightarrow C$ will follow similarly. So suppose $D(A, B \wedge C)$. Then there is a φ such that $D(A, \varphi(B \wedge C))$. As B is a subformula of $\varphi(B \wedge C)$ and so is C, we have D(A,B) and D(A,C). On the other hand, suppose D(A,B) and D(A,C). Then there are G, H such that B is a subformula of G, C is a subformula of H, and E(A,G) and E(A,H). So by (a) and (8), $E(A, G \wedge H)$, and so by (f), $E(A, (G \wedge H)^{\dagger})$. All the propositional variables of both B and C appear in $(G \wedge H)^{\dagger}$. By (4), (6), and (8), we can rearrange and reassociate the conjuncts in $(G \wedge H)^{\dagger}$ to get some K with $(B \wedge C)^{\dagger}$ a subformula of K and E(A,K). So $D(A, (B \wedge C)^{\dagger})$, and hence by (f), $D(A, B \wedge C)$. ∎

To complete the proof of the theorem, let s be a union set-assignment such that D(A,B) iff $s(A) \supseteq s(B)$. It remains to show that E(A,B) iff $s(A) = s(B)$.

If E(A,B), then D(A,B) and D(B,A) by the symmetry of E. Hence $s(A) = s(B)$.

If $\mathsf{s}(A) = \mathsf{s}(B)$, then $\mathsf{D}(A,B)$ and $\mathsf{D}(B,A)$. Hence there are formulas G and H such that A is a subformula of G, and B is a subformula of H, and $\mathsf{E}(A,H)$ and $\mathsf{E}(B,G)$. So by (f), $\mathsf{E}(A,H^{\dagger})$ and $\mathsf{E}(B,G^{\dagger})$, and as in the last part of the proof of Lemma 11 we can find some P, Q such that $\mathsf{E}(A,B^{\dagger}\wedge P)$ and $\mathsf{E}(B,A^{\dagger}\wedge Q)$. So by (f), $\mathsf{E}(A^{\dagger},B^{\dagger}\wedge Q)$ and $\mathsf{E}(B^{\dagger},A^{\dagger}\wedge Q)$, and thus by (7), $\mathsf{E}(A^{\dagger},B^{\dagger})$. So by (f) again, we have $\mathsf{E}(A,B)$. ■

4. An axiom system for Eq

Eq, a logic of equality of referential contents
in $L(\neg,\rightarrow,\wedge)$

$E(A,B) \equiv_{Def} A\rightarrow(B\rightarrow B)$

$A\leftrightarrow B \equiv_{Def} (A\rightarrow B)\wedge(B\rightarrow A)$

axiom schemas

1. $E(A,A)$
2. $E(A,B) \rightarrow E(B,A)$
3. $(E(A,B) \wedge E(B,C)) \rightarrow (E(A,C) \wedge E(A,B))$
4. $(E(A,B) \wedge E(C,C))\rightarrow E(A\wedge C, B\wedge C)$
5. $E(A,\neg A)$
6. $E(A\wedge B, B\wedge A)$
7. $E(A\rightarrow B, A\wedge B)$
8. $E(A\wedge(B\wedge C), (A\wedge B)\wedge C)$
9. $(E(A,B\wedge C) \wedge E(B,A\wedge D)) \rightarrow ((E(A,B) \wedge E(C,C)) \wedge E(D,D)$
10. $E(A, A\wedge A)$
11. $(A\wedge B) \rightarrow (B\wedge A)$
12. $(E(A,B) \wedge \neg(A\wedge\neg B)) \leftrightarrow (A\rightarrow B)$
13. $A\leftrightarrow\neg\neg A$
14. $(\neg(A\wedge B)\wedge B) \rightarrow (\neg A\wedge B)$
15. $(\neg(A\wedge\neg B)\wedge\neg(C\wedge B)) \rightarrow (\neg(A\wedge C)\wedge E(B,B))$
16. $\neg(\neg A\wedge(A\wedge B))$

rules $\dfrac{A,\ A\rightarrow B}{B}$ $\dfrac{A,\ B}{A\wedge B}$ $\dfrac{A\wedge B}{B}$

Let $\vdash_{\mathbf{Eq}}$ stand for the consequence relation of this system.

The system is not independent. For instance, axiom 6 follows from axioms 11 and 12; probably axiom 16 is superfluous. Axioms 1–10 serve to characterize the Eq-relations, and a better proof of Theorem 9 might also lead to a smaller list of axioms. The reason I've chosen these axioms is to make it as easy as possible to prove the Strong Completeness Theorem for **Eq** by modifying the proof for Dependence Logic in §A.7.

Theorem 12 **(*Strong Completeness for* Eq)** $\Gamma \vdash_{\mathbf{Eq}} A$ iff $\Gamma \vDash_{\mathbf{Eq}} A$

There's a certain oddity about these axioms that isn't evident in our other axiom systems. The second conjunct in the antecedent of axiom 4, the second and third conjuncts in the consequent of axiom 9, and the second conjunct in the consequent of axioms 3, 14, and 15 all seem "superfluous" and arbitrary. They are there only to ensure equality of contents of antecedent and consequent. In axiom 15, for instance, any other **Eq**-tautology built from the connectives and B would work just as well as E(B,B). It may be that a syntactic presentation of **Eq** by analytic tableaux would be more natural (cf. *Carnielli, 1987*).

5. Some syntactic observations

The Deduction Theorem fails for **Eq** while the material implication form of it holds, just as for **D** (§A.8). But for the same reasons I gave for Dependence Logic in §A.6 and for Relatedness Logic in §III.H (pp.74–75), I think it inappropriate to axiomatize **Eq** by using only the rule of material detachment and primitives $\neg, \wedge$, and E.

We have that $\mathbf{Eq} \subseteq \mathbf{PC}$ because models in which every wff has the same content, and hence in which the connectives are evaluated classically, are allowed. But Exportation fails for **Eq**, for $[(A \wedge \neg A) \wedge (B \wedge \neg B)] \rightarrow (A \rightarrow B)$ is an **Eq**-tautology, while $[(A \wedge \neg A) \wedge (B \wedge \neg B) \wedge A] \rightarrow B$ fails. Importation, though, holds.

Can we reduce the list of primitives of the language of **Eq**? As in classical logic, $\neg$ isn't definable from $\wedge$ and $\rightarrow$ since any wff built solely from p_0 using the latter connectives is evaluated as true in every model in which p_0 is assigned T. Also, $\rightarrow$ is not definable from $\neg$ and $\wedge$ as it isn't truth-functional. I believe that $\wedge$ is not definable from $\neg$ and $\rightarrow$ in **Eq**, but I haven't been able to prove that. If so, it would be a significant difference from **D**, **Dual D**, **S**, and **R** .

The proof of Theorem 9 characterizing Eq-relations suggests that if quantification over propositions were allowed we could define a connective D(A,B) from $\neg, \rightarrow, \wedge$ such that in any **Eq**-model, $\langle v, s \rangle \vDash D(A,B)$ iff $s(A) \supseteq s(B)$. But without quantification that's not possible, as the connectives $\neg, \rightarrow, \wedge$ in **Eq**-models do not distinguish between $s(A) \supsetneq s(B)$ and $s(A) \cap s(B) = \varnothing$.

It's tempting to believe that $\mathbf{Eq} = \mathbf{D} \cap \mathbf{Dual\ D}$. That is the case for wffs which involve only the connectives $\neg$ and $\wedge$ or are first degree implications (see the next section). But we don't have $\mathbf{Eq} \subseteq (\mathbf{D} \cap \mathbf{Dual\ D})$ since $[A \rightarrow (B \rightarrow B)] \rightarrow [B \rightarrow (A \rightarrow A)]$ is an **Eq**-tautology but is not in **D** or **Dual D**. I don't know whether the reverse inclusion holds, $\mathbf{Eq} \supset (\mathbf{D} \cap \mathbf{Dual\ D})$.

E. A Syntactic Comparison of D, Dual D, Eq, and S

Some logicians explain the notion of relevance in logic in terms of criteria for variable sharing, for instance, *Anderson and Belnap, 1975*, and *Kielkopf, 1977* . Fine, *1977*, compares that syntactic view of relevance with a semantic one. I think that it is a mistaken approach to questions of meaning to rely primarily on syntactic criteria. However, after semantics have been established it is interesting to see what criteria of that sort apply. Let's do that for the logics we've studied, where you can check the observations by using the appropriate completeness theorems.

The following conditions are necessary for $A \to B$ to be a theorem of the logic listed:

i. There is at least one propositional variable that appears in both A and B.
Subject Matter Relatedness Logic, **S**

ii. Every propositional variable that appears in B appears in A.
Dependence Logic, **D**

iii. Every propositional variable that appears in A appears in B.
Dual Dependence Logic, **Dual D**

iv. The same propositional variables appear in both A and B.
Equality of Referential Contents Logic, **Eq**

We also have that

In **S** if (i) fails,
In **D** if (ii) fails,
In **Dual D** if (iii) fails
In **Eq** if (iv) fails

then: If $\vdash C \to (A \to B)$, there must be an occurrence of $\to$ in C.

We need the provisos here because in each of these logics $\vdash A \to (A \to A)$.

We define $A \to B$ to be a *first degree implication* or *first degree entailment* if $\to$ does not appear in either A or B.

For *first degree entailments*:

$\vdash A \to B$ iff $A \to B$ is a **PC**-tautology and
- for **S**, (i) holds.
- for **D**, (ii) holds.
- for **Dual D**, (iii) holds.
- for **Eq**, (iv) holds.

These equivalences fail if $A \to B$ isn't a first degree implication by considering either Importation or Exportation for each logic. However, as each of these logics is contained in **PC**, we have for each:

If $\vdash A \to B$, then $A \to B$ is a **PC**-tautology.

These syntactic criteria are more coarse than you might suppose: the tautologies of Nonsymmetric Relatedness Logic, **R**, satisfy the same criteria as those of **S**.

F. Content as Logical Consequences

1. The consequences of a proposition

Stimulated by the discussion in *Hanson, 1980,* Roger Maddux suggested looking at the content of a proposition as its consequences in classical logic in the framework of dependence-style semantics.

Let's suppose we start with some fixed assignment of propositions to the propositional variables of our formal language. If we were then to interpret $\neg$ and $\wedge$ classically, and equate implication with containment of contents so that $A \rightarrow B$ holds iff the **PC**-logical consequences of A contain those of B, then we'll get back **PC**. That's just the Deduction Theorem.

But that isn't what we want for implication because the truth-values of the propositions matter. Yet we don't want to consider truth-values in determining the content of the propositions for that would trivialize content: every true proposition is a consequence of every proposition, and every false one implies every other in classical logic.

Rather, we begin with an assignment of propositions:

I

$$L(p_0, p_1, \dots \neg, \rightarrow, \wedge)$$
$$\downarrow$$
$$\{p_0, p_1, \dots, \text{complex propositions formed from these using } \neg, \rightarrow, \wedge, \vee\}$$

The content of p is then the collection of semi-formal propositions in this model which are **PC**-logical consequences of p. In this way we can apply the logic where we don't already know the truth-values of the propositions.

For $A \rightarrow B$ to be true we should then have that the consequences of A contain those of B, and not both A is true and B is false. This is dependent implication. If we take negation and conjunction to be classical we have a logic. Let's proceed directly to the Fully General Abstraction.

2. Classically-Dependent Logic, DPC

A *dependent*-**PC** *model* is

II

$$L(p_0, p_1, \dots \neg, \rightarrow, \wedge, \vee)$$
$$\downarrow \quad v,\ s, \text{ and truth-tables}$$
$$\{T, F\}$$

Here v is any function, $v: PV \rightarrow \{T, F\}$. The unique set assignment is

$s(A) = \{ B : A \vdash_{PC} B \}$

And v is extended to all wffs by the dependence truth-conditions: $\neg$ and $\wedge$ are classical, and $\rightarrow$ is evaluated by

$$\mathsf{v}(A \rightarrow B) = \mathsf{T} \text{ iff not (both } \mathsf{v}(A) = \mathsf{T} \text{ and } \mathsf{v}(B) = \mathsf{F}) \text{ and } \mathsf{s}(A) \supseteq \mathsf{s}(B)$$

Classically-Dependent Logic, **DPC**, is the logic presented semantically by these models.

What about the intuition that the logic which takes content to be consequences in classical logic must be classical logic itself? It's certainly true that for any wff A which has either no occurrences of '$\rightarrow$' or which is a first degree entailment, A is a dependent-**PC** tautology iff A is a **PC**-tautology. But if p and q are distinct variables then $q \rightarrow (p \rightarrow q)$ is a **PC**-tautology, but not a dependent-**PC** one: $\mathsf{s}(p) \not\supseteq \mathsf{s}(q)$, so the consequent is always false, yet we may have that q is true. Thus **PC** $\not\subseteq$ **DPC**. More strikingly, **DPC** $\not\subseteq$ **PC** as $\neg(p \rightarrow q)$ is a dependent-**PC** tautology if p and q are distinct. For remember, there is only one set assignment for the fully abstract models of type **II**. If we were to define **DPC** for models of type **I** we could have many different set assignments, one for each realization:

$$\mathsf{s}(A) = \{ B : (\text{the realization of } A) \vdash_{\mathbf{PC}} (\text{the realization of } B) \}$$

In that case $\neg(p \rightarrow q)$ would not be a tautology, for we could have the same proposition assigned to both p and q, so that $\mathsf{s}(p)$ would be the same as $\mathsf{s}(q)$.

Maddux has proposed examining the sequence of logics $\mathbf{DPC_n}$ where $\mathbf{DPC_1} =$ **DPC** and $\mathbf{DPC_{n+1}}$ is defined as was **DPC** except taking $\mathsf{s}(A) = \{B : A \vDash_{\mathbf{DPC_n}} B\}$ (we can't use syntactic consequence unless we axiomatize the logics). Is $\mathbf{DPC_{n+1}} \subseteq \mathbf{DPC_n}$? If so, what's $\bigcap_n \mathbf{DPC_n}$?

VI Modal Logics
– S4, S5, S4Grz, T, B, K, QT, MSI, ML, G, G* –

One aspect of a proposition which can be important in reasoning is whether it is possible or impossible, necessary or not. These qualities are modalities of propositions, sometimes signaled by verbs in the subjunctive mood, such as 'could' or 'would'. In this chapter I will discuss the main ideas of modal logic in terms of possible world semantics of Kripke. In giving set-assignment semantics I will raise the question whether modal logics, like relevance logics, are based on a notion of connection of meanings of propositions.

Roger Maddux and I first began a study of modal logics and set-assignment semantics with an analysis of **S4** that provided a basis for me to consider other modal logics. That logic still seems to me the most straightforward to analyze and I begin the technical analyses with our joint work in §F.

Proofs of the completeness theorems for Kripke semantics for some of the modal logics discussed here can be found in the Appendix to this chapter. In the Summary of Logics at the end of this book I summarize all the various conditions for modal logic set-assignment semantics. For a short course on only classical modal logics and Kripke semantics you can read §A, §B, §D, then the Appendix, Remark 14.a of §G.1, and §K.1.

Another approach to modeling modalities uses many-valued logics which I present in Chapter VIII. Prior, *1955*, discusses that approach and surveys the history of the notions of possibility and necessity, for which you can also consult *Aune, 1968.*

For arguments for and against the (metaphysical) existence of possible worlds see *David Lewis, 1973*, and *Kripke, 1972*, both of which I have drawn on in the following discussions.

There are a number of books which present a general survey of modal logics and more advanced mathematical methods for dealing with them: *Chellas, 1980, Hughes and Cresswell, 1984, Segerberg, 1971*, and *Lemmon, 1977*. The last two have historical outlines of the modern development of classical modal logic. I have followed *Boolos, 1979,* for much of the technical development here.

A. Implication, Possibility, and Necessity

1. Strict implication vs. material implication

In 1912 C. I. Lewis initiated in modern formal logic a debate that dates back to the Greeks: which implications are sound. The logic of *Principia Mathematica*

(*Whitehead and Russell, 1910–1913*) formalized Philonian implication:

> Philo said that the conditional is true whenever it is not the case that its antecedent is true and its consequent false.
>
> Sextus Empiricus, *Against the Mathematicians*, VIII, 112
> translated in *Mates, 1953*, p.97

Lewis rejected that analysis: 'If roses are red then sugar is sweet' isn't true, for there is nothing in the fact that roses are red that precludes sugar being not sweet. It could have been that roses are red and sugar is not sweet.

> But Diodorus says that a conditional is true whenever it neither ever was nor is possible for the antecedent to be true and the consequent false, which is incompatible with Philo's thesis. For, according to Philo, such a conditional as 'If it is day, then I am conversing' is true when it is day and I am conversing, since in that case its antecedent, 'It is day,' is true and its consequent, 'I am conversing,' is true; but according to Diodorus it is false. For it is possible for its antecedent, 'It is day,' to be true and its consequent, 'I am conversing,' to be false at some time, namely, after I have become quiet.
>
> Sextus Empiricus, *Against the Mathematicians*, VIII, 113
> translated in *Mates, 1953*, p.98

Lewis, *1912*, distinguished these two analyses, calling the first *material implication* and the second *strict implication.* Though nowadays we often distinguish which analysis we are using by employing different symbols, $\supset$ for the former, and $\rightarrow$ or $-\!\!3$ for the latter, these are both formalizations of the same natural language connective, 'if ... then ... '. The arguments about which implications are sound are real debates: one single common notion is being analyzed, refined, in different ways. Therefore, the formal language I'll use for modal logics is $L(p_0, p_1, \ldots \neg, \rightarrow, \wedge)$, at least until §D. The formalization of 'or' is normally given in terms of $\neg$ and $\wedge$ as in classical logic

2. Possible worlds

The basic semantic idea of modal logic is that, though conjunction and negation may be evaluated classically, $A \rightarrow B$ is true iff it is impossible that A be true and B be false. Or to put it another way

(1) $A \rightarrow B$ is false iff it's possible that A could be true and B false

Is 'If Caesar crossed the Rubicon then Shakespeare wrote *Macbeth*' true? By this criterion we want to say 'no', for we believe that Shakespeare could have failed to write *Macbeth*. To analyze an implication we survey the possibilities, the ways things could have been.

We can suggestively say that a way things could have been is a *possible world*. That way of talking comes from Leibniz (see *Mates, 1986*) and was imported into formal semantics for modal logics by Kripke, *1959*.

A possible world is something we imagine could be the case. It may be a metaphysical "fact" that a certain possible world "exists," but what is needed for us to reason together is that we agree in general how we will reason about our descriptions of the ways things could be. Then we may withhold judgment and hypothesize that a particular description describes a case that is possible or one that is not possible, and consider the consequences of each.

In particular we must agree on how we are to reason about any one particular description. In any possible world we are interested only in the truth-values of the propositions which specify it when we invoke (1), so it seems that classical logic is appropriate. Hence,

(2) We agree to use classical logic in reasoning about any particular description of the way things could have been.

A convenient abbreviation of (2) is

In every possible world classical logic holds.

That way of saying things sounds suddenly full of metaphysical import, which is a danger of using the terminology of 'possible worlds'.

When we describe a way things could have been we usually only specify the truth-values of a few propositions, namely the ones in the implication we're analyzing. However, we may wish to specify infinitely many, for example, all instances of all the physical laws we now know. In either case, to employ formal methods we must first be given an assignment of propositions $p_0, p_1, \ldots$ to the propositional variables of the formal language. Then we can formally identify a possible world with an assignment of truth-values to the p_i's. That is, since we're using classical logic within the world, a possible world is a model of classical logic. Looked at syntactically this is equivalent to saying that relative to the modes of reasoning we've adopted, **PC**, a possible world is a complete and consistent set of sentences.

Now we can be fully precise about what we will take to be a possible world:

Possible Worlds Abstraction Any model of classical logic is a possible world.

With this agreement we have a standard for our imaginings. A possible world is no longer (just) a psychologically acceptable description. Indeed, a description of the way things could be which you offer might be psychologically unacceptable to me, such as conceiving of two lines through a point both parallel to a third. Yet if you can establish that the description you put forward is indeed a model of classical logic, that is a (**PC**-) consistent and complete set of sentences, then by our agreement I must accept that it is a possible world.

I will return to this discussion in §C.2 below.

Finally, we have

The Classical Modal Logic Abstraction The only properties of a proposition which matter to (this) logic are its form, its truth-value, and whether it is true or false in any given possible world.

3. Necessity

In our new terminology to say that A is possible is to say that there is some possible world of which A is true. Possibility, then, amounts to a sort of existential quantification over possible worlds.

The other side of the coin of possibility is necessity. To say that a proposition is necessary is to say that it couldn't be false. That is, it's not possible that it is false: in every possible world, in any way that we could imagine things to be, it's true. So necessity is like a universal quantification over possible worlds. Because of our agreement to use classical logic on every description we're committed to all classical tautologies being necessary truths. For instance, we may not (or perhaps cannot) imagine that the law of excluded middle fails.

Now we can restate our analysis of implications:

(3) $A \rightarrow B$ is true iff it is not the case that there is some possible world in which A is true and B false.

Abbreviating $\neg(A \wedge \neg B)$ as $A \supset B$, which is material implication, (3) can be restated as follows.

(4) $A \rightarrow B$ is true iff $A \supset B$ is necessary.

4. Different notions of necessity: accessibility relations

The notions of possibility and necessity which we've been discussing so far have been determined by requiring that a particular set of sentences, namely the classical logical tautologies, hold in every possible world: the logical laws are necessary. This is *logical necessity*, or more properly, *classical* logical necessity. There are many other notions of necessity and possibility which are important in philosophy and science (see *Aune, 1967*) and we can consider those, too. We won't, however, change our agreement that the classical logical laws are necessary.

Time dependent necessity corresponds approximately to how we reason about future events. At this time I can wonder who will be President of the U.S.A. in 1993, and consider various possibilities. Suppose I assume that it is Ms. X. Then I can consider, relative to that assumption, that is relative to the possible world in which the year 1993 is as I've stipulated, whether Mr. Y will be President in 1997. All the possible future worlds I imagine will be exactly like the one we live in up to this moment. All the possible worlds subsequent to the one in which Ms. X is President in 1993 will be exactly like that one up to 1993. *Time dependent necessity* or *inevitability at time* t is a notion of necessity *relative to the possible world under consideration.* Right now it's possible that Mr. Y will be President in 1993. Relative to the world in which Ms. X is President in 1993, call it w, it is necessary that Mr. Y is not President in 1993. We might say that a world in which Mr. Y is President in 1993 is *accessible* to the present world but not accessible to w. Accessibility is a relation between possible worlds, and for time dependent necessity world z is accessible relative to world w if w is specified to be at time t and z is exactly like w up to time t. But what does 'exactly like' mean here? To begin with, both worlds must satisfy the classical laws of logic. And in addition any other facts we've specified about w at time t must also hold in world z. We may implicitly assume, say, that world w satisfies the same physical laws as our world, but we must explicitly specify the sentences that are to be true in w at time t if we are to use our formal models. We may also assume that a stipulation of the way things could be in 1993 presupposes a history of how things were before then, but if we do not explicitly stipulate that history we cannot take it into account in our formal reasoning. One thing that must always be specified for time dependent analyses is the time at which the world occurs.

What can we say about this accessibility relation for time dependent necessity? Certainly it is reflexive and transitive. It's also anti-symmetric: if z is accessible to w, then z will occur later than w and hence w cannot be exactly like z at that later time unless $w = z$.

For this notion of necessity let's consider how we will evaluate the truth-value of $A \to B$. Notating 'z is accessible to w' as wRz, we have

(5) $A \to B$ is true at world w iff $A \supset B$ is necessary relative to w
iff at every possible world z such that wRz, $A \supset B$ is true

And then

(6) $A \to B$ is true iff $A \to B$ is true at every possible world w

Recall that I said in passing that we might require world z to satisfy the same physical laws as world w. That's *physical necessity* and corresponds roughly to how we might reason in science. I might postulate a physical law A and consider two worlds w and z satisfying all the laws we now know and in w A is true, in z it's false. So w and z are accessible to our present world, yet inaccessible to each other.

Or in a science fiction vein, I might consider a world w in which gravitational attraction is proportional to the inverse cubed of the distance between objects. With respect to physical necessity, neither that world nor any world accessible to it is accessible to our present one. For *physical necessity*, a world z is accessible to world w, wRz, if all the physical laws specified to hold in w hold in z. To evaluate an implication with respect to physical necessity we would again use (5) and (6).

B. The General Form of Kripke Semantics for Modal Logics

1. The formal framework

We can now formalize the notions we've been discussing. Recall that our formal language is $L(p_0, p_1, \ldots \neg, \rightarrow, \wedge)$, and we define

$$A \supset B \equiv_{Def} \neg(A \wedge \neg B)$$

Suppose we have an assignment of propositions p_0, p_1, ... to p_1, p_2, ... We said that a possible world is a model of classical logic. So it is determined by the propositions true in it. That is, we can characterize it as a collection of p_i's, namely the ones true in it, or through their temporary names in PV.

A *model* is $\langle W, R, e \rangle$ where W is a *nonempty* set whose elements are called *possible worlds,* e is an *evaluation,* $e: W \rightarrow \text{Sub PV}$, and R is a binary relation on W called the *accessibility relation between possible worlds*. We can think of the elements w of W as markers, names of possible worlds. The evaluation e says which propositions (via their temporary names in PV) are true in w, namely $e(w)$. We inductively define $w \vDash A$, read 'A is true at world w', or 'A is true in world w', or 'w validates A', where $\nvDash$ means '$\vDash$ does not hold' and is read as 'is false at' or 'does not validate'.

(7) $w \vDash p$ iff $p \in e(w)$

$w \vDash A \wedge B$ iff $w \vDash A$ and $w \vDash B$

$w \vDash \neg A$ iff $w \nvDash A$

$w \vDash A \rightarrow B$ iff for all z such that wRz, not both $z \vDash A$ and $z \nvDash B$

The first three clauses state that a formal possible world is a model of classical logic. An equivalent formulation of the last condition is:

$w \vDash A \rightarrow B$ iff for all z such that wRz, $z \vDash A \supset B$

We can define $\langle W, R, e, w \rangle$ to be a *model with designated world* w. Then $\langle W, R, e, w \rangle \vDash A$ means $w \vDash A$. This gives an analysis of implication based on necessity and possibility relative to a particular world w. Finally A is true in model $\langle W, R, e \rangle$, written $\langle W, R, e \rangle \vDash A$, iff for all $w \in W$, $\langle W, R, e, w \rangle \vDash A$.

Note: In the literature models are often presented using $e^*: PV \to \text{Sub}\, W$ where the first step of the inductive definition (7) is: $w \models p$ iff $w \in e^*(p)$. This is mathematically equivalent to the definition above: given $e: W \to \text{Sub}\, PV$, define $e^*: PV \to \text{Sub}\, W$ via $e^*(p) = \{w : p \in e(w)\}$; then $\langle W, R, e^* \rangle \models A$ iff $\langle W, R, e \rangle \models A$. The reverse translation I'll leave to you. I'll discuss the intuition behind this alternate presentation of models in §C.

If we have a particular notion of necessity in mind we may wish to model it structurally with a particular class of accessibility relations. We'll call a pair $\langle W, R \rangle$ a *frame*, and say that A is *valid in* $\langle W, R \rangle$, written $\langle W, R \rangle \models A$, iff for all e, $\langle W, R, e \rangle \models A$. If we specify a class of frames, F, we can say that

(8) A is *valid* relative to the notion of necessity corresponding to F means that for all $\langle W, R \rangle \in F$ and all e, $\langle W, R, e \rangle \models A$.

We usually say that a frame or model has a property, for example reflexivity, if R does. However, when we say that $\langle W, R \rangle$ is *finite* we mean that W is finite.

It is now standard to assume a Fully General Abstraction for Kripke semantics and I will do so.

The Fully General Abstraction for Kripke Semantics for Classical Modal Logics

Any set W together with any relation R on W and any function $e: W \to \text{Sub}\, PV$ can comprise a model $\langle W, R, e \rangle$.

In general it is possible to restrict W to be countable, but there are cases where that would make the mathematics significantly more difficult, so I will not make that assumption. However, we may restrict the cardinality of W to be no greater than that of the cardinality of all subsets of the natural numbers.

For a particular notion of necessity we can and do assume a Fully General Abstraction which states that any relation R on W which satisfies the appropriate structural conditions for the notion of necessity under consideration is suitable to use in a model. For example, for a logic of time dependent necessity as described in §B.3 we might take the class of models to be all those $\langle W, R, e \rangle$ such that R is reflexive, transitive, and anti-symmetric (cf. §G.2).

Example As examples of the use of these semantics, let's investigate whether the "paradoxes" of classical logic and of strict implication are valid.

Recall that the "paradoxes" of classical logic are $\neg p \to (p \to q)$ and $p \to (p \to q)$. I'll show that $\neg p \to (p \to q)$ is not valid by exhibiting a model and a world in that model in which it is not true.

Let $W = \{w, z, y\}$ where wRz and zRy, but no other elements are related. Let p be false in z and true in y, and q false in y; we don't need to worry about any other assignments, as we'll see in Lemma 1. Diagrammatically we have:

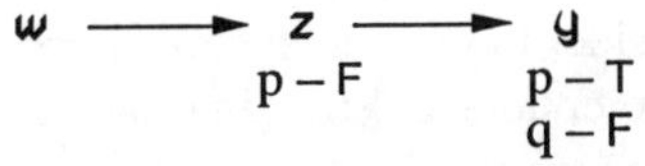

$\neg p \to (p \to q)$ is true in w iff $\neg p \supset (p \to q)$ is true in every world related to w. The only world related to w is z, and there p is false and $\neg p$ is true. Thus $p \to q$ must be true in z for the whole wff to be true. But zRy, and in y, $p \supset q$ is false. So $p \to q$ is false in z. Hence $\neg p \to (p \to q)$ is false in w and so is not valid.

I'll leave to you to show that $p \to (p \to q)$ is not valid.

The "paradox" of strict implication $(p \wedge \neg p) \to q$ is valid, for in any world $(p \wedge \neg p) \supset q$ is true. Similarly, $p \to (q \to q)$ is valid. These are necessary (classical) truths.

Recall that I remarked at the end of §A.2 that when we describe a way things could have been we usually specify the truth-values of only a few propositions, namely the ones in the implication we're analyzing. That's what we did in the formal Kripke semantics in Example 1; I'll now show that's justified. In doing so I'll also demonstrate that the only worlds involved in establishing the truth or falsity of a proposition at a world w are those which we can reach in a finite number of steps from that world.

Given $\langle W, R \rangle$, define *the transitive closure of* R to be R^*, the smallest transitive relation that contains R. That is, wR^*z iff there is a sequence $w = y_1, y_2, \ldots, y_n = z$ such that for each i, $y_i \, R \, y_{i+1}$.

Lemma 1 Let $\langle W, R, e, w \rangle$ be a model with designated world w and $\langle Z, R, e, w \rangle$ be that model restricted to $Z = \{w\} \cup \{z : z \in W \text{ and } wR^*z\}$. That is, in $\langle Z, R, e, w \rangle$ the accessibility relation R and evaluation e are restricted to Z. Then

a. For all A, $\langle W, R, e, w \rangle \vDash A$ iff $\langle Z, R, e, w \rangle \vDash A$.

b. For any e^*, if for all p in A and all $z \in W$, $p \in e^*(z)$ iff $p \in e(z)$, then $\langle W, R, e, w \rangle \vDash A$ iff $\langle Z, R, e, w \rangle \vDash A$.

Proof: I'll show that for all $y \in Z$ and all A, $\langle W, R, e, y \rangle \vDash A$ iff $\langle Z, R, e^*, w \rangle \vDash A$, by induction on the length of A.

It is immediate if the length of A is 1. Suppose the lemma true for all wffs of length $\leq$ n and that A has length n+1. If A is $\neg B$ or $B \wedge C$ the proof is straightforward. So suppose A is $B \to C$. In that case, $\langle W, R, e, y \rangle \vDash B \to C$ iff for all $z \in W$ such that yRz, $\langle W, R, e, z \rangle \vDash B \supset C$. If $y \in Z$ and yRz then wR^*z, so $z \in Z$. So we have by induction,

$$\langle W, R, e, y \rangle \vDash B \to C \text{ iff for all } z \in Z \text{ such that } yRz,\ \langle W, R, e, z \rangle \vDash B \supset C$$
$$\text{iff for all } z \in Z \text{ such that } yRz,\ \langle Z, R, e^*, z \rangle \vDash B \supset C. \quad \blacksquare$$

2. Possibility and necessity in the formal language

In the rest of this chapter we're going to consider logics which correspond to some easily specifiable class of frames with two exceptions in §J.3. In order to characterize syntactically the wffs which are tautologous relative to a particular class of frames it would be useful to have an expression in our formal language corresponding to the metalogical notion of necessity, that is, true in all accessible possible worlds. Fortunately we have one. We set

$$\Box A \equiv_{\text{Def}} \neg A \rightarrow A$$

Then

(9) $\quad w \vDash \Box A$ iff for all z, if wRz then $z \vDash A$

We usually read $\Box A$ as 'It is necessarily the case that A', or for short, 'Necessarily A', or 'It is necessary that A'. Notice that we can now say

(10) $\quad w \vDash A \rightarrow B$ iff $w \vDash \Box(A \supset B)$

which corresponds to (4) above.

Instead of this definition of $\Box A$ we could use $(A \rightarrow A) \rightarrow A$. Nothing that follows depends on which of these definitions we choose, for every logic we examine here contains all instances of $[(A \rightarrow A) \rightarrow A)] \supset (\neg A \rightarrow A)$ and $(\neg A \rightarrow A) \supset [(A \rightarrow A) \rightarrow A]$.

Example Consider the evaluation of $\Box p \supset p$ using (9). This is true at a world w iff not both $\Box p$ is true and p is false at w. But that can happen:

$$\begin{array}{ccc} w & \longrightarrow & z \\ p - F & & p - T \end{array}$$

for $\Box p$ is true iff p is false in every world accessible to w, and w need not be accessible to itself. However, if w is accessible to itself then $\Box p \supset p$ is true at w. So $\Box p \supset p$ is not valid, though it holds in any model in which R is reflexive.

Another example is $\Box(p \wedge q) \supset (\Box p \wedge \Box q)$. Suppose $\Box(p \wedge q)$ is true at a world w. Then in every world z related to w, $p \wedge q$ is true. So in every such z, both p and q are true. Hence, $\Box p$ is true at w and $\Box q$ is true at w, and so $\Box(p \wedge q) \supset (\Box p \wedge \Box q)$ is true at w. Thus $\Box(p \wedge q) \supset (\Box p \wedge \Box q)$ is a tautology.

Given our definition of $\Box A$ we can define

$$\Diamond A \equiv_{\text{Def}} \neg \Box \neg A$$

Then

$w \vDash \Diamond A$ iff there is some z such that wRz and $z \vDash A$

Thus $\Diamond$A being true at a world corresponds to 'A' being possible. So $\Diamond$A is usually read as 'It is possibly the case that A', or for short, 'Possibly A', or 'It is possible that A'.

Note that in every $\langle W, R, e, w \rangle$, $w \vDash \Box A$ iff $w \vDash \neg\Diamond\neg A$, which corresponds to how we first described what it meant for A to be necessary. Alternatively, we could have defined $\Diamond$A as $\neg(A \rightarrow \neg A)$ and then taken $\Box$A to be defined as $\neg\Diamond\neg A$.

Historically, syntactic characterizations of modal logics preceded the semantic analysis of Kripke, and almost all the syntactic characterizations used $\Box$ or $\Diamond$ as a primitive connective (see §D). But are we justified in regarding 'It is necessary that ...' and 'It is possible that ...' as propositional connectives? If so they're certainly quite different from the kind we have used so far in that until now we have not needed to use 'is the case that' or an adverbial modification to form a new proposition from a given one. The only use we've made so far of 'is the case that' is to standardize a variety of uses of 'not', none of which originally involved that phrase.

On the face of it, it would appear to be a confusion of the formal language and the metalanguage to use these as connectives. When we say, 'It is possibly the case that roses are red' we are talking *about* the proposition 'roses are red', not using it. So iterations would not involve the proposition we began with, and seem to make no sense: 'It is necessarily the case that it is possibly the case that roses are red'. The situation seems to be the same as when we rejected viewing R as a connective for relatedness logic (§III.B, pp.64–65, and §III.H, pp.77–78).

There has been a long debate on this question, however, with many logicians arguing that there is no use-mention confusion involved in taking $\Box$ and $\Diamond$ as connectives. Haack, *1973*, Chapter 10, gives a good introduction to those debates; Copeland, *1978*, reviews the arguments and shows that if the metalevel distinctions are taken seriously and an infinite hierarchy of languages is produced, then the confusions are comparable to collapsing these languages which is in some sense technically harmless (cf. Remarks 14, p.170 below). From the point of view I've presented here you're free to view $\Box$ and $\Diamond$ as either sentence connectives or solely as syntactical abbreviations useful for metalogical investigations. Modal logics are justified as logics based on particular abstractions of 'if ... then ... '.

C. A General Form for Set-Assignment Semantics for Modal Logics

1. Semantics

It's often said by modal logicians that the content of a proposition is the possible worlds in which it is true. That way of putting it is misleading, for it suggests that possible worlds exist before we stipulate them in a language. Even were that a metaphysical matter of fact, as David Lewis (*1973,* Chapter 4) argues, it would be beyond our formal methods as our models only allow us to characterize possible worlds in terms of the sentences true in them. However, once we have a model as above we can indeed identify a proposition with the worlds in which it's true, as I pointed out in the Note, p.152. Set-assignment semantics are based on this idea, using the Classical Modal Logic Abstraction.

Given a proposition A we take a set $\mathsf{s}(A)$ to be its content, viewing the elements of $\mathsf{s}(A)$ as descriptions of the ways things could be if A is true. Because we choose to use classical logic in reasoning about any particular description we must have the following, where $\overline{\mathsf{s}(A)}$ denotes the complement of $\mathsf{s}(A)$.

M1. $\mathsf{s}(A \wedge B) = \mathsf{s}(A) \cap \mathsf{s}(B)$

M2. $\mathsf{s}(\neg A) = \overline{\mathsf{s}(A)}$

And because we want to respect the equivalence of $A \rightarrow B$ and $\Box(A \supset B)$ as in (4) and (10), we want

M3. $\mathsf{s}(A \rightarrow B) = \mathsf{s}(\Box(A \supset B))$

Any further structural requirements we impose on contents of propositions will depend on which particular notion of necessity we are modeling.

What should the truth-conditions be? We've already agreed that $\neg$ and $\wedge$ are to be interpreted classically. How should we analyze $A \rightarrow B$?

Let's say that two propositions are *compossible*, or *compatible* if we can conceive of a description, a possible world, in which both are true. So A and B are compossible iff $\mathsf{s}(A) \cap \mathsf{s}(B) \neq \varnothing$. They are incompatible otherwise. Then

$A \rightarrow B$ is true iff it's not possible that both A is true and B is false
iff A and $\neg B$ are incompatible
iff $\mathsf{s}(A) \cap \mathsf{s}(\neg B) = \varnothing$

which models the criteria quoted from Diodorus at the beginning of the chapter (p.147). It is equivalent to:

(11) $A \rightarrow B$ is true iff in every world in which A is true, B is not false
iff $\mathsf{s}(A) \subseteq \mathsf{s}(B)$

Need we also require that if A is true, so is B? That seems a *sine qua non* of implication, for without it we cannot use *modus ponens*: from A and $A \rightarrow B$ conclude B. So we take

(12) $A \rightarrow B$ is true iff $\mathsf{s}(A) \subseteq \mathsf{s}(B)$ and not both A is true and B is false

A *modal semantics for* (*of*) *implication* is a set-assignment model or class of models $\langle \upsilon, \mathsf{s} \rangle$ where:

$\upsilon : PV \rightarrow \{T, F\}$

$\mathsf{s} : \text{Wffs} \rightarrow \text{Sub } \mathsf{S}$

s satisfies M1–M3, and
v is extended to all wffs by the classical tables for $\neg$ and $\wedge$, and (12) for $\rightarrow$

Note that these are the dual dependence truth-conditions.
Corresponding to the Fully General Abstraction for Kripke semantics, we have:

The Fully General Abstraction for Set-Assignment Semantics for Classical Modal Logics Any valuation v and set-assignment s satisfying M1–M3 can comprise a modal semantics for implication.

For particular notions of implication associated with various modal logics we will assume a Fully General Abstraction which states that any modal semantics of implication which satisfies the appropriate structural criteria is a model for that logic.
In these models we have the following tables for $\Box A$ and $\Diamond A$, as demonstrated in Lemma 2 below.

(13)

A	s(A) = S	$\Box$A
any value	fails	F
T	holds	T
F		F

(14)

A	s(A) $\neq \varnothing$	$\Diamond$A
T	fails	T
F		F
T	holds	T
F		T

If we were to take $\neg, \wedge, \Box$ *as primitives instead*, defining $A \rightarrow B$ as $\Box(A \supset B)$ as I'll discuss in §D, *then table* (13) *replaces* (12) *in the definition of modal semantics for implication* and (12) may be derived.

We may, however, use a notion of necessity in which the actual world in which we are evaluating $A \rightarrow B$ is not accessible to itself; then the truth-value of $A \rightarrow B$ would not depend on the truth or falsity of A or of B in the actual world. For instance, one might argue that it doesn't matter whether ex-President Nixon is or is not a dog in evaluating the subjunctive conditional, 'If Nixon were a dog then he would bark too much'. In logics modeling a notion of possibility in which a world may not be accessible to itself the appropriate table for '$\rightarrow$' is (11).

A *weak modal semantics* is a set-assignment model or class of models $\langle$v, s$\rangle$ where:

s satisfies M1–M3, and
v is extended to all wffs by the classical tables for $\neg$ and $\wedge$, and (11) for $\rightarrow$

Recall from §IV.B.1, pp.92–93, that (11) is called *the weak table for the conditional*, and '$\rightarrow$' evaluated by (11) is called *weak implication*.

In these models we have the following tables for $\Box A$ and $\Diamond A$ (see Corollary 3 below).

(15)

A	$s(A)=S$	$\Box A$
any value	fails	F
	holds	T

(16)

A	$s(A)\neq\varnothing$	$\Diamond A$
any value	fails	F
	holds	T

If $\neg, \wedge, \Box$ *are taken as primitive then table* (15) *replaces* (11), which is then derivable.

For logics which have modal semantics of implication *modus ponens*,

$$\frac{A,\ A\rightarrow B}{B}$$

preserves truth in a model and hence the logic is closed under it. But for logics which have only weak modal semantics *modus ponens* can fail to preserve truth in a model. What we do have is that *material detachment*,

$$\frac{A,\ \neg(A\wedge\neg B)}{B}$$

preserves truth in a model.

Here are some further observations we'll need later.

Lemma 2 If $<v, s>$ is a modal semantics for implication in $L(\neg, \rightarrow, \wedge)$ or in $L(\neg, \wedge, \Box)$ then

a. *i.* $s(A\supset B) = \overline{s(A)} \cup s(B)$

ii. $s(A\supset B) = S$ iff $s(A)\subseteq s(B)$

iii. If $\mathbf{PC}\vDash A$, then $v(A)=T$ and $s(A)=S$.

Here **PC** is formulated in the language of $\neg$ and $\wedge$, as explained in §D below.

b. $v(\Box A)=T$ iff $v(A)=T$ and $s(A)=S$

c. $v((A\rightarrow B)\leftrightarrow\Box(A\supset B))=T$

d. $v(\Diamond A)=T$ iff $v(A)=T$ or $s(A)\neq\varnothing$

e. $v(\Box(B\supset C)\supset(\Box B\supset\Box C))=T$

f. $v(\Box A\supset A)=T$

g. For all A and B: $s(A\rightarrow B)\subseteq \overline{s(A)}\cup s(B)$ iff for all A, $s(\Box A)\subseteq s(A)$.

h. If $\mathbf{PC}\vDash A$, then $v(\Box A)=T$.

Proof: The proofs are not difficult but are useful in establishing familiarity with these semantics. I'll do them on the assumption that the language is $L(\neg, \rightarrow, \wedge)$.

Part (a.i) can be proved by induction on the length of a proof of A. (It says that with respect to $\neg$ and $\wedge$ our set-assignment semantics are Boolean algebra models of **PC**.)

b. $v(\Box A)=T$ iff $v(\neg A\rightarrow A)=T$
iff $s(\neg A)\subseteq s(A)$ and $v(\neg A\supset A)=T$

iff $\overline{s(A)} \subseteq s(A)$ and $v(A) = T$
iff $s(A) = S$ and $v(A) = T$

c. By M3 this reduces to showing that $v(A \rightarrow B) = T$ iff $v(\Box(A \supset B)) = T$.
$v(A \rightarrow B) = T$ iff $v(A \supset B) = T$ and $s(A) \subseteq s(B)$
iff $v(A \supset B) = T$ and $s(A \supset B) = S$
iff $v(\Box(A \supset B)) = T$ by (b)

d. $v(\Diamond A) = T$ iff $v(\neg\Box\neg A) = T$
iff $v(\Box\neg A) = F$
iff $s(\neg A) \neq S$ or $v(\neg A) = F$
iff $s(A) \neq \varnothing$ or $v(A) = T$

e. You can establish this by repeated use of (a) and (b).
f. This follows from (b).
g. If for all A and B we have $s(A \rightarrow B) \subseteq \overline{s(A)} \cup s(B)$, then

$$s(\Box A) = s(\neg A \rightarrow A) \subseteq \overline{s(A)} \cup s(A) = s(A)$$

If $s(\Box A) \subseteq s(A)$ for all A, then by M3,

$$s(A \rightarrow B) = s(\Box(A \supset B)) \subseteq s(A \supset B) = \overline{s(A)} \cup s(B)$$

h. This follows from (a.i) and (b). ■

Corollary 3 If $<v,s>$ is a weak modal semantics in either $L(\neg, \rightarrow, \wedge)$ or $L(\neg, \wedge, \Box)$ and we delete '$v(A)=T$' in (b) and (d), then all the above hold except for (f).

We can establish the *functional completeness of* $\{\neg, \rightarrow, \wedge\}$ for any logic characterized by modal semantics of implication. First we define $A \vee B$ classically as $\neg(\neg A \wedge \neg B)$; note that $s(A \vee B) = s(A) \cup s(B)$. Then we can define a connective which represents the relation governing the only nonclassical truth-table:

$$M(A,B) \equiv_{\text{Def}} (A \vee B) \rightarrow A$$

which satisfies $<v,s> \vDash M(A,B)$ iff $s(A) \subseteq s(B)$. The proof of functional completeness then follows just as for Relatedness Logic (§III.H, pp.78–79). For logics characterized by weak modal semantics it is even easier, for we already have that $<v,s> \vDash A \rightarrow B$ iff $s(A) \subseteq s(B)$.

2. The aptness of set-assignment semantics for modal logics: connections of meanings in modal logics

No one is likely to deny that the logical impossibility of $(\neg p \wedge q)$ is a *necessary* condition of q's deducibility from p, but it has been suggested that it is not a *sufficient* condition on the ground that a further condition of q's deducibility from p is that there should be some connection of "content" or "meaning"

> between p and q. It is, however, extremely difficult, if not impossible, to state this additional requirement in precise terms; and to insist on it seems to introduce into an otherwise clear and workable account of deducibility a gratuitously vague element which will make it impossible to determine whether a given formal system is a correct logic of entailment or not.
>
> *Hughes and Cresswell, 1968*, p.336-337

Thus Hughes and Cresswell defend modal logic, reading A→B as 'A entails B' or what they take to be equivalent, 'B is deducible from A'. But I believe that modal logic is based on just such a vague notion of connection of meaning as Hughes and Cresswell wish to exclude from logic, and that modal set-assignment semantics for implication aptly reflect that.

On a modal logician's terms a proposition is to be identified with the possible worlds in which it is true. To understand a proposition, then, is to be able to conceive of the various possible ways in which it could be true.

> The word 'world' has been used by a number of logicians ... and seems to be the most convenient one, but perhaps some such phrase as 'conceivable or evisageable state of affairs' would convey the idea better.
>
> *Hughes and Cresswell, 1968*, p.75

To say that A→B is true is to say that we cannot envisage a state of affairs in which A is true and B is false. That is, the appropriate connection of meaning between A and B obtains. Just because this connection of meaning can be given a rigorous mathematical treatment based on the semantics of classical logic does not mean that in any *application* it is less vague than, say, the referential content of a proposition. Consider how Hughes and Cresswell motivate accessibility relations.

> We can conceive of various worlds which would differ in certain ways from the actual one (a world without telephones, for example). But our ability to do this is at least partly governed by the kind of world we live in: the constitution of the human mind and the human body, the languages which exist or do not exist, and many other things, set certain limits to our powers of conceiving. We could then say that a world, w_1, is accessible to a world, w_2, if w_2 is conceivable by someone living in w_1, and this will make accessibility a relation between worlds as we want it to be.
>
> *Hughes and Cresswell, 1968*, p.77

I can conceive of a world in which Richard Nixon is a dog, yet I cannot conceive of a world which Putnam describes as Twin Earth (*1975, 2*, p.223). Am I wrong? Certainly there is no wrong or right to it. If you and I are to reason together using the modal notions of necessity and possibility we need to agree in general on the structural rules we'll allow our imaginations to follow, and in particular on specific conceivable states of affairs.

A platonist might object to the way I've presented modal logic, saying that conceiving and imagining have nothing to do with it. Propositions exist as abstract objects, and a possible world is just as real as those. Classical modal logic is the right way to reason about those worlds, those possible states of affairs, which are fixed for all time and independent of us and our language. The difficulty with that view is the same one I have with abstract propositions: we have no direct access to these possible worlds, so how in any application of modal logic are we to proceed? All the arguments concerning connections of meaning and imagining that I've put forward would apply to how a platonist is to use his modal logic. The platonist might counter by saying that it's truth and reality he's studying, not how we deal with it.

Set-assignment semantics for modal logics bring out the structural way in which connections of meanings between propositions function in modal logic. They should not be seen as replacing Kripke semantics, but as bringing out the similarities between modal logic and many other logics, setting all of these within a general semantic framework. Moreover, some important modal logics such as **G*** (§K.3) have no Kripke semantics, yet we can characterize them with set-assignment semantics placing them alongside others in the general framework.

Two further observations about set-assignment semantics give some idea of their explanatory capacity. If when we say that the content of a proposition is the possible worlds in which it is true we also mean to include the actual world we inhabit, then the content of a proposition incorporates its truth-value, unlike, say, Dependence Logic (Chapter V). For logics based on that notion formulations (11) and (12) should be equivalent, and so they are.

Moreover, it is correct to speak of the truth or falsity of a modal proposition: we are interested in whether, say, 'Roses are red $\rightarrow$ sugar is sweet' is *true*. And this is a classical conception of truth in which every proposition is true or false but not both. That is something which is obscured by the Kripke semantics but is the basis of the set-assignment ones.

D. On the Syntactic Characterizations of Modal Logics

Each modal logic in this chapter will be presented syntactically, with the exception of **MSI** in §J.3. In this section I will lay out the general format of those presentations and some of the assumptions behind them.

It is currently standard practice to take $\Box$ rather than $\rightarrow$ as a primitive in axiomatizing modal logics. This is contrary to the motivation I have given, but I will nonetheless conform to it so that you may more easily compare this text with others. Every logic in this chapter (except in §J.3) will be presented in the language $L(p_0, p_1, \ldots \neg, \wedge, \Box)$ using axiom schemas.

The following definitions are adopted throughout.

(17) $A \supset B \equiv_{Def} \neg(A \wedge \neg B)$

$A \equiv B \equiv_{Def} (A \supset B) \wedge (B \supset A)$

$A \to B \equiv_{Def} \Box(A \supset B)$

$\Diamond A \equiv_{Def} \neg\Box\neg A$

Kripke semantics are then understood as using (9) in the definition of 'true at world w.' The last clause of (7) which defines $w \vDash A \to B$ is then a derived fact.

Though I will not use it anywhere, it is also customary to define

$A \vee B \equiv_{Def} \neg(\neg A \wedge \neg B)$

As mentioned in §C.1, for logics which have weak modal semantics but not modal semantics for implication *modus ponens*,

$$\frac{A, A \to B}{B}$$

can fail to preserve truth in a model. What is used instead is *material detachment*,

$$\frac{A, A \supset B}{B}$$

Because of the choice of primitives and for uniformity, it is now standard to use the rule of material detachment when presenting modal logics syntactically, establishing for some logics that the rule of *modus ponens* also holds.

If we were to present all the work in this chapter using the language of $\neg$, $\to$, $\wedge$, *then the only changes necessary would be to define* $\Box A$ *as* $\neg A \to A$ *and add to each logic the axiom schema*: $\Box(A \supset B) \equiv (A \to B)$. This is exactly what is needed in Theorem 44, p. 187, to establish the completeness of the Kripke semantics. The set-assignment semantics will remain unchanged: in each of these logics we can prove $(A \to B) \leftrightarrow \Box(A \supset B)$, so we can also include condition M3. It would be a good project to give natural axiomatizations of these logics in that language.

In §II.K.6 I gave an axiomatization of **PC** using only the primitives $\neg$ and $\wedge$. If we allow any formula of the language $L(\neg, \wedge, \Box)$ to be an instance of A, B, or C in those schema, then we have *an axiomatization of* **PC** *based on* $\neg$ *and* $\wedge$ *in the language of modal logic*; and similarly if $\to$ is taken as primitive instead of $\Box$.

In this chapter every logic **L** *satisfies*

a. $\mathbf{PC} \subseteq \mathbf{L}$
by which we mean that every theorem of **PC** based on $\neg$ and $\wedge$ in the language of modal logic is in **L**.

b. **L** is closed under the rule of material detachment, $\dfrac{A, A \supset B}{B}$.

Thus for each logic we will have: if $A \vDash_{\mathbf{PC}} B$ then $\vdash_{\mathbf{L}} A \supset B$.

c. **L** contains the *distribution* schema $\Box(A \supset B) \supset (\Box A \supset \Box B)$.

Many modal logics use an additional rule of inference:

d. **L** is closed under the rule of *necessitation*, $\dfrac{A}{\Box A}$.

The rule of necessitation is intended to apply only to the theorems of the logic and perhaps should be written:

$$\frac{\vdash A}{\vdash \Box A}$$

It reflects the assumption that the laws of logic are not only true, but necessarily true. In terms of the possible worlds motivation, if A is true in all possible worlds, as every law of logic is, then it is necessary, that is, $\Box A$ is true. It is not intended that we take the axiom system for **L** and use it to define a syntactic consequence relation $\Gamma \vdash_{\mathbf{L}} B$, for it is usually claimed that $A \rightarrow B$ already formalizes the notion of A *entails* B, where by that is meant that B *is deducible from* A. Here 'deducible' does not mean with respect to any one particular formal system but the informal notion of consequence.

Therefore, the syntactic presentation of each of the logics in this chapter will be as a collection of theorems, under the usual assumption that PV and Wffs are completed infinite totalities. I will write $\mathbf{L} \vdash A$ for $\vdash_{\mathbf{L}} A$ throughout. In §C of the appendix to this chapter I discuss consequence relations and strong completeness theorems for some of these logics.

Any logic **L** which satisfies each of (a), (b), (c), and (d) is called *normal*. The logic which is characterized by exactly (a), (b), (c), and (d) is called **K** , the smallest normal modal logic, as discussed in §J.1. Any modal logic which contains **K** and which satisfies at least (a), (b) and (c) is called *quasi-normal*. Every logic considered in this chapter is quasi-normal, with the exception of those in §J.3.

In any quasi-normal logic we can prove an additional distribution schema; the proof of that and of two rules that hold in normal systems is a good introduction to the syntactic methods of modal logics.

Lemma 4 *a.* If $\vdash_{\mathbf{K}} A \equiv B$ then $\vdash_{\mathbf{K}} \Box A \equiv \Box B$.

b. If $\vdash_{\mathbf{K}} A \equiv B$ then $\vdash_{\mathbf{K}} \Diamond A \equiv \Diamond B$.

c. $\vdash_{\mathbf{K}} \Box(A \wedge B) \equiv (\Box A \wedge \Box B)$

Proof: a. We have $\mathbf{PC} \vdash (A \equiv B) \supset (A \supset B)$ and hence $\vdash_{\mathbf{K}} (A \equiv B) \supset (A \supset B)$, so an application of the rule of material detachment gives $\vdash_{\mathbf{K}} A \supset B$. (In general I will abbreviate an argument like this by saying simply 'by **PC**'.) Now by necessitation, $\vdash_{\mathbf{K}} \Box(A \supset B)$ and thus using the distribution axioms, $\vdash_{\mathbf{K}} \Box A \supset \Box B$. Similarly, $\vdash_{\mathbf{K}} \Box B \supset \Box A$, so by **PC**, $\vdash_{\mathbf{K}} \Box A \equiv \Box B$.

b. By **PC** we have $\vdash_{\mathbf{K}} \neg B \supset \neg A$. So by necessitation and using the distribution axioms, $\vdash_{\mathbf{K}} \Box\neg B \supset \Box\neg A$, and so by **PC**, $\vdash_{\mathbf{K}} \neg\Box\neg A \supset \neg\Box\neg B$. That is, $\vdash_{\mathbf{K}} \Diamond A \supset \Diamond B$; and similarly $\vdash_{\mathbf{K}} \Diamond B \supset \Diamond A$.

c. By **PC** we have $\vdash_{\mathbf{K}} (A \wedge B) \supset A$, so by necessitation $\vdash_{\mathbf{K}} \Box((A \wedge B) \supset A)$, and thus using the distribution axioms, $\vdash_{\mathbf{K}} \Box(A \wedge B) \supset \Box A$. Similarly, $\vdash_{\mathbf{K}} \Box(A \wedge B) \supset \Box B$. So by **PC**, $\vdash_{\mathbf{K}} \Box(A \wedge B) \supset (\Box A \wedge \Box B)$.

We also have that $\vdash_{\mathbf{K}} \Box(B \supset (A \wedge B)) \supset (\Box B \supset \Box(A \wedge B))$ is a distribution axiom. By **PC** we have $\vdash_{\mathbf{K}} A \supset (B \supset (A \wedge B))$ so using necessitation and distributing we have $\vdash_{\mathbf{K}} \Box A \supset (\Box(B \supset (A \wedge B)))$, and then by **PC**, $\vdash_{\mathbf{K}} \Box A \supset (\Box B \supset \Box(A \wedge B))$. So by **PC** (Importation), $\vdash_{\mathbf{K}} (\Box A \wedge \Box B) \supset \Box(A \wedge B)$. ■

Chellas, *1980*, Chapter 4.1, demonstrates further distribution schemas which hold in normal modal logics.

E. An Outline of the Chapter: Converting Kripke Semantics to Set-Assignment Semantics

As described in the last section, each modal logic will first be presented syntactically as a collection of theorems. Then a class of Kripke frames will be given which is complete for the logic. The method for proving that Kripke semantics are complete for a logic is very different from completeness proofs for the logics in the previous chapters. I will present it in the appendix to this chapter, which you may prefer to read first.

I won't try to give an intuitive reading of the notion of necessity modeled by each of the logics. For that I suggest you consult *David Lewis, 1973,* or *Hughes and Cresswell, 1968.*

Finally, complete set-assignment semantics for the logic will be given. The general method for producing set-assignment semantics from complete Kripke semantics proceeds along the following lines. First we derive a list of conditions on modal or weak modal semantics which ensure that the set-assignment semantics are sound for **L**. To arrange that they are complete we consider how to proceed if we are given a wff A such that $\nvdash_{\mathbf{L}} A$. In that case there is some $\langle \mathsf{W}, \mathsf{R} \rangle$ in the class of Kripke frames which are complete for this logic, and some $w \in \mathsf{W}$ and evaluation e such that $\langle \mathsf{W}, \mathsf{R}, \mathsf{e}, w \rangle \nvDash A$. We define a set-assignment model $<v, s>$ by setting $v(p) = \mathsf{T}$ iff $w \vDash p$, and for all A, $s(A) = \{ z : w\mathsf{R}z$ and $z \vDash A\}$. We then try to prove that for all A, $v(A) = \mathsf{T}$ iff $w \vDash A$. To do that we first check that $<v, s>$ satisfies the original list of conditions we imposed for the soundness of the semantics, and then add other conditions as needed to make the equivalence go through.

Unfortunately, I cannot see a mechanical way to follow this strategy. I will show how to implement it for a number of well-known modal logics and in doing so will survey the standard theory of modal logics as well as variations on the method.

F. S4

in collaboration with **Roger Maddux**

S4

in $L(\neg, \wedge, \Box)$

axiom schemas

PC axioms

$\Box(A \supset B) \supset (\Box A \supset \Box B)$

$\Box A \supset A$

$\Box A \supset \Box\Box A$

rules material detachment

necessitation

We write $\mathbf{S4} \vdash A$ if A is a theorem of this axiom system.

Recall that a relation R on W is *reflexive* if for all $w \in W$, wRw; it is *transitive* if for all $w, y, z \in W$, if wRy and yRz then WRz.

Theorem 5 $\mathbf{S4} \vdash A$ iff for every $\langle W, R \rangle$ which is reflexive and transitive, $\langle W, R \rangle \vDash A$. Moreover, **S4** is decidable.

The first part of the theorem is proved in §A of the appendix to this chapter. The second part is proved in §B of the appendix by showing that the class of finite reflexive, transitive models is complete for **S4**. Hughes and Cresswell, *1968*, Chapters 5 and 6, give a different decision method by justifying that the "obvious" falsification procedure works (cf. the decision procedure for **PC**, §II.G.5, pp.35–36). For example, we can show that $\Box A \supset \Box\Box A$ is valid in all reflexive transitive frames by supposing that it is not valid. In that case, some $\langle W, R, e, w \rangle \nvDash \Box A \supset \Box\Box A$. Thus $w \vDash \Box A$ and $w \nvDash \Box\Box A$. So for all z, if wRz then $z \vDash A$, and yet for some z, wRz and $z \nvDash \Box A$. So for some z and x, wRz and zRx and $x \nvDash A$. But R is transitive, so wRx, a contradiction as all such x validate A. Hence no such model exists and $\Box A \supset \Box\Box A$ is valid.

Let's now consider how to obtain complete set-assignment semantics for **S4**. We know from Lemma 2.a, e, f that modal semantics for implication validate the **PC** axioms as well as the first two modal schemas of **S4**. To arrange that $\Box A \supset \Box\Box A$ is validated we require: if $s(A) = S$ then $s(\Box A) = S$. This condition will ensure that our set-assignment semantics respect the rule of necessitation if we also require that for every axiom schema A, $s(A) = S$. We already have this for the **PC** axioms by Lemma 2.a. For the other schemas which are of the form $B \supset C$ we require equivalently, by Lemma 2.a, that $s(B) \subseteq s(C)$.

We say that $<v, s>$ is an **S4**-*model* if

$<v,s>$ is a modal semantics for implication

and M4. $s(\Box(A \supset B)) \subseteq s(\Box A \supset \Box B)$

M5. If $s(A) = S$ then $s(\Box A) = S$.

M6. $s(\Box A) \subseteq s(A)$

M7. $s(\Box A) \subseteq s(\Box\Box A)$

By Lemma 2.g, M6 can be replaced by $s(A \to B) \subseteq \overline{s(A)} \cup s(B)$. In the presence of M6, M7 is equivalent to $s(\Box\Box A) = s(\Box A)$.

Lemma 6 If **S4**$\vdash$E then for every **S4**-model $<v,s>$, $v(E) = T$ and $s(E) = S$.

Proof: We have that $v(E) = T$ and $s(E) = S$ for every axiom E by Lemma 2 and M4–M7, as pointed out above. So we show that if the lemma holds for the hypotheses of one of the rules of the system, then it holds for the conclusion.

If $s(A \supset B) = S$, $v(A \supset B) = T$, $s(A) = S$, and $v(A) = T$, then $v(B) = T$ and by Lemma 2.a, $s(B) = S$.

If $v(A) = T$ and $s(A) = S$, then $v(\Box A) = T$. And by M5, $s(\Box A) = S$. ■

Lemma 7 ***a.*** For every Kripke model $\langle W, R, e, w \rangle$ for **S4**, there is an **S4**-model $<v,s>$ such that both $v(E) = T$ iff $w \vDash E$ and $s(A) = \{z : wRz \text{ and } z \vDash A\}$.

b. If **S4**$\nvdash$E then for some **S4**-model $<v,s>$, $v(E) = F$ and $s(E) \neq S$.

Proof: a. Let $\langle W, R, e, w \rangle$ be a model for **S4**. Set

$S = \{ z : wRz \}$

$s(A) = \{ z : wRz \text{ and } z \vDash A \}$

By Lemma 1, since R is transitive we need only consider $z \in S$ in evaluating whether w validates A. So it is no loss of generality to assume $W = S$, and then $s(A)$ = the worlds in which A is true.

We have that s satisfies M1–M4, M6, and M7 easily: M4, M6, and M7 follow because they state that the corresponding axioms hold at every world $z \in S$.

Lastly we consider M5. Suppose some $z \notin s(\Box A)$. Then for some y, zRy and $y \nvDash A$. But since R is transitive, we have wRy and $y \nvDash A$, hence $y \notin s(A)$ so $s(A) \neq S$.

Now set $v(p) = T$ iff $w \vDash p$. Extend v to all wffs by the truth-conditions for modal semantics of implication. Then $<v, s>$ is an **S4**-model. We now show that $v(A) = T$ iff $w \vDash A$. The proof is by induction on the length of A, the only interesting step being when A is of the form $\Box B$.

If $v(\Box B) = T$ then $s(B) = S$ and $v(B) = T$. So $w \vDash B$, and for all z such that wRz, $z \vDash B$. So $w \vDash \Box B$.

If $v(\Box B) = \mathsf{F}$ then $s(B) \neq S$ or $v(B) \neq \mathsf{T}$. If the latter then by induction $w \nvDash B$. If the former then for some z, wRz and $z \nvDash B$. In either case, by the reflexivity of R, $w \nvDash \Box B$.

b. If $\mathbf{S4} \nvdash E$ then by Theorem 5 there is some Kripke model with designated world $\langle W, R, e, w \rangle$ where R is reflexive and transitive such that $w \nvDash E$. For such a model the construction above yields an **S4**-model $< v, s >$ such that $v(E) \neq \mathsf{T}$ and, by the reflexivity of R, $s(E) \neq S$. ■

From Lemmas 6 and 7 we have

Theorem 8 **(*Completeness of the Set-Assignment Semantics for* S4)**

$\mathbf{S4} \vdash A$ iff for every **S4**-model $< v, s >$, $v(A) = \mathsf{T}$
iff for every **S4**-model $< v, s >$, $s(A) = S$

Given a Kripke model with designated world $\langle W, R, e, w \rangle$ we have derived an **S4**-model which validates exactly the same wffs. We now show by the methods of *Jónsson and Tarski, 1951*, that we can derive the reverse correspondence as well.

Theorem 9 For every **S4**-model $< v, s >$ there is a reflexive, transitive Kripke model $\langle W, R, e, w \rangle$ such that for all A, $w \vDash A$ iff $v(A) = \mathsf{T}$.

Proof: We will construct the Kripke model by giving an evaluation $e : \mathrm{PV} \to W$ which we noted in §B is equivalent to giving one from W to PV. Let

$W = S \cup \{w\}$ for some object $w \notin S$

$$e(p) = \begin{cases} s(p) \cup \{w\} & \text{if } v(p) = \mathsf{T} \\ s(p) & \text{if } v(p) = \mathsf{F} \end{cases}$$

xRy iff (for all A, if $x \in s(\Box A)$ then $y \in s(A)$) or $x = w$

First, R is reflexive: wRw by definition, and by M6 if $x \in s(\Box A)$ then $x \in s(A)$, so xRx. Second, R is transitive: suppose xRy and yRz. If $x = w$ then xRz. If not, suppose $x \in s(\Box A)$. Then by M7, $x \in s(\Box\Box A)$, so since xRy, $y \in s(\Box A)$ and so $z \in s(A)$. Hence xRz.

Now we will show by induction on the length of wffs that for $x \neq w$, $x \vDash A$ iff $x \in s(A)$. The only interesting case is when A is $\Box B$.

If $x \in s(\Box B)$ then for every y, if xRy then $y \in s(B)$ and hence by induction $y \vDash B$; so $x \vDash \Box B$.

If $x \notin s(\Box B)$, then by M6 $x \notin s(B)$, so by induction $x \nvDash B$, and since R is reflexive, $x \nvDash \Box B$.

Finally, we show by induction on the length of A that $w \vDash A$ iff $v(A) = \mathsf{T}$. Again, the only nontrivial step is if A is $\Box B$. We have

$w \models \Box B$ iff for all z, if wRz then $z \models B$
iff $w \models B$ and for all $z \neq w$, $z \in s(B)$, as demonstrated above
iff $v(B) = T$ and $s(B) = S$, by induction
iff $v(\Box B) = T$. ■

Note that this provides another proof of (the contrapositive of) Lemma 6.

We have not been able to find a correspondence such as this between set-assignment models and Kripke models of the appropriate frame for other modal logics.

A condition satisfied by every **S4**-model $<v,s>$ which arises from a Kripke model by the construction of Lemma 7 is

P. If $v(A) = T$ then $s(A) \neq \varnothing$.

There are, however, $<v,s>$ models satisfying M1–M7 which do not satisfy P. For instance, let $S = \{1\}$ and define $s(A) = \{1\}$ iff $\neg p_0 \vdash_{\mathbf{PC}} A^*$, where A^* is A with every occurrence of every p_j replaced by p_0 and $\Box A$ by $\neg A \supset A$ (in the language $L(p_0, p_1, \ldots \neg, \rightarrow, \wedge)$ this would amount to reading $\rightarrow$ as the **PC**-connective). Note that in **PC**, $(\Box A)^*$, $(\Diamond A)^*$, $\Diamond A^*$, and $\Box A^*$ are all semantically equivalent to A^*, and for every A, A^* is semantically equivalent to either p_0 or $\neg p_0$. If we take the model in which $v(p_i) = T$ for all i, then P is not satisfied. (We are grateful to an anonymous referee for contributing this example.)

But given any $<v,s>$ modal semantics for implication satisfying M1–M7 there is another modal semantics for implication $<v^*, s^*>$ satisfying M1–M7 and P such that for all A, $v^*(A) = v(A)$: derive a Kripke model with designated world as in the last construction for Theorem 9 and then construct $<v^*, s^*>$ from that as in Lemma 7. Thus it is a matter of taste whether to add P to the list of conditions for **S4**. By adding more conditions we get a better reading of $s(A)$ as the worlds in which A is true. For instance, P says that if A is true then there's some world in which it's true. On the other hand, fewer conditions simplify the metamathematical proofs. If condition P is not included then note that our conditions involve only s, and hence the assignment of truth-values to the propositional variables is independent of s.

Algebraic Aside Given any pair s, S satisfying M1–M7, let $W = \{s(A): A \text{ is a wff}\}$. Then consider $M = \langle W, \cup, \cap, \overline{}, \varnothing, S, ^\circ \rangle$ where $s(A)^\circ = s(\Box A)$. We can easily establish that M is an **S4** modal algebra (see, e.g., *Lemmon, 1967*). We may view the truth-value assignment v as a 2 element modal algebra. In the construction of Theorem 9 we've shown that given any modal algebra, the "amalgamation" of it with the 2 element model according to the truth-conditions for modal semantics of implication is isomorphic to an **S4** modal algebra. And conversely, any **S4** modal algebra is so representable.

G. Two Extensions of S4

1. S5, logical necessity

S5
in $L(\neg, \wedge, \Box)$

axiom schemas

PC axioms

$\Box(A \supset B) \supset (\Box A \supset \Box B)$

$\Box A \supset A$

$\Box A \supset \Box\Box A$

$\Diamond A \supset \Box\Diamond A$

rules material detachment
necessitation

Recall that a relation R on W is *symmetric* if for all $w \in W$, if wRz then zRw. We call a reflexive, symmetric, transitive frame an *equivalence* frame.

Theorem 10 $S5 \vdash A$ iff for every equivalence frame $\langle W, R \rangle$, $\langle W, R \rangle \models A$.

A proof of this can be found in Appendix §A.

To obtain complete set-assignment semantics for **S5** we will add further conditions to those for **S4**-models. To begin with, our models should respect the rule of necessitation, so we will take $s(\Diamond A \supset \Box\Diamond A) = S$. This follows from $s(\Diamond A) \subseteq s(\Box\Diamond A)$ but we can use the stronger condition:

M8. $s(\Diamond A) = s(\Box\Diamond A)$

There are modal semantics of implication satisfying M1–M8 in which $\Diamond A \supset \Box\Diamond A$ fails. For instance, the model we looked at in the discussion of condition P in the last section. If you consider Lemma 2.a, d you'll see that mixing conditions for v and s will allow us to validate this last schema.

M9. If $v(A) = T$ or $s(A) \neq \varnothing$, then $s(\Diamond A) = S$.

We define $<v,s>$ to be an **S5**-*model* if it is a modal semantics for implication and satisfies M1–M9.

Lemma 11 If $S5 \vdash E$ then for every **S5**-model $<v,s>$, $v(E) = T$ and $s(E) = S$.

Proof: As for Lemma 6 for **S4** using the discussion above. ■

Lemma 12 If $\mathbf{S5} \nvdash E$ then for some **S5**-model $<v,s>$, $v(E) = F$ and $s(E) \neq S$.

> ***Proof:*** Suppose $\mathbf{S5} \nvdash E$. Then there is some Kripke model $\langle W, R, e, w\rangle$ where R is an equivalence relation such that $\langle W, R, e, w\rangle \nvDash E$. Choose such a model. As in the proof of Lemma 7 for **S4**, define $s(A) = \{ z: wRz \text{ and } z \vDash A\}$ and set $v(p) = T$ iff $w \vDash p$. The proof follows just as for the **S4** case except that we must now show that conditions M8 and M9 are fulfilled.
>
> M8. If $z \in s(\Box\Diamond A)$ then $z \vDash \Box\Diamond A$, so for all y such that zRy, there is some x such that yRx and $x \vDash A$. So for some x, zRx and $x \vDash A$ by the transitivity of R, so $z \in s(\Diamond A)$. For the other containment, if $z \in s(\Diamond A)$ then as $z \vDash \Diamond A$, for some y, zRy and $y \vDash A$. But for all x, y for which zRx and zRy we have xRy as R is an equivalence relation. Therefore for all x such that zRx, there is some y such that xRy and $y \vDash A$. So $z \in s(\Box\Diamond A)$.
>
> The proof of M9 then follows the proof that $v(A) = T$ iff $w \vDash A$. If $v(A) = T$ or $s(A) \neq \varnothing$ then for some x (possibly w itself), wRx and $x \vDash A$. Hence, as R is an equivalence relation, if wRz then zRx and so $z \in s(\Diamond A)$. Thus $s(\Diamond A) = S$. ■

Theorem 13 $\mathbf{S5} \vdash A$ iff for every **S5**-model $<v,s>$, $v(A) = T$
iff for every **S5**-model $<v,s>$, $s(A) = S$.

Remarks 14: ***a.*** Given a Kripke model $\langle W, R, e, w\rangle$ for **S5**, the only elements of W that contribute to determining whether $w \vDash A$ are those which are related to w, for R is reflexive and transitive (see Lemma 1). This is also true for **S4**. But for **S5** we also have symmetry, so relative to any particular world in the model, accessibility is universal. It could be argued that this corresponds to *logical necessity*: no restrictions are put on our imaginings of states of affairs except that they obey the classical laws of logic. In § VIII.E.2 we exploit this to give a characterization of **S5** as a many-valued logic.For this particular notion the difficulty of interpreting iterated modalities vanishes: all modalities can be collapsed to just one occurrence of $\Box$ or $\Diamond$. The following schemas are valid in **S5** as you can check using Theorem 10.

$$\Box\Diamond A \equiv \Diamond A$$
$$\Diamond\Box A \equiv \Box A$$
$$\Diamond\Diamond A \equiv \Diamond A$$
$$\Box\Box A \equiv \Box A$$

Recalling that $\Diamond A$ was defined as $\neg\Box\neg A$, we have that any formula $\lambda_1 \lambda_2 \ldots \lambda_n A$ where each λ_i is $\Box$, $\Diamond$, or $\neg$, is equivalent to one of A, $\neg A$, $\Box A$, $\neg\Box A$, $\Diamond A$, or $\neg\Diamond A$.

b. We can replace condition M9 by condition P which we considered for **S4**-models (p.168) plus

Y. If $s(A) \neq \varnothing$ then $s(\Diamond A) = S$.

For modal semantics for implication, M1–M8, and M9 are together equivalent to M1–M8, P and Y.

Bas van Fraassen, *1967*, uses an analysis similar to our motivation to develop semantics for **S5** which are very much like the set-assignment semantics presented in this section.

2. S4Grz

S4Grz
in $L(\neg, \wedge, \Box)$

axiom schemas

PC axioms

$\Box(A \supset B) \supset (\Box A \supset \Box B)$

$\Box A \supset A$

$\Box A \supset \Box\Box A$

$(\Box(\Box(A \supset \Box A) \supset A)) \supset A$

rules material detachment
necessitation

Note that the new schema is: $((A \rightarrow \Box A) \rightarrow A) \supset A$. I discuss the motivation for **S4Grz** in §K.1 below. The initials 'Grz' are ascribed to this system because of the work of Grzegorczyk, *1967*.

We say that R is *anti-symmetric* if yRz and zRy together imply that $y = z$. We say that a frame $\langle W, R \rangle$ is a *finite weak partial order* if W is finite and R is reflexive, transitive, and anti-symmetric.

Theorem 15

S4Grz $\vdash A$ iff for every $\langle W, R \rangle$ which is a finite weak partial order, $\langle W, R \rangle \models A$.

For a proof of this see *Segerberg, 1971*, pp.96–103, or *Boolos, 1979,* Chapter 13 (note that both Segerberg and Boolos use '$\rightarrow$' for what I call '$\supset$'). As Segerberg notes, 'finite' is essential here.

For the set-assignment semantics, we ensure that our models respect necessitation for consequences of the new axiom schema by requiring:

M10. $s(\Box(\Box(A \supset \Box A) \supset A)) \subseteq s(A)$

And we ensure that the new axiom is true in every model by requiring:

M11. If $v(A) = F$ then $s(\Box(A \supset \Box A)) \nsubseteq s(A)$.

We say that $<v, s, S>$ is an **S4Grz**-*model* if S is finite and $<v, s>$ is a modal semantics for implication satisfying M1–M7, M10, and M11.

Theorem 16 **S4Grz** $\vdash$ A iff for every **S4Grz** -model $<v, s>$, $v(A) = T$
iff for every **S4Grz** -model $<v, s>$, $s(A) = S$.

Proof: It's straightforward to show that these models are sound for **S4Grz** by the same methods as for **S4** (Lemma 6).

To show that the semantics are complete the proof follows as for **S4** (Lemma 7) except that we now have to show that the model $<v, s>$ we construct satisfies M10, which is immediate, and M11, which I'll do now.

Suppose $v(A) = F$. Then $w \nvDash A$. Since we have

$$\langle W, R, e, w\rangle \vDash (\Box(\Box(A \supset \Box A) \supset A)) \supset A$$

we must have $w \nvDash \Box(\Box(A \supset \Box A) \supset A)$. Hence there is a z such that wRz and $z \nvDash \Box(A \supset \Box A) \supset A$. This can only be if $z \nvDash A$ and $z \vDash \Box(A \supset \Box A)$. Thus $z \in s(\Box(A \supset \Box A))$ and $z \notin s(A)$. ■

H. Two Normal Logics

1. T

T
in $L(\neg, \wedge, \Box)$
axiom schemas
PC axioms
$\Box(A \supset B) \supset (\Box A \supset \Box B)$
$\Box A \supset A$
rules material detachment
necessitation

Theorem 17 T $\vdash$ A iff for every $\langle W, R\rangle$ which is reflexive, $\langle W, R\rangle \vDash A$.

The proof of this can be found in §A of the Appendix.

Let's reflect on what changes need to be made from the set-assignment semantics for **S4**. If we want to produce a model $<v, s>$ from a Kripke model $\langle W, R, e, w\rangle$ as in Lemma 7, we will see that if we define $S = \{z : wRz\}$ and $s(A) = \{z : wRz \text{ and } z \vDash A\}$, then $s(A) = S$ does not imply that $s(\Box A) = S$. Yet something like that is going to be required to ensure that the rule of necessitation is respected by our models.

So why don't we take $S = W$, or $S =$ the transitive closure of $\{z : wRz\}$, and then $s(A) = \{z : z \vDash A\}$? In that case necessitation is respected via condition M6, but we could have $w \vDash \Box A$ and $s(A) \neq S$, so $w \vDash \Box A$, yet $v(A)$ would be F. We did not have this problem with the previous modal logics because R was transitive in the models we were modifying there, or equivalently $\Box A \supset \Box\Box A$ was a schema of the logic, and so M5 was valid in the model derived from the Kripke model.

We can, however, combine these two approaches to defining S by taking, in essence, two content sets for every A, both $s(A) = \{z : z \vDash A\}$ and $s(A) = \{z : wRz \text{ and } z \vDash A\}$. We do this by first taking a set C, to correspond to W, and then designating a subset $S \subseteq C$, which will correspond to $\{z : wRz\}$. For each A we assign $t(A) \subseteq S$, corresponding to $\{z : z \vDash A\}$, and then $s(A) = t(A) \cap S$. In this way s will respect necessitation, but only for the theorems of **T**.

We say that s is a **T**-*set-assignment* if there are t, C, and $S \subseteq C$, where S is called the *designated subset*, such that

$$t : \text{Wffs} \to \text{Sub}\, C \text{ and } t \text{ satisfies M1–M6, and for all } A,\ s(A) = t(A) \cap S.$$

A pair $<v, s>$ is a **T**-*model* if s is a **T**-set-assignment and v is extended to all wffs by the truth-conditions for modal semantics of implication.

Note that every **T**-model is a modal semantics for implication, as you can check. A **T**-set-assignment may not satisfy M5. But we have the following.

Lemma 18 If $\mathbf{T} \vdash E$ then for every **T**-model $<v,s>$, $v(E) = T$ and $s(E) = S$.

Proof: Let $<v,s>$ be a **T**-model with t, C, and S as in the definition above. Just as in the proof of Lemma 6 for **S4**, if $\mathbf{T} \vdash E$ then $t(E) = C$ since t satisfies M5. Hence for every E such that $\mathbf{T} \vdash E$, we have $s(E) = S$. The proof now follows as for Lemma 6. ■

Thus the effect of using the designated subset is to relax condition M5 to apply only to the theorems of **T**.

Theorem 19 If $\mathbf{T} \nvdash E$ then for some **T**-model $<v,s>$, $v(E) = F$ and $s(E) \neq S$.

Proof: If $\mathbf{T} \nvdash E$ then there is Kripke model $\langle W, R, e, w\rangle$ where R is reflexive such that $\langle W, R, e, w\rangle \nvDash E$. Fix one and take $C = W$ and $t(A) = \{z : z \vDash A\}$. Then t satisfies M1–M6. Take $S = \{z : wRz\}$ and set $s(A) = t(A) \cap S$. Define $v : PV \to \{T, F\}$ by $v(p) = T$ iff $w \vDash p$, extending v to all wffs by the truth-conditions for modal semantics for implication. Then $<v,s>$ is a **T**-model and the proof that $w \vDash A$ iff $v(A) = T$ follows as for **S4**. ■

Theorem 20 $\mathbf{T} \vdash A$ iff for every **T**-model $<v,s>$, $v(A) = T$
iff for every **T**-model $<v,s>$, $s(A) = S$.

For the class of models we've used here the truth-value assignment v can be given independently of the set-assignment s. But as for **S4** the class of **T**-models which satisfy condition P is also complete.

The class of set-assignments for **T**-models is somewhat complicated by the use of a designated subset with the result that this class is not simply presented by the criteria of §IV.D. We can shift that complication to the truth-conditions for the models if we prefer. The following class of models $<\mathsf{v},\mathsf{s}>$ is also complete for **T**: let $\mathsf{s}:\text{Wffs}\rightarrow\mathsf{C}$ satisfy M1–M6, and $\mathsf{S}\subseteq\mathsf{C}$, with v extended to all wffs by the classical evaluation of $\neg$ and $\wedge$ and $\mathsf{v}(\Box A)=\mathsf{T}$ iff $\mathsf{s}(A)\cap\mathsf{S}=\mathsf{S}$. (In $L(\neg,\rightarrow,\wedge)$ this is replaced by: $\mathsf{v}(A\rightarrow B)=\mathsf{T}$ iff $(\mathsf{s}(A)\cap\mathsf{S})\subseteq(\mathsf{s}(B)\cap\mathsf{S})$ and ($\mathsf{v}(A)=\mathsf{F}$ or $\mathsf{v}(B)=\mathsf{T}$).) These semantics are simply presented. The same modification will work for **B** in the next section.

2. B

B
in $L(\neg,\wedge,\Box)$

axiom schemas

PC axioms

$\Box(A\supset B)\supset(\Box A\supset\Box B)$

$\Box A\supset A$

$A\supset\Box\Diamond A$

rules material detachment
necessitation

The schema $A\supset\Box\Diamond A$ is called the 'Brouwerian axiom'. See *Hughes and Cresswell, 1968,* p.58, for a history of this system and its name. Note that **B** is an extension of **T**.

Theorem 21 $\mathbf{B}\vdash A$ iff for every $\langle\mathsf{W},\mathsf{R}\rangle$ which is reflexive and symmetric, $\langle\mathsf{W},\mathsf{R}\rangle\models A$.

The proof of this can be found in §A of the Appendix.

To give set-assignment semantics for **B** we use designated subsets as for **T** and mix conditions for v and s as for **S5**.

We say that s is a **B**-*set-assignment* if there are t, C and $\mathsf{S}\subseteq\mathsf{C}$ such that $\mathsf{t}:\text{Wffs}\rightarrow\text{Sub}\,\mathsf{C}$ and t satisfies M1–M6 as well as

M12. $\mathsf{t}(A)=\mathsf{t}(\Box\Diamond A)$

and for all A, $\mathsf{s}(A)=\mathsf{t}(A)\cap\mathsf{S}$. A pair $<\mathsf{v},\mathsf{s}>$ is a **B**-*model* if s is a **B**-set-assignment, v uses the truth-conditions for modal semantics of implication, and together they satisfy

M13. If $\mathsf{v}(A)=\mathsf{T}$ then $\mathsf{s}(\Diamond A)=\mathsf{S}$.

Note that every **B**-model is a modal semantics of implication.

Theorem 22 $\mathbf{B} \vdash A$ iff for every **B**-model $<v, s>$, $v(A) = T$ and $s(A) = S$.

Proof: For soundness we show that for every **B**-model $<v, s>$, $v(A \supset \Box\Diamond A) = T$. Suppose $v(A) = T$. Then $v(\Diamond A) = T$ and also by M13, $s(\Diamond A) = S$. Hence $v(\Box\Diamond A) = T$.

The rest of the proof is as for **T**. The only new point to verify is that the model constructed as in Lemma 19 satisfies M13. That's done in the following manner after proving $v(A) = T$ iff $w \models A$. If $v(A) = T$ then $w \models A$. Hence for any z for which wRz we have zRw, so $z \models \Diamond A$. Hence $s(\Diamond A) = S$. ∎

We also have a complete class of models for **B** if we replace M13 by conditions P and Y.

J. The Smallest Logics Characterized by Various Semantics

1. K

K
in $L(\neg, \wedge, \Box)$
axiom schemas
PC axioms
$\Box(A \supset B) \supset (\Box A \supset \Box B)$
rules material detachment
necessitation

Theorem 23 $\mathbf{K} \vdash A$ iff for every $\langle W, R \rangle$, $\langle W, R \rangle \models A$.
That is, **K** is the smallest logic characterized by Kripke frames.

This is proved in §A of the Appendix.

Any logic which is characterized by a class of modal semantics of implication must contain the schema $(A \wedge (A \rightarrow B)) \supset B$. But **K** doesn't: consider $\langle W, R, e, w \rangle$ where $W = \{w, z\}$, wRz but not wRw, and $w \models p$, $w \not\models q$, $z \models p$ and $z \models q$. Then $\langle W, R, e, w \rangle \not\models (p \wedge (p \rightarrow q)) \supset q$. Accordingly we use weak modal semantics for **K** coupled with a designated subset approach as for **T**.

We say that s is a **K**-*set-assignment* if there are t, C, and $S \subseteq C$ such that

t: Wffs $\rightarrow$ Sub C satisfying M1–M5 and $s(A) = t(A) \cap S$

A pair $<v, s>$ is a **K**-*model* if s is a **K**-set-assignment and v uses the truth-conditions of *weak modal semantics*.

Note that every **K**-model is a weak modal semantics.

Theorem 24 $\mathbf{K} \vdash A$ iff for every **K**-model $<v,s>$, $v(A) = T$ and $s(A) = S$.

The proof is as for **T** using the version of Lemma 2 for weak modal semantics (Corollary 3).

Hence **K** *is the smallest logic which has both Kripke semantics and weak modal semantics.*

So far as I can tell, weak modal semantics satisfying only M1–M4 will validate the closure of **PC** under only one application of necessitation and, thereafter, any number of applications of material detachment.

2. QT and quasi-normal logics

What is the smallest logic which has both Kripke semantics and modal semantics of implication? By Theorem 23 it must extend **K** and be closed under material detachment. By Lemma 2.f we know that it must contain the schema $\Box A \supset A$. So let's look at the smallest logic satisfying these conditions.

QT
in $L(\neg, \wedge, \Box)$
is the closure of $\mathbf{K} \cup \{\Box A \supset A\}$ under the rule of material detachment.

For Kripke semantics we have that $\langle W, R, e, w\rangle \vDash \Box A \supset A$ iff wRw. If we require of our class of Kripke frames that R be reflexive, that is zRz for all z, then we get **T**. Putting a global condition on $\langle W, R\rangle$ corresponds to closing the logic under the rule of necessitation. However, we may require only a local condition on R by taking a frame with a designated world, $\langle W, R, w\rangle$. Then $\langle W, R, w\rangle \vDash A$ is defined to mean that all e, $\langle W, R, e, w\rangle \vDash A$.

In §A of the appendix to this chapter, I prove that

$\mathbf{QT} \vdash A$ iff for every $\langle W, R, w\rangle$ such that wRw, $\langle W, R, w\rangle \vDash A$.

Define $<v, s>$ to be a **QT**-*model* if s is a **K**-set-assignment and v uses the truth-conditions for *modal semantics for implication.*

Note that every **QT**-model is a modal semantics for implication.

Theorem 25 $\mathbf{QT} \vDash A$ iff for every **QT**-model $<v,s>$, $v(A) = T$.

The proof follows as for **T**. We cannot claim that if $\mathbf{QT} \vDash A$ then for every **QT**-model $<v,s>$, $s(A) = S$. Consider the model $\langle W, R, e, w\rangle$ for **QT** where $W = \{w, z\}$, $w \vDash p$, $z \nvDash p$ and wRw, wRz, zRw, but not zRz. Then $z \nvDash \Box p \supset p$. So the model $<v, s>$ derived from this Kripke model as in the proof of Theorem 19 will not satisfy $s(\Box p \supset p) = S$.

Modal logics which extend **K** are called *quasi-normal*, hence the name **QT**, *Quasi*-**T**, for this logic which has the same axioms as **T** and is quasi-normal but not normal. Similarly, we can define **QS4** as the closure of

$$\mathbf{K} \cup \{\Box A \supset A\} \cup \{\Box A \supset \Box\Box A\}$$

under material detachment, and **QS5** and **QB** analogously. Kripke semantics and set-assignment semantics can be given for these along the lines of the ones we've given for **QT**. Segerberg, *1971*, has an exposition of quasi-normal logics; Blok and Köhler, *1983*, treat them algebraically.

3. The logic characterized by modal semantics of implication

MSI
in $L(\neg, \wedge, \Box)$
is $\{A : v(A) = T$ in every modal semantics of implication$\}$

Let $\mathbf{PC}^{\Box}$ be the closure of **PC** in the language $L(\neg, \wedge, \Box)$ under the rule of necessitation. Based on Lemma 2 and the fact that **MSI** is closed under both *modus ponens* and material detachment, I conjecture that **MSI** is the same as the following logic.

ML
in $L(\neg, \wedge, \Box)$
axiom schemas
$\mathbf{PC}^{\Box}$
$\Box(A \supset B) \supset (\Box A \supset \Box B)$
$\Box A \supset A$
rule material detachment

First note that $\mathbf{ML} \subseteq \mathbf{QT}$ and $\mathbf{ML} \subseteq \mathbf{MSI}$. Also, $\mathbf{ML} \vdash (A \wedge (A \rightarrow B)) \supset B$: because $\mathbf{ML} \vdash \Box(A \supset B) \supset (A \supset B)$, and, as $\mathbf{PC} \subseteq \mathbf{ML}$, $\mathbf{ML} \vdash (A \wedge \Box(A \supset B)) \supset B$. Thus **ML** is closed under *modus ponens*. I do not know if we get the same logic if we replace the schema $\Box A \supset A$ by the rule of *modus ponens*.

The natural language for **MSI** and **ML** is really $L(\neg, \rightarrow, \wedge)$, and we can take **MSI** to be defined in it in the obvious way. Then we can define **ML** for these primitives as follows.

ML
in $L(\neg, \rightarrow, \wedge)$

$\Box A \equiv_{Def} \neg A \rightarrow A$

$A \leftrightarrow B \equiv_{Def} (A \rightarrow B) \wedge (B \rightarrow A)$

axiom schemas

PC$^{\Box}$

$\Box(A \supset B) \supset (\Box A \supset \Box B)$

$(A \wedge (A \rightarrow B)) \supset B$

$\Box(A \supset B) \leftrightarrow (A \rightarrow B)$

rule material detachment

For **ML** in $L(\neg, \rightarrow, \wedge)$ note that **ML** $\subseteq$ **MSI**. And **ML** $\vdash \Box A \supset A$: we have $\vdash (\neg A \wedge (\neg A \rightarrow A)) \supset A$ and by **PC**, $\vdash (A \wedge (\neg A \rightarrow A)) \supset A$, so by **PC** $\vdash (\neg A \rightarrow A) \supset A$; that is, $\vdash \Box A \supset A$.

We can axiomatize **MSI** in this language as we did **D**, §V.A.7, using the defined connective M of §C above. That method, however, does not lead to very natural axioms.

We may translate from $L(\neg, \rightarrow, \wedge)$ to $L(\neg, \wedge, \Box)$ by translating $\neg$ and $\wedge$ homophonically, and taking $(A \rightarrow B)^* = \Box(A \supset B)$. This preserves semantic consequence for **MSI**. The translation in the other direction is homophonic for $\neg$ and $\wedge$, and takes $(\Box A)^{\dagger} = \neg(A^{\dagger}) \rightarrow A^{\dagger}$, and also preserves semantic consequence. That is,

$\Gamma \vdash_{\mathbf{MSI}} A$ in $L(\neg, \rightarrow, \wedge)$ iff $\Gamma^* \vdash_{\mathbf{MSI}} A^*$ in $L(\neg, \wedge, \Box)$

$\Gamma \vdash_{\mathbf{MSI}} A$ in $L(\neg, \wedge, \Box)$ iff $\Gamma^{\dagger} \vdash_{\mathbf{MSI}} A^{\dagger}$ in $L(\neg, \rightarrow, \wedge)$

Indeed, *the same mappings of languages are translations of every other modal logic of this chapter to itself*, in particular **ML**.

K. Modal Logics Modeling Notions of Provability

1. □ read as 'It is provable that'

> Deducibility and provability are strange notions, and different though their properties may be from those of implication and necessity, the symbolism of modal logic turns out to be exceedingly useful notation for representing the forms of sentences of formal theories that have to do with the notions of deducibility, provability, and consistency, and the techniques devised to study systems of modal logic disclose facts about these notions that are of great interest.
>
> *Boolos, 1979*, p.4

A good survey of the relation of modal logics to notions of provability in arithmetic can be found in *Boolos, 1980 B*. In this section I'll give a brief synopsis of the main connections, presupposing some familiarity with classical first-order logic, and then turn to semantic analyses of the modal logics **G** and **G***.

Let **PA** denote Peano Arithmetic, the first-order theory of arithmetic with induction (see, for example, *Epstein and Carnielli*, Chapter 23, or *Boolos, 1979*, p.35). We may Gödel number the formulas of the language, denoting by $[\![A]\!]$ the Gödel number of A. Then we may define in the language of **PA** a predicate *Bew* (for the German 'Beweisbar' = provable) which corresponds to provability under that Gödel numbering. That is, for any natural number m,

$$\mathrm{Bew}(m) \text{ holds iff } m = [\![A]\!] \text{ and } \mathbf{PA} \vdash A$$

A *realization* is a map $\varphi : \mathrm{PV} \rightarrow$ sentences of the language of **PA**. The *provability translation* of modal sentences *under realization* φ is the map from the sentences of the language of modal logic to those of the language of **PA** defined inductively by:

$$p^{\varphi} = \varphi(p)$$

$$(\neg A)^{\varphi} = \neg(A^{\varphi})$$

$$(A \wedge B)^{\varphi} = A^{\varphi} \wedge B^{\varphi}$$

$$(\Box A)^{\varphi} = \mathrm{Bew}([\![A^{\varphi}]\!])$$

That is, we read $\Box A$ as 'it is provable that'. If we take $L(\neg, \rightarrow, \wedge)$ to be the language of modal logic then the last part of the translation is replaced by:

$$(A \rightarrow B)^{\varphi} = \mathrm{Bew}([\![A^{\varphi} \supset B^{\varphi}]\!])$$

Two modal logics are closely connected to such translations. The first is **G**, which is $\mathbf{K} \cup \{\Box(\Box A \supset A) \supset \Box A\}$ closed under material detachment and necessitation. The other, called **G***, is the closure of $\mathbf{G} \cup \{\Box A \supset A\}$ under the rule of material detachment.

Thus **G*** is to **G** as **QT** is to **K**. We'll find that the set-assignment semantics of **G*** are related to those of **G** just as those of **QT** are to **K**: they use the same set-assignments while the former are modal semantics of implication and the latter are weak modal semantics.

The connection of **G** and **G*** to provability in arithmetic is established by the following two theorems of Solovay, which can be found in *Boolos, 1979*.

Theorem 26 $\mathbf{G} \vdash A$ iff every provability translation of A is a theorem of **PA**. That is, for all φ, $\mathbf{PA} \vdash A^{\varphi}$

If $\mathbf{G} \vdash A$ then $\mathbf{G} \vdash \Box A$, as **G** is closed under the rule of necessitation. Hence **G** *can be viewed as the sentences which express provable principles of provability.*

Theorem 27 $\mathbf{G}^* \vdash A$ iff every provability translation of A is true in the standard model of **PA**. That is, if **N** represents the natural numbers then for all φ, $\mathbf{N} \models A^{\varphi}$.

Hence **G*** *can be viewed as the sentences which express correct principles of provability.*

The logic **S4Grz** of §G.2 can also be viewed as expressing provability principles in terms of the following translations.

A *provability and truth translation* of modal sentences under realization φ is the map defined by:

$$p^{\dagger} = \varphi(p) \wedge \mathrm{Bew}([\![\varphi(p)]\!])$$

$$(\neg A)^{\dagger} = \neg A^{\dagger}$$

$$(A \wedge B)^{\dagger} = A^{\dagger} \wedge B^{\dagger}$$

$$(\Box A)^{\dagger} = \mathrm{Bew}([\![A^{\dagger}]\!]) \wedge A^{\dagger}$$

We can construe this as reading $\Box A$ as 'it is provable and true that'.

Similarly we can define a map # as we defined † except we take $p^{\#} = \varphi(p)$.

Theorem 28 *a.* $\mathbf{S4Grz} \vdash A$ iff every provability-and-truth translation of A is a theorem of **PA**.

b. $\mathbf{S4Grz} \vdash A$ iff for every map #, $\mathbf{PA} \vdash A^{\#}$.

A proof of (a) can be found in *Goldblatt, 1978,* and of (b) in *Boolos, 1979* and *1980 B*. Thus **S4Grz** can be viewed as the sentences which express provable principles of provability-and-truth. In § VII.C we'll see that the intuitionist propositional logic can be construed as expressing correct principles of provability and truth similar to these.

2. G

G

in $L(\neg, \wedge, \Box)$

axiom schemas

PC axioms

$\Box(A \supset B) \supset (\Box A \supset \Box B)$

$\Box(\Box A \supset A) \supset \Box A$

rules material detachment

necessitation

A relation is *anti-reflexive* if no element is related to itself. A frame $\langle W, R \rangle$ is a *finite strict partial order* if W is finite and R is transitive, anti-reflexive, and anti-symmetric.

Theorem 29 $\mathbf{G}\vdash A$ iff for every $\langle W,R\rangle$ which is a finite strict partial order, $\langle W,R\rangle \vDash A$.

For a proof of this see *Boolos, 1979*.

We say that $<v,s>$ is a **G**-*model* if $<v,s>$ is a *weak modal semantics* and s satisfies M1–M5 as well as

M14. $s(\Box(\Box A \supset A)) \subseteq s(\Box A)$

M15. $s(\Box A) \subseteq s(A)$ iff $s(A) = S$.

Note that in **G**-models we allow the truth-value assignment v to be independent of s.

Lemma 30 If $\mathbf{G}\vdash E$ then for every **G**-model $<v,s>$, $v(E) = T$ and $s(E) = S$.

Proof: The proof follows as for Lemma 6 except that we use the version of Lemma 2 for weak modal semantics (Corollary 3). The only new case is to show that $v(\Box(\Box A \supset A) \supset \Box A) = T$. If $v(\Box(\Box A \supset A)) = T$, then $s(\Box A \supset A) = S$. Hence $s(\Box A) \subseteq s(A)$. So by M15, $s(A) = S$, and then since we are using weak modal semantics of implication, $v(\Box A) = T$. ■

Lemma 31 If $\mathbf{G}\nvdash E$ then for some finite **G**-model $<v,s>$, $v(E) = F$ and $s(E) \neq S$.

Proof: If $\mathbf{G}\nvdash E$ then there is some $\langle W,R,e,w\rangle$ such that $w \nvDash E$ and $\langle W,R\rangle$ is a finite strict partial order. Choose one and set $S = \{ z : wRz\}$ and $v(p) = T$ iff $w \vDash p$. Extend v to all wffs by the truth-conditions of weak modal semantics.

It's easy to show that $<v,s>$ satisfies M1–M4 and M14. Using the transitivity of R we can show that it satisfies M5, as we did for Lemma 7. We need to show that $<v,s>$ satisfies M15.

Suppose $s(A) \neq S$. Then for some x, $x \nvDash A$. If $x \vDash \Box A$ we are done. If not, then for some y, xRy and $y \nvDash A$. Again, if $y \vDash \Box A$ we are done. If not we can continue; as W is finite this process must terminate in an end point, that is, a z with no world related to it, such that $z \nvDash A$. But since z is an end point it vacuously validates $\Box A$. Hence $s(\Box A) \nsubseteq s(A)$.

$w \longrightarrow x \longrightarrow y \dashrightarrow z$

A is F   A is F   A is F

□A is F

Finally we show that $v(A) = T$ iff $w \vDash A$ by induction on the length of A. The only interesting case is if A is $\Box B$. Then $v(\Box B) = F$ iff $s(B) \neq S$, which is iff there is some z, wRz and $z \nvDash B$, which is iff $w \nvDash \Box B$. ■

Combining Lemmas 30 and 31 we have:

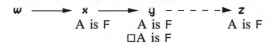

Theorem 32 $\mathbf{G}\vdash A$ iff for every **G**-model $<v,s>$, $v(A) = \mathsf{T}$
iff for every **G**-model $<v,s>$, $s(A) = \mathsf{S}$
iff for every finite **G**-model $<v,s>$, $v(A) = \mathsf{T}$.

The proof of Lemma 30 shows that the restricted class of **G**-models which satisfy $s(\Box A) \neq \varnothing$ is also complete for **G**. We make one last observation about **G** before we turn to **G***.

Lemma 33 $\mathbf{G}\nvdash \Box A \supset A$

Proof: Suppose to the contrary that $\mathbf{G}\vdash \Box A \supset A$. Then by necessitation, $\mathbf{G}\vdash \Box(\Box A \supset A)$. Hence via the second axiom schema, $\mathbf{G}\vdash \Box A$ and so $\mathbf{G}\vdash A$. But the same argument works to establish that $\mathbf{G}\vdash \neg A$. That's a contradiction by Theorem 29. ■

3. G*

G*
in $L(\neg, \wedge, \Box)$
is the closure of $\mathbf{G} \cup \{\Box A \supset A\}$ under the rule of material detachment.

By Lemma 33, $\mathbf{G}^* \neq \mathbf{G}$. The same argument shows that **G*** is not closed under necessitation, that is, it is not normal. Thus **G*** is quasi-normal and is to **G** as **QT** is to **K**.

No characterization of **G*** in terms of a class of Kripke models is known. It won't work to modify the Kripke semantics given for **G** in the last section by adding the requirement that each model be stipulated with respect to a designated world w for which wRw. Such a model may no longer validate **G**; for example, if $w \nvDash p$ yet for all other $z \in W$, $z \vDash p$, then $w \nvDash \Box(\Box p \supset p) \supset \Box p$.

Boolos, *1980 A*, has given semantics for **G*** in terms of a notion of a formula being "eventually true" at a world in a Kripke model. I'll present those as modified to apply to the language in which we're working.

Let $\langle W, R, e \rangle$ be a Kripke model. For this model we have already defined $w \vDash B$ for every for $w \in W$. We now define *for every natural number* j, $w \vDash^{(j)} B$ by induction on j and the length of B.

$w \vDash^{(j)} p$ iff $w \vDash p$

$w \vDash^{(j)} \neg B$ iff $w \nvDash^{(j)} B$

$w \vDash^{(j)} B \wedge C$ iff $w \vDash^{(j)} B$ and $w \vDash^{(j)} C$

$w \vDash^{(j)} \Box B$ iff $w \vDash \Box B$ and for all $k < j$, $w \vDash^{(k)} B$

Define

$w \models^* A$ iff for some i, for all $j \geq i$, $w \models^{(j)} A$

We read $w \models^* A$ as A *is eventually true at* w, or w *eventually validates* A.

Finally define: $\langle W, R, e \rangle \models^* A$ iff for all $w \in W$, $w \models^* A$. And $\langle W, R \rangle \models^* A$ iff for all e, $\langle W, R, e \rangle \models^* A$.

Theorem 34 $\mathbf{G}^* \vdash A$ iff for every $\langle W, R \rangle$ which is a finite strict partial order, $\langle W, R \rangle \models^* A$.

This is proved in *Boolos, 1980.*

Before turning to set-assignment semantics for **G*** we need some observations about these new semantics.

Lemma 35 *a.* $w \models^{(0)} B$ iff $w \models B$

b. $w \models^{(j+1)} \Box B$ iff $w \models^{(j)} \Box B$ and $w \models^{(j)} B$

c. If $w \models^* \Box B$ then for all j, $w \models^{(j)} B$.

Proof: I'll only do (c) and leave the rest to you. If $w \models^* \Box B$, then for any j there is some $k > j$, such that $w \models^{(k)} \Box B$. Hence by definition, for all $i < k$, $w \models^{(i)} B$, so $w \models^{(j)} B$. ■

Set-assignment semantics for **G*** can be given as a simple modification of those for **G**: we use the same set-assignments but replace the weak modal truth-conditions by those for modal semantics for implication, just as we did in modifying the semantics of **K** to get those for **QT**.

We say that $<v, s>$ is a **G***-*model* if $<v, s>$ is a modal semantics of implication and s satisfies M1–M5, M14, and M15.

It's straightforward to show that every **G***-model validates **G*** by recalling that every modal semantics for implication validates $\Box A \supset A$ (Lemma 2.f).

Lemma 36 If $\mathbf{G}^* \nvdash E$ then for some **G***-model $<v, s>$, $v(E) = F$.

Proof: If $\mathbf{G}^* \nvdash E$, then by Theorem 34 there is some $\langle W, R, e, w \rangle$ such that $\langle W, R \rangle$ is a finite strict partial order and $\langle W, R, e, w \rangle \not\models E$. Choose one.

As R is transitive we may assume without loss of generality that $W - \{w\} = \{ z : wRz \}$. We may also assume, by relabeling if necessary, that no natural number is in W.

Define

$S = (W - \{w\}) \cup \{ j : j \geq 0 \}$

$s(A) = \{ z : z \neq w \text{ and } z \models A \} \cup \{ j : w \models^{(j)} A \}$

Take $v(p) = T$ iff $w \models p$, and extend v to all wffs by the truth-conditions for modal semantics of implication. Note that we use '$z \models$' and not '$z \models^*$' in the definition of $s(A)$.

We now show that $v(E) = T$ iff $w \vDash^* E$, and that $<v, s>$ is a **G***-model.

The proof that $v(E) = T$ iff $w \vDash^* E$ is by induction on the length of E and is easy except when E is $\Box B$. If $v(\Box B) = T$ then $v(B) = T$ and $s(B) = S$. So for every z such that wRz, $z \vDash B$ and for all j, $w \vDash^{(j)} A$. Hence by Lemma 35.b, $w \vDash^* \Box B$.

If $v(\Box B) = F$, then $v(B) = F$ or $s(B) \neq S$. If the latter, then for some z, wRz and $z \nvDash B$, so $w \nvDash \Box B$, so $w \nvDash^* \Box B$. If the former, then by induction $w \nvDash^* B$, so for some j, $w \nvDash^{(j)} B$. By Lemma 35.c, $w \nvDash^* \Box B$.

We now proceed to show that $<v, s>$ is a **G***-model. It's easy to verify M1 and M2.

M5. If $s(A) = S$ then for all z such that wRz, $z \vDash A$ and for all j, $w \vDash^{(j)} A$. Hence $w \vDash \Box A$. And so by induction using Lemma 35.a, b, for all j, $w \vDash^{(j)} \Box A$. And also for all z, $z \vDash \Box A$. So $s(\Box A) = S$.

To show that M3, M4, and M14 hold, we'll prove that if $\mathbf{G} \vdash A$ then $s(A) = S$. If $\mathbf{G} \vdash A$, then $(W - \{w\}) \subseteq s(A)$. To show that every j is in $s(A)$ we induct first on j and then on the length of a proof of A. It is immediate for $j = 0$. Suppose it's true for j. I'll leave to you that $w \vDash^{(j+1)} \Box(B \supset C) \supset (\Box B \supset \Box C)$. For an axiom of the form $\Box(\Box B \supset B) \supset \Box B$, suppose $w \vDash^{(j+1)} \Box(\Box B \supset B)$. Then $w \vDash^{(j)} \Box B \supset B$ and by Lemma 35.b, $w \vDash^{(j)} \Box(\Box B \supset B)$, hence by induction (on this schema) $w \vDash^{(j)} \Box B$ and also $w \vDash^{(j)} B$. Thus $w \vDash^{(j+1)} \Box B$ by Lemma 35.c. It's now straightforward to complete the proof for consequences of the axioms.

Finally we establish M15 by showing that if $s(A) \neq S$ then $s(\Box A) \nsubseteq s(A)$. Suppose $s(A) \neq S$. If there is some z such that wRz and $z \nvDash A$, then proceed as in the proof of Lemma 30 to get that $s(\Box A) \nsubseteq s(A)$. Otherwise, suppose that for all z such that wRz, $z \vDash A$, so that $w \vDash \Box A$. Let j be minimal such that $j \notin s(A)$, that is, $w \nvDash^{(j)} A$. Then for all $k < j$, $w \vDash^{(k)} A$. Hence $w \vDash^{(j)} \Box A$, so $j \in s(\Box A)$. ■

Theorem 37 $\mathbf{G^*} \vdash A$ iff for every **G***-model $<v, s>$, $v(E) = T$.

Appendix: Completeness Theorems for Kripke Semantics

A. Completeness Theorems

In this section I'll prove the completeness theorems for Kripke semantics for **K**, **T**, **B**, **S4**, and **S5** that were cited in the body of the chapter. The proofs follow *Boolos, 1979,* using canonical models. Unless specified otherwise, whenever I refer to a logic **L** in this appendix it is to be understood as one of these five, although the results and methods apply to many other logics as you can read in *Chellas, 1980.*

We first show that the Kripke semantics proposed for the various logics are sound. We say that a relation R is *euclidean* if for all w, y, z, if wRy and wRz, then yRz.

Lemma 38 $\langle W,R\rangle \vDash \Box A \supset A$ iff R is reflexive

$\langle W,R\rangle \vDash \Box A \supset \Box\Box A$ iff R is transitive

$\langle W,R\rangle \vDash A \supset \Box\Diamond A$ iff R is symmetric

$\langle W,R\rangle \vDash \Diamond A \supset \Box\Diamond A$ iff R is euclidean

If $\langle W,R\rangle$ is reflexive and euclidean, then R is an equivalence relation.

Proof: For the first equivalence,

$\langle W,R\rangle \vDash \Box A \supset A$ iff for all e, w, $\langle W,R,e,w\rangle \vDash \Box A \supset A$
iff for all e, w, if $w \vDash \Box A$ then $w \vDash A$
iff for all e, w, if for all z such that wRz, we have $z \vDash A$, then $w \vDash A$

If R is reflexive, then clearly $\langle W,R\rangle \vDash \Box A \supset A$. If R is not reflexive, then for some w it is not the case that wRw. Define an evaluation e such that for all $p \neq p_1$, $p \in e(z)$ for all z, and $p_1 \in e(z)$ iff $z \neq w$ (this will work even if $W = \{w\}$). Then $w \vDash \Box p_1$. But $w \nvDash p_1$, so $\langle W,R\rangle \nvDash \Box A \supset A$.
I'll leave the other parts to you (cf. p.165). ■

A class of frames C is *sound* for a modal logic $\mathbf{L}$ means that if $\mathbf{L} \vdash A$, then for all $\langle W,R\rangle \in C$, $\langle W,R\rangle \vDash A$.

Theorem 39 The following classes of frames are sound for the respective logics:

K all frames

T all reflexive frames

B all reflexive and symmetric frames

S4 all reflexive and transitive frames

S5 all equivalence frames

Proof: It's just a matter of checking it for the first case and then the others follow by Lemma 38. ■

I'll now show that for each of the five logics there is one particular model, called a canonical model, in which exactly the theorems of that logic hold. The worlds of that model will be complete and consistent sets of wffs, using definitions of those notions which essentially coincide with those for **PC**. This makes sense if you recall that we can view each world as a **PC**-model and that **PC**-models can be correlated to complete and consistent sets of wffs.

For the reasons given in §D above I won't define '$\Gamma \vdash A$' in this section.

For the notion of consistency relative to a logic **L** we define

Γ is **L**-$\wedge$-*inconsistent* if for some $B_1, \ldots, B_n$ in Γ, $\vdash_{\mathbf{L}} \neg(B_1 \wedge \cdots \wedge B_n)$
Γ is **L**-$\wedge$-*consistent* otherwise.

We can use the ambiguous notation for the conjunction because $\mathbf{PC} \subseteq \mathbf{L}$ (see §D), so if, for example, $\vdash_{\mathbf{L}} \neg(B_1 \wedge (B_1 \wedge B_3))$, then any result of associating or permuting the B_i's in the formula is also a theorem of **L**. Note that n may be 1.
We define Γ to be *complete* iff for all A, one of A, $\neg$A is in Γ.

Lemma 40 If Γ is **L**-$\wedge$-consistent, then one of $\Gamma \cup \{\neg A\}$ or $\Gamma \cup \{A\}$ is **L**-$\wedge$-consistent. Hence, if Γ is **L**-$\wedge$-consistent and complete, then for each A, exactly one of A, $\neg$A is in Γ.

Proof: If $\Gamma \cup \{A\}$ is **L**-$\wedge$-inconsistent then since Γ is **L**-$\wedge$-consistent there must be $B_1, \ldots, B_n$ such that $\vdash_{\mathbf{L}} \neg(A \wedge B_1 \wedge \cdots \wedge B_n)$. If $\Gamma \cup \{\neg A\}$ is also **L**-$\wedge$-inconsistent there must be $C_1, \ldots, C_n$ in Γ such that $\vdash_{\mathbf{L}} \neg(\neg A \wedge C_1 \wedge \cdots \wedge C_n)$. Hence by **PC** (see §D), $\vdash_{\mathbf{L}} \neg(B_1 \wedge \cdots \wedge B_n \wedge C_1 \wedge \cdots \wedge C_n)$, and hence Γ is **L**-$\wedge$-inconsistent. ■

Theorem 41 If Σ is **L**-$\wedge$-consistent then there is some **L**-$\wedge$-consistent and complete Γ such that $\Sigma \subseteq \Gamma$.

Proof: Let the wffs of the language be ordered as $A_0, A_1, \ldots$. Define

$$\Gamma_0 = \Sigma$$

$$\Gamma_{n+1} = \begin{cases} \Gamma_n \cup \{A_n\} & \text{if this is } \mathbf{L}\text{-}\wedge\text{-consistent} \\ \Gamma_n \cup \{\neg A_n\} & \text{otherwise} \end{cases}$$

Using the previous lemma, $\Gamma = \bigcup_n \Gamma_n$ is complete and **L**-$\wedge$-consistent. ■

Lemma 42 If Γ is **L**-$\wedge$-consistent and complete, then:
a. $\mathbf{L} \subseteq \Gamma$ and Γ is closed under material detachment.
b. If $A \vdash_{\mathbf{PC}} B$ then: if $\Box A \in \Gamma$, then $\Box B \in \Gamma$; if $\neg\Box B \in \Gamma$ then $\neg\Box A \in \Gamma$.

Proof: ***a.*** Suppose $\vdash_{\mathbf{L}} A$. Then if $A \notin \Gamma$ we have $\neg A \in \Gamma$. But as $\vdash_{\mathbf{L}} \neg(\neg A)$ (by **PC**) we have a contradiction on the **L**-$\wedge$-consistency of Γ. So $A \in \Gamma$. If A and $A \supset B$ are in Γ, then if $B \notin \Gamma$ we must have $\neg B \in \Gamma$. But $\vdash_{\mathbf{PC}} \neg(A \wedge (A \supset B) \wedge \neg B)$ and that contradicts the **L**-$\wedge$-consistency of Γ; so $B \in \Gamma$.
b. This follows from Lemma 4 (p.163) and part (a). ■

The *canonical model* for a modal logic **L** is $\langle \mathcal{W}_{\mathbf{L}}, \mathcal{R}_{\mathbf{L}}, e_{\mathbf{L}} \rangle$ where

$$\mathcal{W}_{\mathbf{L}} = \{\Gamma : \Gamma \text{ is } \mathbf{L}\text{-}\wedge\text{-consistent and complete}\}$$

$\Gamma\, R_L\, \Delta$ iff for all A, if $\Box A \in \Gamma$ then $A \in \Delta$

$e_L(\Gamma) = \{p : p \in \Gamma\}$

Note that W_L is uncountable.

Lemma 43 Given $\Gamma \in W_L$, if for all $\Delta \in W_L$ such that $\Gamma\, R_L\, \Delta$ we have $B \in \Delta$, then $\Box B \in \Gamma$.

Proof: Let $\Sigma = \{A : \Box A \in \Gamma\}$. If $\Sigma \cup \{\neg B\}$ were **L**-$\wedge$-consistent then there would be some complete and **L**-$\wedge$-consistent $\Delta \supseteq \Sigma \cup \{\neg B\}$; but $B \notin \Delta$ yet $\Gamma\, R_L\, \Delta$, a contradiction. So $\Sigma \cup \{\neg B\}$ is **L**-$\wedge$-inconsistent.

So for some $B_1, \ldots, B_n$ in Σ, either $\vdash_L \neg(B_1 \wedge \cdots \wedge B_n)$ or $\vdash_L \neg(\neg B \wedge B_1 \wedge \cdots \wedge B_n)$. In either case via **PC**, $\vdash_L \neg B \supset \neg(B_1 \wedge \cdots \wedge B_n)$ and hence $\vdash_L (B_1 \wedge \cdots \wedge B_n) \supset B$. And so by necessitation and the distribution axioms and Lemma 4, $\vdash_L (\Box B_1 \wedge \cdots \wedge \Box B_n) \supset \Box B$, whence by **PC** (Exportation) $\vdash_L \Box B_1 \supset (\Box B_2 \supset \cdots \supset (\Box B_n \supset \Box B)) \cdots)$. As each $B_i \in \Sigma$, we have by definition each $\Box B_i \in \Gamma$, so by Lemma 42, $\Box B \in \Gamma$. ■

Theorem 44 ***a.*** $\langle W_L, R_L, e_L, \Gamma \rangle \vDash A$ iff $A \in \Gamma$.

b. $\vdash_L A$ iff $\langle W_L, R_L, e_L \rangle \vDash A$.

Proof: a. Given $\langle W_L, R_L, e_L \rangle$ we proceed for all Γ by induction on the length of A. It's true for the propositional variables. Then,

$\Gamma \vDash \neg A$ iff $\Gamma \nvDash A$
 iff $A \notin \Gamma$ by induction
 iff $\neg A \in \Gamma$ by the completeness of Γ

$\Gamma \vDash A \wedge B$ iff $\Gamma \vDash A$ and $\Gamma \vDash B$
 iff $A, B \in \Gamma$ by induction
 iff $(A \wedge B) \in \Gamma$ by Lemma 42, as $\mathbf{PC} \subseteq \mathbf{L}$

$\Gamma \vDash \Box A$ iff for every Δ, if $\Gamma\, R_L\, \Delta$ then $\Delta \vDash A$
 iff for every Δ, if $\Gamma\, R_L\, \Delta$ then $A \in \Delta$ by induction
 iff $\Box A \in \Gamma$ by the previous lemma

and we have proved part (a).

Note that if we take $L(\neg, \rightarrow, \wedge)$ as our language, we have to use the axiom schema $\Box(A \supset B) \equiv (A \rightarrow B)$ and argue, as above,

$\Gamma \vDash A \rightarrow B$ iff $\Gamma \vDash \Box(A \supset B)$
 iff $\Box(A \supset B) \in \Gamma$
 iff $(A \rightarrow B) \in \Gamma$

b. By Lemma 42, for each $\Gamma \in W_L$, $\mathbf{L} \subseteq \Gamma$, so if $\vdash_L A$ then $\langle W_L, R_L, e_L \rangle \vDash A$. If $\nvdash_L A$ then $\{\neg A\}$ is **L**-$\wedge$-consistent. By Lemma 41 there is an **L**-$\wedge$-consistent complete Γ such that $\neg A \in \Gamma$. Hence $\langle W_L, R_L, e_L, \Gamma \rangle \not\models A$, so $\langle W_L, R_L, e_L \rangle \not\models A$. ∎

We now have to verify that the accessibility relation of each canonical model has the appropriate properties. We cannot use Lemma 38 since in Theorem 44 we considered only one model from each canonical frame $\langle W_L, R_L \rangle$.

Lemma 45 If **L** contains the schema:

a. $\Box A \supset A$ then $\langle W_L, R_L \rangle$ is reflexive.

b. $\Box A \supset \Box\Box A$ then $\langle W_L, R_L \rangle$ is transitive.

c. $A \supset \Box\Diamond A$ then $\langle W_L, R_L \rangle$ is symmetric.

d. $\Diamond A \supset \Box\Diamond A$ then $\langle W_L, R_L \rangle$ is euclidean.

Proof: a. Suppose $\vdash_L \Box A \supset A$. Then for all A, if $\Box A \in \Gamma$ then by Lemma 42, $A \in \Gamma$, so $\Gamma R_L \Gamma$.

b. If $\Gamma R_L \Delta$ and $\Delta R_L \Sigma$, we need to show that $\Gamma R_L \Sigma$. This is if and only if for all A, if $\Box A \in \Gamma$ then $A \in \Sigma$. But if $\Box A \in \Gamma$ then $\Box\Box A \in \Gamma$ by assumption using Lemma 42, so $\Box A \in \Delta$, so $A \in \Sigma$.

c. Assume $\Gamma R_L \Delta$. To show $\Delta R_L \Gamma$ we need for all B that if $\Box B \in \Delta$ then $B \in \Gamma$. Suppose $\Box B \in \Delta$ and $B \notin \Gamma$. Then $\neg B \in \Gamma$. So by the assumption of this part we get $\Box\Diamond\neg B \in \Gamma$, so $\Diamond\neg B \in \Delta$. That is, $\neg\Box\neg\neg B \in \Delta$, so by Lemma 42.b, $\neg\Box B \in \Delta$, which is a contradiction on the **L**-$\wedge$-consistency of Δ. Hence $\Delta R_L \Gamma$.

d. Suppose $\Gamma R_L \Delta$ and $\Gamma R_L \Sigma$. We want $\Delta R_L \Sigma$. Suppose not and some $\Box B \in \Delta$, yet $B \notin \Sigma$. Then $\Box B \notin \Gamma$, so $\neg\Box B \in \Gamma$. Thus $\Diamond\neg B \in \Gamma$ by using Lemma 42.b. So by the assumption of this part, $\Box\Diamond\neg B \in \Gamma$. Hence $\Diamond\neg B \in \Delta$, so $\neg\Box B \in \Delta$ and $\Box B \notin \Delta$, a contradiction. ∎

A class C of frames is *complete* for a modal logic **L** means that $L \vdash A$ iff for all $\langle W, R \rangle \in C$, $\langle W, R \rangle \models A$

Theorem 46 **(*Completeness of Kripke Semantics*)** The following classes of frames are complete for the respective logics:

K all frames

T all reflexive frames

B all reflexive and symmetric frames

S4 all reflexive and transitive frames

S5 all equivalence frames

Proof: This follows from Theorem 39, Theorem 44, and Theorem 45. ∎

Recall that **QT** is characterized as the closure of $\mathbf{K} \cup \{\Box A \supset A\}$ under material detachment. We can adapt the proofs above to **QT**.

Theorem 47 $\mathbf{QT} \vdash A$ iff for every $\langle W, R, w \rangle$ such that wRw, $\langle W, R, w \rangle \models A$.

Proof: It's easy to show that if $\mathbf{QT}\vDash A$ then for every $\langle W, R, e, w\rangle$ such that wRw we have $w\vDash A$.

Now suppose $\mathbf{QT}\nvdash A$. First, $\mathbf{QT}\cup\{\neg A\}$ is $\mathbf{QT}$-$\wedge$-consistent and hence is contained in some $\mathbf{QT}$-$\wedge$-consistent and complete set of wffs Σ. Choose such a Σ. Now let $\langle W, R, e\rangle$ be the canonical model for **K**. Set $M = \langle W, R, e, \Sigma\rangle$. Since $(\Box B \supset B) \in \Sigma$ we have $M\vDash \Box B \supset B$. Hence if $\Box B \in \Sigma$ then $B \in \Sigma$, so $\Sigma R \Sigma$. And since $\neg A \in \Sigma$, $M \nvDash A$. ■

B. Decidability and the Finite Model Property

In this section we'll see that for each of the logics **K**, **T**, **B**, **S4**, and **S5** we can add the word 'finite' to the description of the class of models characterizing it in Theorem 46. From this we'll be able to deduce that each of these logics is decidable.

Given a collection of wffs Γ, we say that Γ *is closed under subformulas* if for all $A \in \Gamma$, if B is a subformula of A then $B \in \Gamma$.

Lemma 48 Let $M = \langle W, R, e\rangle$ be any Kripke model and Γ any collection of wffs closed under subformulas. Then there is a model $M^* = \langle W^*, R^*, e^*\rangle$ such that:

i. For all $A \in \Gamma$, $M \vDash A$ iff $M^* \vDash A$.

ii. For every $w^*, z^* \in W^*$ with $w^* \neq z^*$ there is some $A \in \Gamma$ such that either $w^* \vDash A$ and $z^* \nvDash A$, or $w^* \nvDash A$ and $z^* \vDash A$.

iii. If R is reflexive, or symmetric, or transitive, then R^* is, too.

iv. If Γ is finite, then M^* is finite.

Proof: Given M and Γ. Define a relation on W:

$$x \approx y \text{ iff for all } A \in \Gamma,\ x \vDash A \text{ iff } y \vDash A$$

This is an equivalence relation. Let x^* denote the equivalence class of x. Define:

$$W^* = \{x^* : x \in W\}$$

$$x^* R^* y^* \text{ iff for all } A \in \Gamma, \text{ if } x \vDash \Box A \text{ then } y \vDash A$$

$$e^*(x^*) = \{p : p \in \Gamma \text{ and } x \vDash p\}$$

$$M^* = \langle W^*, R^*, e^*\rangle$$

By the definition of $\approx$, both R^* and e^* are well-defined, that is, they do not depend on the choice of representative of the equivalence class. Note that if xRy then $x^* R^* y^*$.

If $x^* \neq y^*$ then there is some $A \in \Gamma$ such that x^* evaluates A differently from y^*. So if Γ is finite then W^* is finite. It remains to show that for all $A \in \Gamma$, $M \vDash A$ iff $M^* \vDash A$. I will show by induction on the length of A that for all $A \in \Gamma$ and $x \in W$, $\langle W, R, e, x\rangle \vDash A$ iff $\langle W^*, R^*, e^*, x^*\rangle \vDash A$.

If A has length 1 it's true by definition. So suppose it's true for all shorter wffs and A is $B \wedge C$. Then

$$\begin{aligned} x \vDash B \wedge C \ &\text{iff} \ x \vDash B \text{ and } x \vDash C \\ &\text{iff} \ x^* \vDash B \text{ and } x^* \vDash C \quad \text{by induction, since } \Gamma \text{ is closed under subformulas} \\ &\text{iff} \ x^* \vDash B \wedge C \end{aligned}$$

If A is $\neg B$ then the proof is similar.

If A is $\Box B$, suppose $x^* \vDash \Box B$. If $x R y$ then $x^* R^* y^*$, so $y^* \vDash B$. Hence by induction as $B \in \Gamma$, $y \vDash B$ and thus $x \vDash \Box B$. In the other direction, suppose $x \vDash \Box B$. If $x^* R^* y^*$ then $y \vDash B$ by the definition of R^*. So by induction $y^* \vDash B$, so $x^* \vDash \Box B$.

Finally, I'll show that R^* inherits the properties of R.

Suppose that R is reflexive. So if $x \vDash \Box A$ then $x \vDash A$, and hence for all x^*, $x^* R^* x^*$.

Suppose R is symmetric and $x^* R^* y^*$. Suppose also that for some A, $y \vDash \Box A$ but $x \nvDash A$. Then $x \vDash \neg A$ and so by the symmetry of R, $x \vDash \Box \Diamond \neg A$. Hence $y \vDash \Diamond \neg A$, and so $y \vDash \neg \Box A$, a contradiction. Hence $x \vDash A$ and so $y^* R^* x^*$.

Suppose R is transitive and $x^* R^* y^*$ and $y^* R^* z^*$. Suppose that for some A, $x \vDash \Box A$. Since R is transitive, $x \vDash \Box\Box A$, so $y \vDash \Box A$, and hence $z \vDash A$. Hence $x^* R^* z^*$. ∎

Theorem 46 and Lemma 48 show that each of **K**, **T**, **B**, **S4**, and **S5** has the *finite model property*: if a wff fails in some model of the logic, then it fails in a finite model. The same proof applies to **QT**. Thus

Theorem 49 The following classes of frames are complete for the respective logics:

K all *finite* frames

T all *finite* reflexive frames

B all *finite* reflexive and symmetric frames

S4 all *finite* reflexive and transitive frames

S5 all *finite* equivalence frames

The class of all $\langle W, R, w \rangle$ such that W is finite and wRw is complete for **QT**.

Theorem 50 **K**, **T**, **B**, **S4**, **S4Grz**, **S5**, and **G** are decidable.

Proof: By 'decidable' we mean that there is an effective procedure which given any A determines whether A is a theorem or not.

Let **L** be any one of these logics. Since each is axiomatized by a finite number of schema we have an effective procedure for listing out all theorems of the logic: list all proofs. We can also effectively list all finite frames which satisfy the appropriate conditions for the logics (Theorem 49 for **K**, **T**, **B**, **S4**, **S4Grz**, **S5**; Theorem 15 for **S4Grz**; Theorem 29 for **G**). And we can effectively check whether a given wff is valid in a finite frame because by Lemma 1 we need only consider the propositional variables in that wff.

Now suppose we are given a wff A. To decide if A is a theorem dovetail these two listing procedures until either we find a proof of A, in which case A is a theorem, or we find a finite model of **L** which invalidates A, in which case A is not a theorem. ■

Hughes and Cresswell, *1968*, Chapters 5 and 6, give a more useful decision procedure for **T**, **S4**, and **S5**.

C. Consequence Relations and the Deduction Theorem

For normal logics there are two ways we can define a syntactic consequence relation depending on whether we allow the rule of necessitation to apply to non-theorems.

1. Without necessitation

Let **L** be one of the five logics **K**, **T**, **B**, **S4**, or **S5**. Define for $\Gamma \subseteq$ Wffs, $\Gamma \vdash_{\mathbf{L}} A$ to mean that there are wffs $B_1, \ldots, B_n = A$ such that each B_i is in **L**, or is in Γ, or is a direct consequence of earlier B_i's by the rule of material detachment. So $\varnothing \vdash_{\mathbf{L}} A$ iff $\mathbf{L} \vdash A$ as defined in the body of the chapter.

This is not our standard definition of proof, for in a derivation we allow B_i to be a theorem and not just an axiom of **L**. Thus this notion of consequence requires us to interleave two proof procedures, one for theorems of **L** and one for consequences of Γ. We need this because without the rule of necessitation we could not prove the theorems of **L**. The result is that this is the **PC** notion of consequence, for

$$\Gamma \vdash_{\mathbf{L}} A \text{ iff } \Gamma \cup \mathbf{L} \vdash_{\mathbf{PC}} A$$

Thus the appropriate notions of completeness and consistency are as for **PC**, and the definition of a theory is standard: Γ is a $\vdash_{\mathbf{L}}$- theory if $\mathbf{L} \subseteq \Gamma$ and Γ is closed under the proof rule of $\vdash_{\mathbf{L}}$, namely, material detachment. Then *the elements of the canonical model are complete consistent* $\vdash_{\mathbf{L}}$*-theories*: these are the canonical possible worlds.

Theorem 51 ***(Strong Completeness with respect to Kripke Models with Designated World)***

For each logic **L** listed below we have:

$\Gamma \vdash_{\mathbf{L}} A$ iff for every $\langle W, R, e, w \rangle$ in the class listed, if $w \models \Gamma$, then $w \models A$.

K all $\langle W, R, e, w \rangle$

T all reflexive $\langle W, R, e, w \rangle$

B all reflexive and symmetric $\langle W, R, e, w \rangle$

S4 all reflexive and transitive $\langle W, R, e, w \rangle$

S5 all equivalence $\langle W, R, e, w \rangle$

Proof: $\Rightarrow$| I'll leave this direction to you.

$\Leftarrow$| Let **L** be one of these logics and suppose that $\Gamma \nvdash_{\mathbf{L}} A$. Then $\Gamma \cup \{\neg A\}$ is **L**-$\wedge$-consistent: if not, then for some $B_1, \ldots, B_n \in \Gamma$ either $\vdash_{\mathbf{L}} \neg(B_1 \wedge \cdots \wedge B_n)$ or $\vdash_{\mathbf{L}} \neg((B_1 \wedge \cdots \wedge B_n) \wedge \neg A)$. In both cases, using the fact that $\mathbf{PC} \subseteq \mathbf{L}$ we get $\vdash_{\mathbf{L}} B_1 \supset (B_2 \supset \cdots \supset (B_n \supset A)) \cdots)$, so $\Gamma \vdash_{\mathbf{L}} A$, a contradiction. Hence there is some complete **L**-$\wedge$-consistent Σ such that $\Gamma \cup \{\neg A\} \subseteq \Sigma$. Choose one.

Define $\mathsf{U} = \{\Delta \in \mathsf{W}_{\mathbf{L}} : \neg A \in \Delta\}$. Then $\mathsf{U} \neq \varnothing$ as $\Sigma \in \mathsf{U}$. So define $\mathsf{M} = \langle \mathsf{U}, \mathsf{R}_{\mathbf{L}}, \mathsf{e}_{\mathbf{L}}, \Sigma \rangle$ where $\mathsf{R}_{\mathbf{L}}$ and $\mathsf{e}_{\mathbf{L}}$ are defined as in the canonical model. Now proceed as in Lemma 43, Theorem 44, and Lemma 45 to prove that M satisfies the appropriate condition stated in the theorem and that for all $\Delta \in \mathsf{U}$, $\Delta \models B$ iff $B \in \Delta$. So we have $\Sigma \models \Gamma$ and $\Sigma \nvDash A$. ∎

We define the *semantic consequence relation for the set-assignment semantics* for each of these logics in the usual way: $\Gamma \models_{\mathbf{L}} A$ iff for every **L**-model $<v,s>$, if $<v,s> \models \Gamma$ then $<v,s> \models A$.

Theorem 52 **(*Strong Completeness for Set-Assignment Semantics*)**
For **L** any one of **K**, **T**, **B**, **S4**, or **S5**, $\Gamma \vdash_{\mathbf{L}} A$ iff $\Gamma \models_{\mathbf{L}} A$.

Proof: $\Rightarrow$| This direction is easy.

$\Leftarrow$| Suppose $\Gamma \nvdash_{\mathbf{L}} A$. Then by the previous theorem, some $\langle \mathsf{W}, \mathsf{R}, \mathsf{e}, w \rangle \models \Gamma$ but $w \nvDash A$. For each of these logics a method was given in the body of the chapter for converting such a model with designated world into an **L**-model $<v,s>$ in which $v(B) = \mathsf{T}$ iff $w \models B$. Then $<v,s> \models \Gamma$ and $v(A) = \mathsf{F}$. ∎

Theorem 53 **(*Material Implication Form of the Deduction Theorem*)**
If **L** is any one of **K**, **T**, **B**, **S4**, or **S5**, then:

$$\Gamma \cup \{A\} \vdash_{\mathbf{L}} B \text{ iff } \Gamma \vdash_{\mathbf{L}} A \supset B$$

$$\Gamma \cup \{A\} \models_{\mathbf{L}} B \text{ iff } \Gamma \models_{\mathbf{L}} A \supset B$$

Proof: The syntactic part comes from our observation above that $\Gamma \vdash_{\mathbf{L}} A$ iff $\Gamma \cup \mathbf{L} \vdash_{\mathbf{PC}} A$; the semantic part is immediate from the definitions. ∎

2. With necessitation

Let **L** be any one of the five logics **K**, **T**, **B**, **S4**, or **S5**, and let $\Gamma \subseteq$ Wffs. Define $\Gamma \vdash_{\mathbf{L}}^{\Box} A$ to mean that there are $B_1, \ldots, B_n = A$ such that each B_i is an axiom of **L**, or is in Γ, or is a direct consequence of earlier B_i's by the rule of material detachment or the rule of necessitation. Thus to proceed on the hypothesis of A is also to assume that A is necessary.

This is the usual notion of syntactic derivation and here we also have $\varnothing \vdash_{\mathbf{L}}^{\Box} A$ iff $\mathbf{L} \vdash A$.

Theorem 54 ***(Strong Completeness with respect to Kripke Frames)***

For each logic **L** listed below we have:

$\Gamma \vdash_{\mathbf{L}}^{\Box} A$ iff for every frame $\langle W, R\rangle$ in the class listed, if $\langle W, R\rangle \vDash \Gamma$, then $\langle W, R\rangle \vDash A$.

K all frames

T all reflexive frames

B all reflexive and symmetric frames

S4 all reflexive and transitive frames

S5 all equivalence frames

On the face of it this notion of consequence seems unnatural: assuming A to be true, why should A be necessary? This is corroborated by two technical consequences of the definition.

First, the two usual forms of the Deduction Theorem fail. For each logic for every A, $A \vdash_{\mathbf{L}}^{\Box} \Box A$, but both $\vdash_{\mathbf{L}} A \supset \Box A$ and $\vdash_{\mathbf{L}} A \rightarrow \Box A$ can fail.

Second, I see no obvious way to modify the set-assignment semantics for these logics to obtain a strong completeness theorem for $\vdash_{\mathbf{L}}^{\Box}$. We might suppose that one additional condition would do: if $v(A) = T$ then $s(A) = S$, corresponding to the idea of necessitation. But that's too strong for it validates $A \supset \Box A$.

However, we do have the following Deduction Theorem for **S4** and **S5**.

Theorem 55 If **L** is either **S4** or **S5**, then

a. $A \vdash_{\mathbf{L}}^{\Box} B$ iff $\vdash_{\mathbf{L}}^{\Box} \Box A \supset B$.

b. $A \vdash_{\mathbf{L}}^{\Box} B$ iff $\vdash_{\mathbf{L}}^{\Box} \Box A \supset \Box B$.

Proof: I'll prove both of these at once.

$\Leftarrow$ Given A we have the following proofs:

a. A, $\Box A$ (by necessitation), $\Box A \supset B$, B

b. A, $\Box A$, $\Box A \supset \Box B$, $\Box B$, $\Box B \supset B$ (since **L** is **S4** or **S5**), B

$\Rightarrow$ Suppose $B_1, \ldots, B_n$ is a proof of B from A. I'll show by induction that for all i, $\vdash_{\mathbf{L}}^{\Box} \Box A \supset B_i$ and $\vdash_{\mathbf{L}}^{\Box} \Box A \supset \Box B_i$.

If $n = 1$, then if B_1 is A we have $\vdash_{\mathbf{L}}^{\Box} \Box A \supset A$ since **L** is **S4** or **S5**; by **PC** we also have $\vdash_{\mathbf{L}}^{\Box} \Box A \supset \Box A$. Otherwise, B_1 is an axiom of **L**, in which case we have $\vdash_{\mathbf{L}}^{\Box} \Box B_1$ by necessitation, and the result follows by **PC**.

Suppose now it's true for n and $B_1, \ldots, B_{n+1}$ is a proof of B from A. Then: i. B_{n+1} is A, or ii. B_{n+1} is an axiom of **L**, or iii. for some $i, j \leq n$, B_j is $B_i \supset B_{n+1}$, or iv. B_{n+1} is $\Box B_i$ for some $i \leq n$. Cases (i) and (ii) were done above. For (iii), by induction $\vdash_{\mathbf{L}}^{\Box} \Box A \supset (B_i \supset B_{n+1})$ and $\vdash_{\mathbf{L}}^{\Box} \Box A \supset B_i$, so by **PC**, $\vdash_{\mathbf{L}}^{\Box} \Box A \supset B_{n+1}$. Also by induction, $\vdash_{\mathbf{L}}^{\Box} \Box A \supset \Box(B_i \supset B_{n+1})$ and $\vdash_{\mathbf{L}}^{\Box} \Box A \supset \Box B_i$, and by distribution, $\vdash_{\mathbf{L}}^{\Box} \Box A \supset (\Box B_i \supset \Box B_{n+1})$, so by **PC**

$\vdash_{\mathbf{L}}^{\Box} \Box A \supset \Box B_{n+1}$. For case (iv), by induction $\vdash_{\mathbf{L}}^{\Box} \Box A \supset \Box B_{n+1}$, and as **L** is **S4** or **S5**, $\vdash_{\mathbf{L}}^{\Box} \Box B_{n+1} \supset \Box\Box B_{n+1}$, so by **PC**, $\vdash_{\mathbf{L}}^{\Box} \Box A \supset \Box\Box B_{n+1}$. ■

Note that this proof can be adapted to show that

$\Gamma, A \vdash_{\mathbf{L}}^{\Box} B$ iff $\Gamma \vdash_{\mathbf{L}}^{\Box} \Box A \supset B$

$\Gamma, A \vdash_{\mathbf{L}}^{\Box} B$ iff $\Gamma \vdash_{\mathbf{L}}^{\Box} \Box A \supset \Box B$

It would be interesting to find a semantic proof of Theorem 55 using Theorem 54.

Deduction Theorems for other modal logics using this notion of syntactic consequence are more complicated. See *Porte, 1982, Surma, 1972,* and *Perzanowski, 1973.*

VII Intuitionism
– Int and J –

In this chapter I discuss Heyting's formalization of intuitionist reasoning and Kripke semantics for it. I rely primarily on the readings of those semantics given by Fitting and Dummett to place this formalization of intuitionism within the general framework of logics of Chapter IV. It is not necessary, however, to have read that chapter to follow the discussion here.

In this chapter we will see for the first time a nonclassical table for negation.

A. Intuitionism and Logic

At the end of the last century mathematicians first began to use completed infinite totalities in mathematical constructions and proofs. Such sets were justified on a platonist conception of mathematics, often related to a formalist reduction of mathematics to logic.

> In particular, in introducing new numbers, mathematics is only obliged to give definitions of them, by which such a definiteness and, circumstances permitting, such a relation to the older numbers are conferred upon them that in given cases they can definitely be distinguished from one another. As soon as a number satisfies all these conditions, it can and must be regarded as existent and real in mathematics.
>
> *Cantor, 1883*, p. 182

There were a number of mathematicians who objected to using nonconstructive proofs or definitions. But Brouwer, *1907* and *1908*, went further and argued that the classical laws of logic which are valid in finite domains do not necessarily apply to (potentially) infinite collections. In the following years he and his colleagues developed a distinct program of mathematics which is now called 'intuitionism'.

The intuitionist believes that mathematics, that is the doing of mathematics, is prior to logic. Formal logic may be interesting and useful but it cannot be the basis of mathematics.

> And in the construction of [all mathematical sets of units which are entitled to that name] neither the ordinary language nor any symbolic language can have any other rôle than that of serving as a nonmathematical auxiliary, to assist the mathematical memory or to enable different individuals to build up the same set.
>
> *Brouwer, 1912*, p. 81

The fundamental concepts of mathematics can be built by us independently of sense experience starting from a "basal" intuition.

> This neo-intuitionism considers the falling apart of moments of life into qualitatively different parts, to be reunited only while remaining separated by time, as the fundamental phenomenon of the human intellect, passing by abstracting from its emotional content into the fundamental phenomenon of mathematical thinking, the intuition of the bare two-oneness. ... Finally this basal intuition of mathematics, in which the connected and the separate, the continuous and discrete are united, gives rise immediately to the intuition of the linear continuum, i.e., of the "between," which is not exhaustible by the interposition of new units and which therefore can never be thought of as a mere collection of units.
>
> *Brouwer, 1912*, p. 80

It is on the question of how to reason about the infinite, or the potentially infinite that the intuitionists disagree with the classical mathematician, for in the realm of the finite they concur that classical logic is appropriate.

The intuitionist argues that the law of excluded middle, $A \vee \neg A$, is not universally valid. For instance, let $\varphi(n)$ be the sentence 'there is a prime pair greater than n,' which means that there is some x such that both $2x+1$ and $2x+3$ are both prime and greater than n. So for example, $\varphi(10)$ is true since both 11 and 13 are prime. At present it is not known whether there is a largest prime pair. The classical mathematician claims that for every natural number n, $\varphi(n) \vee \neg\varphi(n)$ is true; the intuitionists reject that, saying that $\varphi(10^{8489726}) \vee \neg\varphi(10^{8489726})$ cannot be justifiably asserted since no method has been given for constructing a prime pair $> 10^{8489726}$ nor a proof that no such pair exists.

A further example of the difference in reasoning between a classical mathematician and an intuitionist is the classical proof that there are irrational numbers a, b such that a^b is rational. First recall that a number is rational if it is of the form p/q where p and q are integers, that a real number is irrational if it is not rational, and that $\sqrt{2}$ is irrational. Now consider $\sqrt{2}^{\sqrt{2}}$. If it is rational then we are done. If not, let $a = \sqrt{2}^{\sqrt{2}}$ and $b = \sqrt{2}$. Then $a^b = (\sqrt{2}^{\sqrt{2}})^{\sqrt{2}} = 2$. Since, classically, either $\sqrt{2}^{\sqrt{2}}$ is rational or $\sqrt{2}^{\sqrt{2}}$ is not rational, we have the result. Intuitionists reject this argument: which pair of irrationals gives the result? To prove $A \vee B$ is, for them, to give a proof of A or a proof of B.

> The solution is to abandon the principle of bivalence, and suppose our statements to be true just in case we have established that they are, i.e., if mathematical statements are in question, when we at least have an effective method of obtaining a proof of them.
>
> *Dummett, 1977,* p.375

Brouwer takes as a basic insight that the validity of $A \vee \neg A$ is identified with the principle that every mathematical problem is solvable (see *Brouwer, 1928*, pp. 41–42, translated in *Bochenski*, p. 295).

Similarly, $\neg\neg A \rightarrow A$ is not accepted as universally valid by the intuitionists. Consider the decimal $b = .b_1 b_2 \ldots b_n \ldots$ where

$$b_n = \begin{cases} 3 & \text{if no string of 7 consecutive 7's appears before the } n^{\text{th}} \text{ decimal} \\ & \text{place in the expansion of } \pi \\ 0 & \text{otherwise} \end{cases}$$

We may prove '$\neg\neg$(b is rational)' by showing that '$\neg$(b is rational)' leads to a contradiction: if b were not rational then it could not be a finite string of 3's, $.33 \ldots 3$. So it would have to be $1/3$, which is a contradiction. But it is, *at present*, not correct to assert 'b is rational' for no method is known to compute numbers p and q such that $b = p/q$.

In explaining the meaning of the logical constants Dummett says

A proof of $A \wedge B$ is anything that is a proof of A and of B.

A proof of $A \vee B$ is anything that is a proof of A or of B. ...

A proof of $A \rightarrow B$ is a construction of which we can recognize that, applied to any proof of A, it yields a proof of B. ...

A proof of $\neg A$ is usually characterized as a construction of which we can recognize that, applied to any proof of A, it will yield a proof of a contradiction.

Dummett, 1977, pp. 12–13

In the explication of '$\neg$' Dummett warns that 'a contradiction' must not be understood to be some statement of the form $B \wedge \neg B$ lest the characterization be circular. Rather it is intended to mean some particular statement, such as '$0 = 1$'; or negation is assumed to be clear when applied to arithmetic equations involving no variables, such as '$47 \times 23 = 1286$' and a contradiction is then understood as $(A \rightarrow B) \wedge \neg B$ where B is such an equation.

With this background let's turn to how intuitionistic reasoning has been formalized as a logic.

B. Heyting's Formalization of Intuitionism

Intuitionistic mathematics is an activity of thought, and every language—even the formalistic—is for it only a means of communication. It is impossible in principle to establish a system of formulae that would have the same value as intuitionistic mathematics, since it is impossible to reduce the possibilities of thought to a finite number of rules that thought can previously lay down. The endeavour to reproduce the most important parts of mathematics in a language of formulae is justified exclusively by the great conciseness and definiteness of this last as compared with customary languages, properties which fit it to facilitate penetration of the intuitionistic concepts and their application in research. ... The relationship between this [formal] system and mathematics is this, that on a determinate interpretation of the constants and under certain restrictions on substitution for variables every formula expresses a correct mathematical proposition. (E.g. in the propositional calculus the variables must be replaced only by senseful [sinnerfülte] mathematical sentences.)

Heyting, 1930, translated in *Bochenski*, pp. 293–294

Heyting presented his formal system syntactically as a collection of theorems. It is now standard to refer to it as the *intuitionist propositional calculus*.

1. Heyting's axiom system Int

Int

in L(¬, →, ∧, ∨)

axiom schemas

I. $A \to (A \wedge A)$

II. $(A \wedge B) \to (B \wedge A)$

III. $(A \to B) \to ((A \wedge C) \to (B \wedge C))$

IV. $((A \to B) \wedge (B \to C)) \to (A \to C)$

V. $A \to (B \to A)$

VI. $(A \wedge (A \to B)) \to B$

VII. $A \to (A \vee B)$

VIII. $(A \vee B) \to (B \vee A)$

IX. $((A \to C) \wedge (B \to C)) \to ((A \vee B) \to C)$

X. $\neg A \to (A \to B)$

XI. $((A \to B) \wedge (A \to \neg B)) \to \neg A$

rules *modus ponens* adjunction

$$\frac{A, A \to B}{B} \qquad \frac{A, B}{A \wedge B}$$

Heyting used the rule of substitution rather than schemas.

Until §E, whenever I write $\vdash$ I will mean $\vdash_{\mathbf{Int}}$.

What is the nature of a project to give formal semantics to this system? By Heyting's quote, with which apparently all intuitionists agree, we cannot hope to fully capture or accurately represent the intuitionists' notion of meaning with formal semantics for a formal language. At best formal semantics can give us a projective knowledge of how intuitionists reason; that is, we can gain enough insight to be able to reason propositionally in agreement with them if we wish. But the same could be said for any logic. What distinguishes the intuitionists is the degree to which they claim the precedence of intuition over logical systems and the extent to which they feel their notions have been misunderstood by classically trained mathematicians and logicians.

2. Kripke semantics for Int

I'll present formal semantics for **Int** along the lines of *Kripke, 1965*. See *Troelstra and van Dalen, 1988,* for a survey of formal semantics for **Int**, and *Dummett, 1977*, pp.213–214, for an historical account.

$\langle W, R, e \rangle$ is a (Kripke) *model* if W is a nonempty set, R is a *reflexive, transitive* relation on W, and $e : \mathrm{PV} \to \mathrm{Sub}\, W$. We call e an *evaluation* and say that the model is *finite* if W is finite. The pair $\langle W, R \rangle$ is a *frame*.

We define a relation $\vDash$, read as 'validates', between elements w of W and wffs, where $\nvDash$ means '$\vDash$ does not hold'.

1. $w \vDash p$ iff for all z such that wRz, $z \in e(p)$
2. $w \vDash A \wedge B$ iff $w \vDash A$ and $w \vDash B$
3. $w \vDash A \vee B$ iff $w \vDash A$ or $w \vDash B$
4. $w \vDash \neg A$ iff for all z such that wRz, $z \nvDash A$
5. $w \vDash A \to B$ iff for all z such that wRz, $z \nvDash A$ or $z \vDash B$

Then $\langle W, R, e \rangle \vDash A$ iff for all $w \in W$, $w \vDash A$.

Here and throughout this chapter *except in* §B.4 *we do not necessarily assume that* PV *and* Wffs *are completed infinite totalities.* Assignments or evaluations such as e above can be understood as meaning that we have a method such that given any variable p_i we can produce a subset of W.

This is the presentation given by Fitting, *1969*. Dummett, *1977*, gives an equivalent formulation by requiring that for each p, $e(p)$ is *closed under* R, or for short is R-*closed*; that is, if $w \in e(p)$ and wRz then $z \in e(p)$. He can then replace condition (1) by: $w \vDash p$ iff $w \in e(p)$. That formulation is equivalent to the one given above.

For either formulation it is not hard to prove the following.

Lemma 1 *a.* For any $\langle W, R, e \rangle$ and $w \in W$, $w \vDash A$ iff for all z such that wRz, $z \vDash A$.
b. If $\vdash_{\mathrm{Int}} A$ then A is valid in every Kripke model.
c. If $\Gamma \vdash_{\mathrm{Int}} A$ then every Kripke model which validates Γ also validates A.

Proof: Part (a) is proved by induction on the length of wffs, part (b) by induction on the length of proofs. Part (c) follows by proving that the collection of wffs validated at any w is closed under deduction. ■

Dummett further classifies a subcollection of frames as *Kripke trees*. For the purposes of this chapter I will take these to be $\langle W, R \rangle$ such that $\langle W, R \rangle$ is a weak partial order (i.e., R is reflexive, transitive and anti-symmetric) and there is an *initial point* $w \in W$ which has no predecessor under R (i.e., for no $z \neq w$ do we have zRw) and which is related to all elements of W. Whenever I refer to a model $\langle W, R, e, w \rangle$ as a *Kripke tree* I will mean that w is the initial point.

How are these semantics supposed to reflect the intuitionists' understanding of logic, particularly Dummett's reading of the connectives? Let's first quote Fitting.

> W is intended to be a collection of ... states of knowledge. Thus a particular w

> in W may be considered as a collection of (physical) facts known at a particular time. The relation R represents (possible) time succession. That is, given two states of knowledge w and z in W, to say $w\mathsf{R}z$ is to say: if we now know w, it is possible that later we will know z. Finally, to say that $w \vDash A$ is to say: knowing w, we know A, or: from the collection of facts w, we may deduce the truth of A.
>
> Under this interpretation condition [4 above] for example, may be interpreted as follows: from the facts w we may conclude $\neg A$ if and only if from no possible additional facts can we conclude A. ... [Lemma 1.a is interpreted as:] If from a certain amount of information we can deduce A, given additional information, we still can deduce A, or if at some time we know A is true, at any later time we still know A is true.
>
> *Fitting, 1969,* p.21

Dummett refers to the points in W as 'states of information' and says that p *is true at* w iff $w \in \mathsf{e}(p)$ (recall that he requires $\mathsf{e}(p)$ to be an R-closed set). He then says:

> Given any set of formulas, the sentence-letters occurring in them represent unanalysed constituent statements: we are considering states of information only in so far as they bear on the verification of these constituent statements. A state of information consists in a knowledge of two things: which of the constituent statements have been verified; and what future states of information are possible. That the constituent statement represented by a sentence letter p has been verified in the state of information represented by a point w is itself represented by the fact that $w \in \mathsf{e}(p)$. That the state of information represented by w may subsequently be improved upon by achieving the state represented by a point z is represented by the fact that $w\mathsf{R}z$. Note that there is no assumption that, at any point, we shall actually every [sic] acquire more information.
>
> The requirement that $\mathsf{e}(p)$ be an [R-closed set] ... corresponds intuitively to the assumption that, once a constituent statement has been verified, it remains verified; i.e., that we do not forget what we have verified.
>
> *Dummett, 1977,* p.182

Formally, these semantics characterize **Int**.

Theorem 2 **(*Completeness of the Kripke Semantics*)**

For any finite collection of sentences Γ,

$\Gamma \vdash_{\mathbf{Int}} A$ iff every finite Kripke tree which validates Γ also validates A.

Dummet, *1977, gives an intuitionistically acceptable proof of this* (taking into account the comments concerning alternate axiomatizations in §5 below): in proving that if $\Gamma \nvdash_{\mathbf{Int}} A$ then A is not valid, he actually produces a finite Kripke tree which validates Γ and invalidates A.

Thus if we confine our attention to finite collections of propositions or wffs then finite models suffice for analysing our logic: about these the intuitionist agrees we can reason classically. And since only finite models are involved we can, if we wish, understand the Fully General Abstraction for these semantics in intuitionistic terms.

Why is Theorem 2 stated only for finite collections? If we allow Γ to be infinite and $\Gamma \nvdash A$ then it is not clear how to proceed intuitionistically to produce a model of Γ which invalidates A. We cannot "survey" all of Γ at once. But even if we reason classically the reduction from the class of all Kripke trees to the class of finite Kripke trees seems to require that we restrict ourselves to finite collections Γ. *Using classical reasoning* we have the following for all collections Γ.

Theorem 3 ***a.*** $\Gamma \vdash_{\mathbf{Int}} A$ iff every Kripke tree which validates Γ also validates A.
b. $\Gamma \vdash_{\mathbf{Int}} A$ iff every Kripke model which validates Γ also validates A.

I give a classical proof of Theorems 2 and 3 in §B.4 below. Only there in this chapter, and in Corollary 5 below, and in the discussion of translations in §C do I use intuitionistically unacceptable reasoning (I hope).

Let's see how these semantics reflect the intuitionists' rejection of the law of excluded middle and the law of double negation. For notation define $\neg^1 A = \neg A$, and for $n \geq 1$, $\neg^{n+1} A = \neg(\neg^n A)$, and $A \leftrightarrow B$ as $(A \rightarrow B) \wedge (B \rightarrow A)$.

Corollary 4 For **Int** we have:

a. $\nvdash A \vee \neg A$

b. $\nvdash \neg\neg A \rightarrow A$

c. $\vdash A \rightarrow \neg\neg A$

d. $\vdash \neg\neg\neg A \rightarrow \neg A$

e. $\vdash \neg A \rightarrow \neg\neg\neg A$

f. For $n \geq 1$, $\vdash \neg^{2n+1} A \leftrightarrow \neg A$ and $\vdash \neg^{2n+2} A \leftrightarrow \neg\neg A$

g. $\vdash \neg\neg(A \rightarrow B) \rightarrow (\neg\neg A \rightarrow \neg\neg B)$

h. $\vdash \neg\neg(A \wedge B) \rightarrow (\neg\neg A \wedge \neg\neg B)$

Proof: I'll exhibit models below in which $A \vee \neg A$ and $\neg\neg A \rightarrow A$ fail. It's a good exercise to show that (b)–(e) and (g), (h) are valid. Part (f) then follows by induction on the number of occurrences of $\neg$.

$A \vee \neg A$ fails:

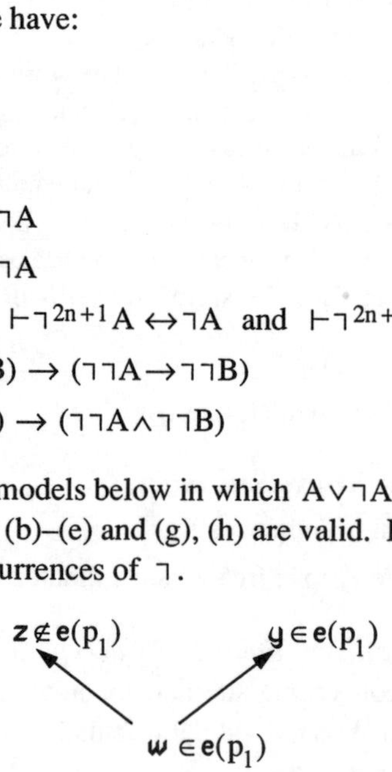

Since wRy and wRz, we have $w \nvDash p_1$ and $w \nvDash \neg p_1$. So $w \nvDash p_1 \vee \neg p_1$.

$\neg\neg A \rightarrow A$ fails: $w \longrightarrow z \longrightarrow y$
$\notin e(p_1)$ $\notin e(p_1)$ $\in e(p_1)$

Since wRy and zRy and $y \vDash p_1$, we have that $w \vDash \neg\neg p_1$. But $w \nvDash p_1$.
Hence $w \nvDash \neg\neg p_1 \rightarrow p_1$. ∎

As mentioned in §A, for an intuitionist a proof of $A \vee B$ is a proof of A or a proof of B. I'll give a demonstration that here using classical reasoning.

Corollary 5 *a.* $A \vee B$ is valid in every finite Kripke tree iff A is valid in every finite Kripke tree or B is valid in every finite Kripke tree.

b. $\vdash A \vee B$ iff $\vdash A$ or $\vdash B$.

Proof: Part (b) follows from part (a) and Theorem 2.

a. $\Leftarrow$ This is immediate.

$\Rightarrow$ I will give an intuitionistically acceptable proof that if both A and B are not valid then $A \vee B$ is not valid. Classically that is equivalent to what we want; it is much more difficult to establish that intuitionistically.

If both A and B are not valid then there is some $\langle W_1, R_1, e_1, w_1 \rangle$ such that $w_1 \nvDash A$, and $\langle W_2, R_2, e_2, w_2 \rangle$ such that $w_2 \nvDash B$. Define $\langle W, R, e, z \rangle$, where $z \notin W_1 \cup W_2$, by:

$W = W_1 \cup W_2$

xRy iff $(x = z)$ or $(x, y \in W_1$ and $xR_1y)$ or $(x, y \in W_2$ and $xR_2y)$

$e(p) = \{z\} \cup e_1(p) \cup e_2(p)$ for all p

Then $\langle W, R, e, z \rangle$ is a finite Kripke tree as you can check. And $z \nvDash A$, since zRw_1 and $w_1 \nvDash A$; and $z \nvDash B$, since zRw_2 and $w_2 \nvDash B$. Hence $z \nvDash A \vee B$. ∎

Finally, using the completeness theorem we can compare **PC** to **Int**.

Corollary 6 **PC** = the closure of **Int** $\cup \{(A \rightarrow B) \rightarrow ((\neg A \rightarrow B) \rightarrow B)\}$ under *modus ponens*.

Proof: In the axiomatization of **PC** in $L(\neg, \rightarrow, \wedge, \vee)$ in §II.K.6, p. 56, all the other schemas are valid in every finite Kripke tree, and hence theorems of **Int**. ∎

In §C.3 we'll see that **PC** can also be characterized as the closure of **Int** $\cup \{\neg\neg A \rightarrow A\}$ or **Int** $\cup \{A \vee \neg A\}$ under *modus ponens*.

The next two sections are devoted to proving the Deduction Theorem and Strong Completeness Theorem for Kripke semantics for **Int**. They and the section following on alternate axiomatizations may be skipped with no loss of continuity, though the completeness proof may give some insight into the nature of the translations in §C.

3. Some syntactic derivations and the Deduction Theorem

The following syntactic derivations are used to prove the Syntactic Deduction Theorem for **Int** and are needed in the completeness proof of the next section.

Lemma 7

a. If $\Gamma \vdash A \to B$ and $\Gamma \vdash B \to C$, then $\Gamma \vdash A \to C$.

b. $\vdash A \wedge B \to A$

c. $\vdash A \wedge B \to B$

d. $\vdash A \to A$

e. $\vdash B \to (A \vee B)$

f. If $\Gamma \vdash A \to B$ and $\Gamma \vdash A \to (B \to C)$, then $\Gamma \vdash A \to C$.

g. If $\Gamma \vdash A \to B$ and $\Gamma \vdash A \to C$, then $\Gamma \vdash A \to (B \wedge C)$.

Proof:

a. i.	$A \to B$	premise
ii.	$B \to C$	premise
iii.	$(A \to B) \wedge (B \to C)$	rule of adjunction on (i) and (ii)
iv.	$((A \to B) \wedge (B \to C)) \to (A \to C))$	axiom IV
v.	$A \to C$	*modus ponens* on (iii) and (iv)
b. i.	$A \to (B \to A)$	axiom V
ii.	$(A \to (B \to A)) \to (A \wedge B \to ((B \to A) \wedge B)$	axiom III
iii.	$(A \wedge B) \to ((B \to A) \wedge B)$	*modus ponens* on (i) and (ii)
iv.	$((B \to A) \wedge B) \to (B \wedge (B \to A))$	axiom II
v.	$(B \wedge (B \to A)) \to A$	axiom VI
vi.	$(A \wedge B) \to A$	by (iii), (iv) and (v) using part (a)
c. i.	$(A \wedge B) \to (B \wedge A)$	axiom III
ii.	$(B \wedge A) \to B$	by part (b)
iii.	$(A \wedge B) \to B$	by (i) and (ii) using part (a)
d. i.	$A \to (A \wedge A)$	axiom I
ii.	$(A \wedge A) \to A$	by part (b)
iii.	$A \to A$	by (i) and (ii) using part (a)
e. i.	$B \to (B \vee A)$	axiom VII
ii.	$(B \vee A) \to (A \vee B)$	axiom VIII
iii.	$B \to (A \vee B)$	by (i) and (ii) using part (a)
f. i.	$A \to (B \to C)$	premise
ii.	$(A \to (B \to C)) \to ((A \wedge B) \to ((B \to C) \wedge B))$	axiom III
iii.	$(A \wedge B) \to ((B \to C) \wedge B)$	*modus ponens* on (i) and (ii)
iv.	$((B \to C) \wedge B) \to (B \wedge (B \to C))$	axiom II
v.	$(B \wedge (B \to C)) \to C$	axiom VI

vi.	$(A \wedge B) \to C$	by (iii), (iv) and (v) using part (a)
vii.	$(B \wedge A) \to (A \wedge B)$	axiom II
viii.	$(B \wedge A) \to C$	by (vi) and (vii) using part (a)
ix.	$(A \to B) \to ((A \wedge A) \to (B \wedge A))$	axiom III
x.	$A \to B$	premise
xi.	$(A \wedge A) \to (B \wedge A)$	*modus ponens* on (ix) and (x)
xii.	$A \to (A \wedge A)$	axiom I
xiii.	$A \to (B \wedge A)$	by (xi) and (xii) using part (a)
xiv.	$A \to C$	by (xiii) and (viii) using part (a)

g. i.	$(A \to B) \to (A \wedge C \to B \wedge C)$	axiom III
ii.	$A \to B$	premise
iii.	$(A \wedge C) \to (B \wedge C)$	*modus ponens* on (i) and (ii)
iv.	$(A \to C) \to ((A \wedge A) \to (A \wedge C))$	axiom III
v.	$(A \to C)$	premise
vi.	$(A \wedge A) \to (A \wedge C)$	*modus ponens* on (iv) and (v)
vii.	$A \to (A \wedge A)$	axiom I
viii.	$A \to (A \wedge C)$	by (vi) and (vii) using part (a)
ix.	$A \to (B \wedge C)$	by (viii) and (iii) using part (a) ■

***Theorem 8* (*The Deduction Theorem for* Int)** $\Gamma \cup \{A\} \vdash_{\mathbf{Int}} B$ iff $\Gamma \vdash_{\mathbf{Int}} A \to B$.

Proof: Let $A_0, A_1, \ldots, A_n = B$ be a proof of B from $\Gamma \cup \{A\}$. I will show by induction that for all $i \leq n$, $\Gamma \vdash A \to A_i$.

Either A_0 is an axiom, or $A_0 \in \Gamma$, or A_0 is A. For the first two we have the result by using axiom V. If A_0 is A then we are done by Lemma 7.d.

Suppose $\Gamma \vdash A \to A_i$ for all $i < k$. Then if A_k is an axiom, is in Γ, or is A we are done as before. Otherwise there are $i, j < k$ such that A_k is obtained from A_i and A_j by one of the rules. In that case we are done by Lemma 7.f,g. ■

4. Completeness theorems for Int

I will prove in this section that the class of finite Kripke trees is strongly complete for **Int**. The methods I will use to establish this, however, are not intuitionistically acceptable.

We make the following definitions:

Σ is a *theory* if $\Sigma \supseteq \mathbf{Int}$ and Σ is closed under *modus ponens* and adjunction.

Σ is *consistent* if for no A do we have $\Sigma \vdash A$ and $\Sigma \vdash \neg A$.

Σ is *full* if Σ is a consistent theory such that for every A and B, if $(A \vee B) \in \Sigma$ then $A \in \Sigma$ or $B \in \Sigma$.

Lemma 9 *a.* Σ is consistent iff for some A, $\Sigma \nvdash A$.

b. If Σ is full then:

i. $\Sigma \vdash A$ iff $A \in \Sigma$

ii. $A \wedge B \in \Sigma$ iff $A \in \Sigma$ and $B \in \Sigma$

iii. $A \vee B \in \Sigma$ iff $A \in \Sigma$ or $B \in \Sigma$

iv. $\Sigma \vdash A \vee B$ iff $\Sigma \vdash A$ or $\Sigma \vdash B$

Proof: a. If $\Sigma \vdash B$ and $\Sigma \vdash \neg B$, then by axiom X for every A, $\Sigma \vdash A$. The converse is immediate.

b. i. $\Sigma \supseteq$ **Int** and is closed under the rules.

ii. This follows from part (a) and Lemma 7.

iii. If $A \vee B \in \Sigma$ then by definition $A \in \Sigma$ or $B \in \Sigma$. If $A \in \Sigma$ then by axiom VII and part (i), $A \wedge B \in \Sigma$. If $B \in \Sigma$ then by Lemma 7 and part (i), $A \vee B \in \Sigma$.

iv. This follows from parts (i) and (iii). ■

The first use of infinitistic intuitionistically unacceptable reasoning in this section occurs in the proof of the next lemma.

Lemma 10 If $\Gamma \nvdash_{\mathbf{Int}} E$ then there is some full $\Sigma \supseteq \Gamma$ such that $E \notin \Sigma$.

Proof: Let $B_1, B_2, \ldots$ be a listing of all wffs. Define

$$\Sigma_0 = \Gamma$$

$$\Sigma_{j+1} = \begin{cases} \Sigma_j \cup \{B_j\} & \text{if } \Sigma_j \nvdash B_j \rightarrow E \\ \Sigma_j & \text{otherwise} \end{cases}$$

$$\Sigma = \bigcup_j \Sigma_j$$

By Lemma 7.d, $\vdash E \rightarrow E$, so for all j, $\Sigma_j \vdash E \rightarrow E$, and hence $E \notin \Sigma$. I'll show by induction that for all j, $\Sigma_j \nvdash E$. It's true for $j = 0$. Suppose it's true for all $i \leq j$. If $\Sigma_{j+1} \vdash E$ then by induction we must have $\Sigma_{j+1} = \Sigma_j \cup \{B_j\}$, but then by the Deduction Theorem, $\Sigma_j \vdash B_j \rightarrow E$, a contradiction. Hence $\Sigma \nvdash E$, so by Lemma 9, Σ is consistent. It remains to show that Σ is full.

To show that Σ is a theory, suppose that $\Sigma \vdash A$. Were $A \notin \Sigma$, then by construction $\Sigma \vdash A \rightarrow E$, so we would have $\Sigma \vdash E$ which is a contradiction. Hence $A \in \Sigma$.

Suppose $A \vee B \in \Sigma$. If $A \notin \Sigma$ and $B \notin \Sigma$ then by the construction $\Sigma \vdash A \rightarrow E$ and $\Sigma \vdash B \rightarrow E$. But then by axiom IX, $\Sigma \vdash A \vee B \rightarrow E$, a contradiction. So $A \in \Sigma$ or $B \in \Sigma$. Hence Σ is full. ■

Lemma 11 If Γ is full, then for all E:

a. $\Gamma \vdash E$ iff for every $\Sigma \supseteq \Gamma$ which is full, $E \in \Sigma$.

b. $\Gamma \vdash \neg E$ iff for every $\Sigma \supseteq \Gamma$ which is full, $E \notin \Sigma$.

Proof: a. If $\Gamma \vdash E$ then for all full $\Sigma \supseteq \Gamma$, $\Sigma \vdash E$ and so by Lemma 9, $E \in \Sigma$. The converse is immediate.

b. $\Rightarrow|$ If $\Gamma \vdash \neg E$ then for all full $\Sigma \supseteq \Gamma$, $\Sigma \vdash \neg E$, so by the consistency of Σ, $E \notin \Sigma$.

$\Leftarrow|$ Suppose that for all full $\Sigma \supseteq \Gamma$, $E \notin \Sigma$. So $\Gamma \cup \{E\}$ is inconsistent, for were it not then taking $\Gamma \cup \{E\}$ for Γ in Lemma 10 we would have a contradiction. Hence $\Gamma \cup \{E\} \vdash \neg E$. So by the Deduction Theorem, $\Gamma \vdash E \to \neg E$, and by Lemma 7, $\Gamma \vdash E \to E$. So using Axiom XI, $\Gamma \vdash \neg E$. ■

Define the *Canonical Model* for **Int** to be:

$\langle \mathsf{W}_{\mathbf{Int}}, \subseteq, \mathsf{e}_{\mathbf{Int}} \rangle$, where

$\mathsf{W}_{\mathbf{Int}} = \{\Gamma : \Gamma \text{ is full}\}$

$\mathsf{e}_{\mathbf{Int}}(\Gamma) = \{p : p \in \Gamma\}$

Note that the canonical model is reflexive, transitive, and anti-symmetric. It is also uncountably infinite.

Lemma 12 *a.* $\langle \mathsf{W}_{\mathbf{Int}}, \subseteq, \mathsf{e}_{\mathbf{Int}}, \Gamma \rangle \models A$ iff $A \in \Gamma$

b. $\langle \mathsf{W}_{\mathbf{Int}}, \subseteq, \mathsf{e}_{\mathbf{Int}} \rangle \models A$ iff $\vdash_{\mathbf{Int}} A$

Proof: a. By induction on the length of A. It is immediate if the length of A is 1. Suppose A has length greater than 1 and it is true for all wffs shorter than A and all Γ. We have the following cases.

A is $B \wedge C$:

$\Gamma \models B \wedge C$	iff $\Gamma \models B$ and $\Gamma \models C$	
	iff $B \in \Gamma$ and $C \in \Gamma$	by induction
	iff $B \wedge C \in \Gamma$	by Lemma 9

A is $B \vee C$:

$\Gamma \models B \vee C$	iff $\Gamma \models B$ or $\Gamma \models C$	
	iff $B \in \Gamma$ or $C \in \Gamma$	by induction
	iff $B \vee C \in \Gamma$	because Γ is full and by Lemma 9

A is $\neg B$:

$\Gamma \models \neg B$	iff for all $\Sigma \supseteq \Gamma$, $\Sigma \not\models B$	
	iff for all $\Sigma \supseteq \Gamma$, $B \notin \Sigma$	by induction
	iff $\Gamma \vdash \neg B$	by Lemma 11
	iff $\neg B \in \Gamma$	by Lemma 9

A is $B \to C$. First note that:

$\Gamma \models B \to C$ iff for all $\Sigma \supseteq \Gamma$, $\Sigma \not\models B$ or $\Sigma \models C$

iff for all $\Sigma \supseteq \Gamma$, $B \notin \Sigma$ or $C \in \Sigma$

Suppose $\Gamma \vDash B \to C$. I'll show that $\Gamma \cup \{B\} \vdash C$ from which, by the Deduction Theorem, $\Gamma \vdash B \to C$ and hence by Lemma 9, $B \to C \in \Gamma$. If $\Gamma \cup \{B\}$ is inconsistent we are done. So suppose $\Gamma \cup \{B\}$ is consistent. If $\Gamma \cup \{B\} \nvdash C$ then by Lemma 10, for some full $\Sigma \supseteq \Gamma \cup \{B\}$, $C \notin \Sigma$ which contradicts the fact that $\Gamma \vDash B \to C$.

Now suppose $B \to C \in \Gamma$. If $\Sigma \supseteq \Gamma$ is full and $B \in \Sigma$ then $C \in \Sigma$ as Σ is closed under deduction. So $\Gamma \vDash B \to C$.

b. This follows by Lemma 10. ■

Note: An *endpoint* of a Kripke model is a z such that for no y do we have zRy. Any endpoint validates **PC** since the evaluation at the endpoint proceeds as in a classical model taking $v(p) = T$ iff $z \in e(p)$. In particular, if Γ is an endpoint of the canonical model then $\Gamma \vDash$ **PC**, so **PC** $\subset \Gamma$. As **PC** $\vdash A \vee \neg A$, for every A, $A \in \Gamma$ or $\neg A \in \Gamma$. That is, the endpoints of the canonical model are **PC**-complete and consistent sets of wffs.

We say that Γ is *closed under subformulas* if given any $A \in \Gamma$, if B is a subformula of A, then $B \in \Gamma$.

Lemma 13 Given any Kripke model $\langle W, R, e\rangle$ and Γ a finite collection of wffs closed under subformulas, then there is a finite anti-symmetric Kripke model $\langle W^*, R^*, e^*\rangle$ such that for every $A \in \Gamma$, $\langle W, R, e\rangle \vDash A$ iff $\langle W^*, R^*, e^*\rangle \vDash A$.

Proof: The proof mimics that of Lemma VI.48.

Define an equivalence relation on W:

$$x \approx y \text{ iff for all } A \in \Gamma,\ x \vDash A \text{ iff } y \vDash A$$

Denote by x^* the equivalence class of x. Define:

$$W^* = \{x^* : x \in W\}$$

$$x^* R^* y^* \text{ iff for all } A \in \Gamma, \text{ if } x \vDash A \text{ then } y \vDash A$$

$$e(x^*) = \{p : p \in \Gamma \text{ and } x \vDash p\}$$

Because of the definition of $\approx$, R^* and e^* are well-defined, R^* is reflexive, transitive, and anti-symmetric, and W^* is finite. Moreover by Lemma 1.a, if xRy then $x^*R^*y^*$.

In order to show that for all $A \in \Gamma$, all $x \in W$, $\langle W, R, e, x\rangle \vDash A$ iff $\langle W^*, R^*, e^*, x^*\rangle \vDash A$, we proceed by induction on the length of A. The only interesting cases are if A is of the form $B \to C$ or $\neg B$.

If A is $B \to C$, suppose $x \vDash B \to C$. By way of contradiction suppose that there is some y^* such that $x^*R^*y^*$ and $y^* \vDash B$ yet $y^* \nvDash C$. Then by induction, since $B, C \in \Gamma$, $y \vDash B$ and $y \nvDash C$. But then $y \nvDash B \to C$, contradicting $x^*R^*y^*$. So $x^* \vDash B \to C$.

Now suppose $x^* \models B \to C$. If $x R y$ and $y \models B$, then $x^* R^* y^*$ and, by induction, $y^* \models B$. But then $y^* \models C$, so by induction $y \models C$. Hence $x \models B \to C$.

If A is $\neg B$, suppose $x \models \neg B$ and $x^* R^* y^*$. If $y^* \models B$ then by induction $y \models B$, contradicting $x^* R^* y^*$. So $y^* \not\models B$, and $x^* \models \neg B$.

If $x^* \models \neg B$ and $x R y$ then $x^* R^* y^*$ and $y^* \not\models B$. So by induction $y \not\models B$. Hence $x \models \neg B$. ∎

Theorem 14 **(*Completeness of the Kripke Semantics for* Int)**

a. $\Gamma \vdash_{\mathbf{Int}} A$ iff every Kripke model which validates Γ also validates A.

b. $\Gamma \vdash_{\mathbf{Int}} A$ iff every Kripke tree which validates Γ also validates A.

c. For finite Γ, $\Gamma \vdash_{\mathbf{Int}} A$ iff every finite Kripke tree which validates Γ also validates A

d. **Int** is decidable.

Proof: a. From left to right is Lemma 1.b.

If $\Gamma \nvdash_{\mathbf{Int}} A$ then consider the submodel of the canonical model, $\langle W_\Gamma, \subseteq, e_{\mathbf{Int}} \rangle$ where $W_\Gamma = \{\Sigma : \Sigma \text{ is full and } \Sigma \supseteq \Gamma\}$. Since $\Gamma \nvdash A$, by Lemma 10, $W_\Gamma \neq \varnothing$; and for all B, $\langle W_\Gamma, \subseteq, e_{\mathbf{Int}} \rangle \models B$ iff $B \in \Gamma$. Hence $\langle W_\Gamma, \subseteq, e_{\mathbf{Int}} \rangle \models \Gamma$ and $\langle W_\Gamma, \subseteq, e_{\mathbf{Int}} \rangle \not\models A$.

b. This follows by part (a) using the methods of Lemma VI.1, p.153, to cull a Kripke tree from the model $\langle W_\Gamma, \subseteq, e_{\mathbf{Int}} \rangle$.

c. This follows from part (b) by Lemma 13.

d. The decidability of **Int** follows from part (c) as in the proof of Theorem VI.50, p.190. ∎

5. An alternate axiomatization of Int

The following axiomatization is due to Dummett, *1977*, p.126.

Int

in $L(\neg, \to, \wedge, \vee)$

axiom schemas

1. $A \to (B \to A)$
2. $A \to (B \to (A \wedge B))$
3. $(A \wedge B) \to A$
4. $(A \wedge B) \to B$
5. $A \to (A \vee B)$
6. $B \to (A \vee B)$
7. $(A \vee B) \to ((A \to C) \to ((B \to C) \to C))$
8. $(A \to B) \to ((A \to (B \to C)) \to (A \to C))$
9. $(A \to B) \to ((A \to \neg B) \to \neg A)$
10. $A \to (\neg A \to B)$

rule $\dfrac{A, A \to B}{B}$

Dummett gives an intuitionistically acceptable proof that his system is characterized by finite Kripke trees. Classically that is enough to establish via Theorem 14 that it is the same as Heyting's. To establish the equivalence with Heyting's in an intuitionistically acceptable manner, first note that Dummett's system contains **Int** via Lemma 1.b. To show the containment in the other direction you can derive in **Int** each (instance of each) schema of Dummett using Lemma 7, the Deduction Theorem, and the observation that $\{A, B\} \vdash_{\mathbf{Int}} C$ iff $A \wedge B \vdash_{\mathbf{Int}} C$.

C. Translations and Comparisons with Classical Logic

In this section we'll see how to relate **Int** to classical systems of logic via translations. Since these comparisons are to systems which are not intuitionistically acceptable, I will make no effort to use intuitionistically acceptable reasoning.

1. Translations of Int into modal logic and classical arithmetic

If you are familiar with Kripke semantics for modal logics (§VI.B.1) you may have noticed that the semantics for **Int** are very similar to those for **S4** and **S4Grz**. Using Theorem 3 (Theorem 14 of the previous section) we can interpret **Int** in **S4** and **S4Grz**. Consider the following map from $L(\neg, \rightarrow, \wedge, \vee)$ to $L(\neg, \wedge, \Box)$.

$$\begin{aligned} p^* &= \Box p \\ (A \wedge B)^* &= A^* \wedge B^* \\ (A \vee B)^* &= A^* \vee B^* \\ (A \rightarrow B)^* &= \Box(A^* \supset B^*) \\ (\neg A)^* &= \Box\neg(A^*) \end{aligned}$$

and $\Gamma^* = \{A^* : A \in \Gamma\}$. Note that $(A \rightarrow B)^* = A^* \rightarrow B^*$.

Theorem 15 ***a.*** $\Gamma \vdash_{\mathbf{Int}} A$ iff $\Gamma^* \vdash_{\mathbf{S4}} A^*$

b. For finite Γ, $\Gamma \vdash_{\mathbf{Int}} A$ iff $\Gamma^* \vdash_{\mathbf{S4Grz}} A^*$.

Proof: Any Kripke model for **Int** can be viewed as a Kripke model for modal logic via the Note on p.152. And $\langle W, R, e, w \rangle \models A$ in the intuitionist semantics iff $\langle W, R, e, w \rangle \models A^*$ in the modal semantics. Part (a) is then a consequence of Theorem 3 and Theorem VI.46 and Theorem VI.51. Part (b) follows by Theorem 2 (Theorem 14) and the characterization of **S4Grz** quoted in §VI.G.2. ■

Gödel, *1933 B*, was the first to interpret **Int** in **S4**, long before formal semantics had been given for either logic. His translation was suggested by reading '$\Box$' as 'it is provable that':

$$p' = p$$
$$(A \wedge B)' = \Box A' \wedge \Box B'$$
$$(A \vee B)' = \Box A' \vee \Box B'$$
$$(A \rightarrow B)' = \Box A' \supset \Box B'$$
$$(\neg A)' = \neg \Box A'$$

or alternatively, $(A \wedge B)' = \Box A' \wedge \Box B'$, and/or $(\neg A)' = \Box \neg \Box A'$.

We can also interpret **Int** in terms of provability in classical arithmetic via the translations of §VI.K.1. We can compose the map $*$ above with any provability-and-truth translation $\#$ by defining $(A \vee B)^\# = A^\# \vee B^\#$ to obtain a map of the language $L(\neg, \rightarrow, \wedge, \vee)$ of **Int** to that of Peano Arithmetic, **PA**:

$$p_i^+ = \alpha_i \wedge \mathrm{Bew}(\llbracket \alpha_i \rrbracket) \quad \text{for some sentence } \alpha_i \text{ of the language of } \mathbf{PA}$$
$$(A \wedge B)^+ = A^+ \wedge B^+$$
$$(A \vee B)^+ = A^+ \vee B^+$$
$$(A \rightarrow B)^+ = (A^+ \supset B^+) \wedge \mathrm{Bew}(\llbracket A^+ \supset B^+ \rrbracket)$$
$$(\neg A)^+ = \neg(A^+) \wedge \mathrm{Bew}(\llbracket \neg A^+ \rrbracket)$$

Combining Theorem 14 with Theorem VI.28 we have:

Theorem 16 **Int**$\vdash$A iff for every translation $+$ as above A^+ is a theorem of Peano Arithmetic

An intuitionist implication in arithmetic asserts the truth and provability of a material implication; an intuitionist negation asserts that the sentence is false and provably so (compare the first quote by Dummett in §A, p. 197).

By a series of observations about the logics **G*** and **S4Grz**, Goldblatt, *1978*, invokes Theorem VI.27 to prove the following stronger fact.

Theorem 17 **Int**$\vdash$A iff for every translation $+$ as above, A^+ is true of the natural numbers.

2. Translations of classical logic into Int

We have that **Int** $\subset$ **PC** as every axiom of **Int** is a **PC**-tautology and the rules are **PC**-valid: all the intuitionists principles are acceptable to the classical logician.

Nonetheless, Gödel, *1933 A*, has shown that **Int** may be viewed as an extension of **PC** if we take the latter to be formalized in $L(\neg, \wedge)$. In order to show that, I need to first establish a translation of **PC** into **Int**, due essentially to Glivenko, *1929*. Let $\neg\neg\Gamma = \{\neg\neg A : A \in \Gamma\}$.

Theorem 18 In $L(\neg, \rightarrow, \wedge, \vee)$,

a. $\Gamma \vdash_{\mathbf{PC}} A$ iff $\neg\neg\Gamma \vdash_{\mathbf{Int}} \neg\neg A$

b. $\vdash_{\mathbf{PC}} \neg A$ iff $\vdash_{\mathbf{Int}} \neg A$

Proof: a. $\Leftarrow$| **Int** $\subset$ **PC** and these systems use the same rule. So if $\neg\neg\Gamma \vdash_{\mathbf{Int}} \neg\neg A$ then $\neg\neg\Gamma \vdash_{\mathbf{PC}} \neg\neg A$. Since for all B, $\vdash_{\mathbf{PC}} \neg\neg B \leftrightarrow B$, we have $\Gamma \vdash_{\mathbf{PC}} A$.

$\Rightarrow$| I will give two quite different proofs, the first semantic, the second (essentially) syntactic.

First proof: We first observe that for any B and any finite Kripke tree $\langle W, R, e \rangle$,

$\langle W, R, e \rangle \models \neg\neg B$ iff for all w, $w \models \neg\neg B$
iff for all w and all z, if wRz then $z \not\models \neg B$
iff for all w and all z for which wRz there is some x such that zRx and $x \models B$

For any endpoint x of $\langle W, R, e \rangle$ (i.e., for no $z \neq x$ do we have xRz) we have that if $\vdash_{\mathbf{PC}} B$ then $x \models B$ because the evaluation at an endpoint proceeds as in the classical model taking $v(p) = T$ iff $x \in e(p)$.

Suppose that we have $\vdash_{\mathbf{PC}} A$. Then for any point w of any finite Kripke tree model and any z such that wRz, there is some endpoint x such that zRx. Since $x \models A$ we have, as observed above, $w \models \neg\neg A$. So $\vdash_{\mathbf{Int}} \neg\neg A$.

Suppose now that $\Gamma \vdash_{\mathbf{PC}} A$. Since the syntactic consequence relation is compact we may assume that Γ is finite. At any endpoint t of a finite Kripke tree model $\langle W, R, e \rangle$, if $t \models \Gamma$ then $t \models A$. So for any point z in the model, if $z \vdash \neg\neg\Gamma$ then for every y such that zRy there is an x such that yRx and $x \models \Gamma$. For such an x there is some endpoint t such that xRt, and by Lemma 1.a, $t \models \Gamma$. Hence $t \models A$, so $z \models \neg\neg A$. Hence by Theorem 2, $\neg\neg\Gamma \vdash_{\mathbf{Int}} \neg\neg A$.

Second proof: We induct on the length of a proof of A, using the axiomatization of **PC** from §II.K.6, p. 56.

If the length is 1, then either A is an axiom of **PC** or else $A \in \Gamma$. If $A \in \Gamma$ then we are done. If A is an instance of an axiom schema of **PC** other than $(A \rightarrow B) \rightarrow ((\neg A \rightarrow B) \rightarrow B)$ then it is also a theorem of **Int** as you can check using Theorem 2. So by Corollary 4.c, $\vdash_{\mathbf{Int}} \neg\neg A$. Using Theorem 2 you can also check that the double negation of that other schema is a theorem of **Int**.

Now suppose that it is true for any wff which has a proof of length $\leq n$ steps, and $A_1, \ldots, A_n, A_{n+1}$ is a proof of A from Γ in **PC**. The last step must be an application of *modus ponens* on $A_i \rightarrow A_{n+1} = A_j$ where $i, j \leq n$. We have by induction that $\neg\neg\Gamma \vdash_{\mathbf{Int}} \neg\neg A_i$ and $\neg\neg\Gamma \vdash_{\mathbf{Int}} \neg\neg(A_i \rightarrow A_{n+1})$. By Corollary 4.g, $\vdash_{\mathbf{Int}} \neg\neg(A_i \rightarrow A_{n+1}) \rightarrow (\neg\neg A_i \rightarrow \neg\neg A_{n+1})$, so by using *modus ponens* twice we have $\neg\neg\Gamma \vdash_{\mathbf{Int}} \neg\neg A_{n+1}$.

b. From right to left is because **Int** $\subset$ **PC**. So suppose $\vdash_{\mathbf{PC}} \neg A$. Then by part (a), $\vdash_{\mathbf{Int}} \neg\neg\neg A$, so by Corollary 4.d, $\vdash_{\mathbf{Int}} \neg A$. ∎

Now we can show that **PC** ⊂ **Int** in the following sense.

Corollary 19 If A is a wff of $L(\neg, \wedge)$ then $\vdash_{\mathbf{PC}} A$ iff $\vdash_{\mathbf{Int}} A$

Proof: From right to left is because **Int** ⊂ **PC**.

Suppose $\vdash_{\mathbf{PC}} A$. If A is a negation then we are done by the previous theorem. So suppose A is not a negation. Then A must be of the form $B_1 \wedge \ldots \wedge B_n$, $n \geq 1$, where each B_i is not a conjunction. Therefore, each B_i must be either a negation or a propositional variable. Since $\vdash_{\mathbf{PC}} A$ we have $\vdash_{\mathbf{PC}} B_i$ for each i. But no variable is a theorem of **PC**. Thus A has the form $\neg C_1 \wedge \ldots \wedge \neg C_n$ where for each i, $\vdash_{\mathbf{PC}} \neg C_i$. But then by the previous theorem we have $\vdash_{\mathbf{Int}} \neg C_i$. So by the rule of adjunction, $\vdash_{\mathbf{Int}} A$. ∎

Corollary 19 also establishes that we cannot define both → and ∨ from ¬ and ∧ in **Int**, for otherwise we would have **Int** = **PC** in $L(\neg, \rightarrow, \wedge, \vee)$. Actually, the four connectives of **Int** are all independent: no one of them can be defined in terms of the other three. I present part of McKinsey's *1939* proof of that in §VIII.G.

However, we may use the definitions of → and ∨ in **PC** to effect a definition of **PC** in $L(\neg, \rightarrow, \wedge, \vee)$ within **Int**, as first observed by Łukasiewicz in *1952*. Define the translation:

$$\begin{aligned} p^{\dagger} &= p \\ (A \wedge B)^{\dagger} &= A^{\dagger} \wedge B^{\dagger} \\ (\neg A)^{\dagger} &= \neg(A^{\dagger}) \\ (A \vee B)^{\dagger} &= \neg(\neg A^{\dagger} \wedge \neg B^{\dagger}) \\ (A \rightarrow B)^{\dagger} &= \neg(A^{\dagger} \wedge \neg B^{\dagger}) \end{aligned}$$

Corollary 20 $\vdash_{\mathbf{PC}} A$ iff $\vdash_{\mathbf{Int}} A^{\dagger}$

Neither this mapping nor the homophonic mapping of **PC** into **Int** of Corollary 19 respects the syntactic (and therefore the semantic) consequence relation: we cannot improve the corollary to $\Gamma \vdash_{\mathbf{PC}} A$ iff $\Gamma \vdash_{\mathbf{Int}} A$, for we have $\neg\neg p \vdash_{\mathbf{PC}} p$. And if we had $\neg\neg p \vdash_{\mathbf{Int}} p$ then by the Deduction Theorem we would have $\vdash_{\mathbf{Int}} \neg\neg p \rightarrow p$ which we know is false. However, the following translation due to Gentzen, *1936*, does preserve consequence.

$$\begin{aligned} (p)^{\circ} &= \neg\neg p \\ (A \wedge B)^{\circ} &= A^{\circ} \wedge B^{\circ} \\ (A \rightarrow B)^{\circ} &= A^{\circ} \rightarrow B^{\circ} \\ (\neg A)^{\circ} &= \neg(A^{\circ}) \\ (A \vee B)^{\circ} &= \neg(\neg A^{\circ} \wedge \neg B^{\circ}) \end{aligned}$$

Theorem 21 The translation $^\circ$ from $L(\neg, \rightarrow, \wedge, \vee)$ to itself is a grammatical translation of **PC** into **Int**: $\Gamma \vdash_{\mathbf{PC}} A$ iff $\Gamma^\circ \vdash_{\mathbf{Int}} A^\circ$.

Proof: Gentzen's proof was entirely syntactic, whereas I will use Theorem 3.

First suppose that $\Gamma^\circ \vdash_{\mathbf{Int}} A^\circ$. Note that $\mathbf{PC} \vdash A \leftrightarrow A^\circ$. So if $\Gamma^\circ \vdash_{\mathbf{Int}} A^\circ$, then as $\mathbf{Int} \subset \mathbf{PC}$, $\Gamma^\circ \vdash_{\mathbf{PC}} A^\circ$. So $\Gamma \vdash_{\mathbf{PC}} A$.

For the other direction we first need a lemma.

Lemma $\vdash_{\mathbf{Int}} \neg\neg(A^\circ) \rightarrow A^\circ$

Proof: We proceed by induction on the length of A. If A is a variable, p, then by Corollary 4.f, $\vdash_{\mathbf{Int}} \neg\neg(\neg\neg p) \rightarrow \neg\neg p$. So suppose A has length greater than 1 and the lemma is true for all shorter wffs.

If A is $\neg B$ then we are done by Corollary 4.f.

If A is $B \wedge C$ then we have by induction $\vdash_{\mathbf{Int}} \neg\neg B^\circ \rightarrow B^\circ$ and $\vdash_{\mathbf{Int}} \neg\neg C^\circ \rightarrow C^\circ$. The lemma then follows by Corollary 4.g.

If A is $B \vee C$ then since $(B \vee C)^\circ = \neg(\neg B^\circ \wedge \neg C^\circ)$ we are done by Corollary 4.f.

Finally we have the case where A is $B \rightarrow C$. By induction we have $\vdash_{\mathbf{Int}} \neg\neg C^\circ \rightarrow C^\circ$. If we can show that for all D,E, $\neg\neg(D \rightarrow E) \vdash_{\mathbf{Int}} D \rightarrow \neg\neg E$, then by the Deduction Theorem we have $\vdash_{\mathbf{Int}} \neg\neg(B^\circ \rightarrow C^\circ) \rightarrow (B^\circ \rightarrow \neg\neg C^\circ)$ and hence by using Lemma 7.a and Axiom V, $\vdash_{\mathbf{Int}} \neg\neg(B^\circ \rightarrow C^\circ) \rightarrow (B^\circ \rightarrow C^\circ)$. So it remains to show that $\neg\neg(D \rightarrow E) \vdash_{\mathbf{Int}} D \rightarrow \neg\neg E$.

Suppose by way of contradiction that there is a finite Kripke tree $\langle \mathsf{W}, \mathsf{R}, \mathsf{e}, w \rangle$ such that $w \vDash \neg\neg(D \rightarrow E)$ and yet $w \nvDash D \rightarrow \neg\neg E$. Then for some z, $w\mathsf{R}z$ and $z \vDash D$, but $z \nvDash \neg\neg E$. Hence for some t, $z\mathsf{R}t$ and $t \vDash \neg E$; so for all v such that $t\mathsf{R}v$, $v \nvDash E$. Take such a v (one must exist as $t\mathsf{R}t$) which is an endpoint of the ordering. By the transitivity of R, $z\mathsf{R}v$ so we have $v \vDash D$ by Lemma 1.a. Yet $v \nvDash E$ so $v \nvDash D \rightarrow E$. Yet since $w \vDash \neg\neg(D \rightarrow E)$ we can show as in the proof of Lemma 18 that $v \vDash D \rightarrow E$ which is a contradiction. This ends the proof of the lemma.

To return to the proof of the theorem, suppose $\Gamma \vdash_{\mathbf{PC}} A$. Then for some $A_1, \ldots, A_n \in \Gamma$, $\{A_1, \ldots, A_n\} \vdash_{\mathbf{PC}} A$. Hence $\{A_1{}^\circ, \ldots, A_n{}^\circ\} \vdash_{\mathbf{PC}} A^\circ$, so $(A_1{}^\circ \wedge \ldots \wedge A_n{}^\circ) \vdash_{\mathbf{PC}} A^\circ$. Thus by Theorem 18, $\neg\neg(A_1{}^\circ \wedge \ldots \wedge A_n{}^\circ) \vdash_{\mathbf{Int}} \neg\neg A^\circ$. By the lemma and Corollary 4.e we have $(A_1{}^\circ \wedge \ldots \wedge A_n{}^\circ) \vdash_{\mathbf{Int}} A^\circ$, hence by Lemma 7, $\{A_1{}^\circ, \ldots, A_n{}^\circ\} \vdash_{\mathbf{Int}} A^\circ$. That is, $\Gamma^\circ \vdash_{\mathbf{Int}} A^\circ$. ■

The mappings of Corollary 19 and Theorem 21 can both be extended to the language of arithmetic (see *Kleene, 1952*, §81). Gödel, *1933 A*, concludes from this that intuitionistic arithmetic is only apparently narrower than classical arithmetic. But it seems to me that to establish Gödel's conclusion we would need to show that the semantics of classical negation and conjunction can be defined in intuitionistic logic, as I discuss in §D.3.b below and §X.B.4.

3. Axiomatizations of classical logic relative to Int

In §B.2, Corollary 6, we saw that

$$\mathbf{PC} = \text{the closure of } \mathbf{Int} \cup \{ (A \to B) \to ((\neg A \to B) \to B) \} \text{ under } \textit{modus ponens}$$

Using Theorem 18 we can give two further axiomatizations of **PC**.

Theorem 22 In $L(\neg, \to, \wedge, \vee)$,

a. **PC** = the closure of **Int** $\cup \{\neg\neg A \to A\}$ under *modus ponens*

b. **PC** = the closure of **Int** $\cup \{A \vee \neg A\}$ under *modus ponens*

Proof: a. $\mathbf{Int} \cup \{\neg\neg A \to A\} \subseteq \mathbf{PC}$ which is closed under *modus ponens*. It remains to show that every theorem of **PC** can be derived from $\mathbf{Int} \cup \{\neg\neg A \to A\}$. Take $\vdash$ to mean 'derivable from using only *modus ponens*.'

Suppose $\mathbf{PC} \vdash A$. Then by Theorem 18, $\neg\neg\,\mathbf{Int} \vdash \neg\neg A$. So $\neg\neg\,\mathbf{Int} \cup \{\neg\neg A \to A\} \vdash A$. Finally $\neg\neg\,\mathbf{Int} \subseteq \mathbf{Int}$ since by Corollary 4.c, $\mathbf{Int} \vdash A \to \neg\neg A$.

b. This follows from part (a) as $\vdash_{\mathbf{Int}} (A \vee \neg A) \to (\neg\neg A \to A)$, which you can show by checking that the wff is valid in every Kripke tree. ■

D. Set-assignment semantics for Int

Dummett is explicit in attributing content to propositions.

> If we take it as a primary function of a sentence to convey information, then it is natural to view a grasp of the meaning of a sentence as consisting in an awareness of its *content*; and this amounts to knowing the conditions under which an assertion made by it is correct.
>
> Dummett, *1977,* p.363

In terms of the formal Kripke semantics and their interpretation given above this would amount to identifying a proposition with those elements of W, that is those states of knowledge or information, in which it is valid. Symbolically, $\mathsf{s}(A) = \{ \mathsf{z} : \mathsf{z} \vDash A \}$. This is what we did for **S4**, and Theorem 15 suggests we try it here.

The basis of that approach is to convert a model $\langle \mathsf{W}, \mathsf{R}, \mathsf{e}, w \rangle$ to a model $<\mathsf{v}, \mathsf{s}>$ such that $\mathsf{s}(A) = \{ \mathsf{z} : \mathsf{z} \vDash A \}$ and $\mathsf{v}(A) = \mathsf{T}$ iff $w \vDash A$. But here we have a complication: $w \nvDash A$ does not imply $w \vDash \neg A$. We may have both $w \nvDash A$ and $w \nvDash \neg A$ corresponding to the intuition that from the information of w we may not be able to deduce either A or $\neg A$. Thus we may not simply take $\mathsf{v}(\neg A) = \mathsf{T}$ iff $\mathsf{v}(A) = \mathsf{F}$. Rather we need to take into account the content of A. I will present set-assignment semantics for **Int** based on these ideas and then discuss their aptness.

1. The semantics

We say that a set-assignment model $<\upsilon, s>$ uses *intuitionist truth-conditions* if:

$\wedge$ and $\vee$ are evaluated classically

$\rightarrow$ is evaluated by the dual dependence table (as for **S4**):

A	B	$s(A) \subseteq s(B)$	$A \rightarrow B$
any values		fails	F
T	T	holds	T
T	F		F
F	T		T
F	F		T

and $\neg$ is evaluated by the table for *intuitionist negation*:

A	$s(A) = \varnothing$	$\neg A$
any value	fails	F
T	holds	F
F		T

That is, $\upsilon(\neg A) = T$ iff $\upsilon(A) = F$ and $s(A) = \varnothing$.

We then say that $<\upsilon, s>$ is an **Int**-*model* if it uses intuitionist truth-conditions and satisfies:

Int 1. $s(A \wedge B) = s(A) \cap s(B)$

Int 2. $s(A \vee B) = s(A) \cup s(B)$

Int 3. $s(\neg A) \cup s(B) \subseteq s(A \rightarrow B)$

Int 4. $s(A) \cap s(A \rightarrow B) \subseteq s(B)$

Int 5. $s(A \rightarrow B) \subseteq s((A \wedge C) \rightarrow (B \wedge C))$

Int 6. $s(A \rightarrow B) \cap s(B \rightarrow C) \subseteq s(A \rightarrow C)$

Int 7. $s(A \rightarrow C) \cap s(B \rightarrow C) = s((A \vee B) \rightarrow C)$

Int 8. $s(A \rightarrow B) \cap s(A \rightarrow \neg B) = s(\neg A)$

Int 9. If $\upsilon(A) = T$ then $s(A) = S$.

Int 10. $s(A \wedge \neg A) = \varnothing$

It is *finite* if S is finite.

We define $\Gamma \vDash_{\mathbf{Int}} A$ for these semantics in the usual way: for every **Int**-model $<\upsilon, s>$ which validates Γ, $\upsilon(A) = T$.

As an example of the use of these set-assignment semantics I'll show that $\neg(A \wedge (A \rightarrow \neg A))$ is true in every **Int**-model $<v,s>$.

We need to show that $s(A \wedge (A \rightarrow \neg A)) = \varnothing$ and $v(A \wedge (A \rightarrow \neg A)) = F$. For the first, if $s(A) = \varnothing$ we are done by Int 1. If $s(A) \neq \varnothing$ then by Int 4, $s(A) \cap s(A \rightarrow \neg A) \subseteq s(\neg A)$ and hence by Int 10 and Int 1, $s(A \wedge (A \rightarrow \neg A)) = \varnothing$. To show that $v(A \wedge (A \rightarrow \neg A)) = F$, if $v(A) = F$ it's immediate. If $v(A) = T$, then $v(\neg A) = F$, so $v(A \rightarrow \neg A) = F$ and we're done. Hence $v(\neg(A \wedge (A \rightarrow \neg A))) = T$.

Lemma 23 If $\Gamma \vdash_{\mathbf{Int}} A$ then $\Gamma \vDash_{\mathbf{Int}} A$.

Proof: It's enough to show that the axioms are valid in **Int**-models and that the rules preserve validity. That's not hard, but it's worth noting which conditions are involved in the verification of the axioms.

I and II follow from Int 1.

III follows from Int 5 and Int 1.

IV follows from Int 6 and Int 1.

V follows from Int 3 and Int 9.

VI follows from Int 1 and Int 4.

VII and VIII follow from Int 2.

IX follows from Int 7, Int 1, and Int 2.

X follows from Int 3, since $v(\neg A) = T$ implies $s(A) = \varnothing$ by the truth-conditions.

XI follows from Int 8 and Int 10. ■

Lemma 24 Given any Kripke tree $\langle W, R, e, w \rangle$ there is an **Int**-model $<v,s>$ such that $v(A) = T$ iff $w \vDash A$ and $s(A) = \{z : z \vDash A\} - \{w\}$.

Proof: a. We need to show that the pair $<v,s>$ described in the lemma is an **Int**-model.

I'll leave to you to show that Int 1–Int 8 and Int 10 are satisfied. To establish Int 9, if $v(A) = T$ then $s(A) = S$, I'll use induction on the length of A. If A is a propositional variable then the result follows by Lemma 1. For conjunctions and disjunctions it is easy. If A is $\neg B$ and $v(\neg B) = T$, then $v(B) = F$ and $s(B) = \varnothing$, so $s(\neg B) = S$. If A is $B \rightarrow C$ and $v(B \rightarrow C) = T$, then $s(B) \subseteq s(C)$ so $s(B \rightarrow C) = S$.

It is easy to establish by induction on the length of wffs that every wff is evaluated by the intuitionist truth-conditions. ■

Theorem 25 **(*Strong Completeness of the Set-Assignment Semantics for* Int)**

$\Gamma \vdash_{\mathbf{Int}} A$ iff $\Gamma \vDash_{\mathbf{Int}} A$

For finite Γ the class of finite **Int**-models is strongly complete.

Proof: The first part is by Lemmas 23 and 24 and Theorem 3 (Theorem 14); the case for finite models uses Theorem 2 (Theorem 14). ■

The only place where nonconstructive reasoning might have entered into the proof of Theorem 25 is in claiming the existence of the Kripke tree and evaluation used in the proof of Lemma 24 . But Dummett explicitly constructs such a model in his proof of Theorem 2, so we can claim that Theorem 25 is intuitionistically acceptable so long as we confine ourselves to finite Γ.

Does every set-assignment model for **Int** arise from a Kripke model by the construction of Lemma 24 ? I suspect not. However, we can pick out those that do with the following condition:

Int K. If $\bigcap\{ s(C) : x \in s(C)\} \subseteq \overline{s(A)} \cup s(B)$, then $x \in s(A \to B)$.

Let I be the class of set-assignment models which use the intuitionist truth-conditions and satisfy Int 1, Int 2, Int 4, Int 8, Int 10, and Int K.

Theorem 26 *a.* If $<v, s>$ is an **Int**-model which arises from a Kripke tree by the construction of Lemma 24, then $<v, s>$ satisfies Int K.

b. Given any $<v, s> \in I$ there is a Kripke model $\langle W, R, e, w \rangle$ such that $v(A) = T$ iff $w \vDash A$ and $s(A) = \{ z : z \vDash A\} - \{w\}$.

c. I is strongly complete for **Int**, and the class of finite models in I is finitely strongly complete.

Proof: Suppose that $<v, s>$ arises from $\langle W, R, e, w \rangle$. And suppose $\bigcap\{ s(C) : x \in s(C)\} \subseteq \overline{s(A)} \cup s(B)$. Then if xRy, by Lemma 1.a, $y \in \bigcap\{ s(C) : x \in s(C)\}$. So if xRy, then $y \in \overline{s(A)}$ or $y \in s(B)$, so $y \nvDash A$ or $y \vDash B$. Hence $x \vDash A \to B$. So $x \in s(A \to B)$.

b. Given $<v, s>$ define:

$W = S \cup \{w\}$ for some object $w \notin S$

$$e(p) = \begin{cases} s(p) \cup \{w\} & \text{if } v(p) = T \\ s(p) & \text{if } v(p) = F \end{cases}$$

xRy iff $y \in \bigcap\{ s(C) : x \in s(C)\}$ or $x = w$

Note that if $x \neq w$, xRy iff for all C, if $x \in s(C)$ then $y \in s(C)$. So R is reflexive and transitive.

I'll now show by induction on the length of A that for $x \neq w$, $x \in s(A)$ iff $x \vDash A$. It's easy to check for propositional variables. Suppose now that it's true for all wffs shorter than A. If A is a conjunction or disjunction the proof is immediate from Int 1 and Int 2.

So suppose that A is $\neg B$. If $x \in s(\neg B)$, then for all y such that xRy, $y \in s(\neg B)$

and hence by Int 10, $y \notin s(B)$. So by induction, $y \nvDash B$. So $x \vDash \neg B$.

If $x \vDash \neg B$, then for all y such that xRy, $y \vDash \neg B$. Hence for all y such that xRy, $y \notin s(B)$ by induction. Thus $\bigcap\{s(C): x \in s(C)\} \subseteq \overline{s(B)} \cup s(\neg B)$. Hence by Int K, $x \in s(B \rightarrow \neg B)$. But $x \in s(B \rightarrow B)$ since $B \rightarrow B$ is a tautology. So by Int 8, $x \in s(\neg B)$.

If A is $B \rightarrow C$, suppose first that $x \in s(B \rightarrow C)$. If xRy then $y \in s(B \rightarrow C)$. So if $y \vDash B$, then by induction $y \in s(B)$. So by Int 4, $y \in s(C)$, and thus $y \vDash C$. Hence $x \vDash B \rightarrow C$.

If $x \vDash B \rightarrow C$, suppose $y \in \bigcap\{s(C): x \in s(C)\}$. Then xRy, so $y \vDash B \rightarrow C$. Hence either $y \nvDash B$ or $y \vDash C$, and so by induction, $y \notin s(B)$, or $y \in s(C)$. Thus by Int K, $x \in s(B \rightarrow C)$.

Now I'll prove by induction on the length of A that $v(A) = T$ iff $w \vDash A$. Recall that by Int 9, if $v(A) = T$ then $s(A) = S$. The only interesting cases in the proof are when A is a negation or conditional.

If A is $\neg B$,

$v(\neg B) = T$ iff $v(B) = F$ and $s(B) = \varnothing$
iff $w \nvDash B$ and for all y such that wRy, $y \notin s(B)$
iff $w \nvDash B$ and for all y such that wRy, $y \nvDash B$
iff $w \vDash \neg B$

If A is $B \rightarrow C$,

$v(B \rightarrow C) = T$ iff $(v(B) = F$ or $v(C) = T)$ and $s(B) \subseteq s(C)$
iff $(w \nvDash B$ or $w \vDash C)$ and for all y such that wRy, $y \notin s(B)$ or $y \in s(C)$
iff $(w \nvDash B$ or $w \vDash C)$ and for all y such that wRy, $y \nvDash B$ or $y \vDash C$
iff $w \vDash B \rightarrow C$

c. This follows by Theorems 2 and 3 using the 1–1 correspondence between Kripke models and models in $\mathcal{I}$ established in Lemma 24 and parts (a) and (b). ■

Note that $\mathcal{I}$ is not simply presented. I would very much like to see a condition that establishes a 1–1 correspondence between set-assignment models and Kripke models that is part of a simple presentation. The correspondence is important for showing that the translation of **Int** into **S4** is semantically faithful (Theorem X.15). I discuss other refinements of the set-assignment semantics in §3.f below.

2. Bivalence in intuitionism: the aptness of set-assignment semantics

Dummett explicitly says that the intuitionist abandons the principle of bivalence for truth. Yet it is not so easy for someone speaking our language to fully escape the Yes–No dichotomy that we all practice and impose on experience. Given any proposition and any particular state of information it is either correct or incorrect to assert A, there being no third way. And given any proposition A and any finite collection of states of information ordered under time either it is always correct to assert A or it is not. *Tertium non datur.*

> It is evident that it is fundamental to the notion of an assertion that it be capable of being either correct or incorrect; and therefore, in so far as assertion is taken to be the primary mode of employment of sentences, it is fundamental to our whole understanding of language that sentences are capable of being true or false, where a sentence is true if an assertion could be correctly made by uttering it, and false if such an assertion would be incorrect.
>
> Dummett, *1973*, p.371

From the viewpoint of the general framework for semantics for propositional logics proposed in Chapter IV, the intuitionist reasons analogously to a modal logician, or a many-valued, or a relevance logician. His notion of the truth of a complex proposition is based on two aspects of its constituent propositions: truth-value and some epistemological mathematical content. The intuitionist disagrees with the classical logician not on the truth-values of the atomic arithmetic formulas, but on the use of the connectives, particularly $\neg$ and $\rightarrow$, which take into account both aspects of the constituent propositions. From this point of view it seems perfectly apt to read T and F as 'true' and 'false' in the tables for **Int**.

We may take Dummett's interpretation of the content of a proposition as the conditions under which it is correct to assert A. Here we must understand $s(A)$ to be a collection of various conditions under each of which it is correct to assert A. So, for example, Int 4 can be read as 'Any condition which justifies my asserting A and which justifies my asserting $A \rightarrow B$ also justifies my asserting B'. Then $v(A) = T$ in a model means that it is always correct to assert A, and in that case $v(\neg A) = F$: we cannot always correctly assert $\neg A$. If, however, $v(A) = F$ we must ask whether there's any state of knowledge or information in our finite model which can verify A, that is, whether $s(A) = \varnothing$ or not. If $s(A) = \varnothing$ then it's always correct to assert $\neg A$ and hence $v(\neg A) = T$. If $s(A) \neq \varnothing$ then sometimes it's correct to assert A; so we can't always assert $\neg A$ and thus $v(\neg A) = F$. Granted this does impose some global (platonic?) point of view on whether we can *ever* verify A. However, we can confine ourselves to finite models: it's no more platonic than the reading Dummett gives in his intuitionistically valid completeness proof of Theorem 2.

You might argue that we should use new symbols here, say C and I for 'correct to assert' and 'incorrect to assert' rather than T and F. But that would obscure the similarity of the duality that the intuitionist imposes on propositions with that which the classical logician does. Both use a proposition to deduce further ones just in case it is true. And the formal logic of each is designed to capture those schemas which are invariably true and which can be used to deduce true propositions for any assignment of propositions. The situation is the same as when I chose to use '$\rightarrow$' to represent whatever notion of one proposition following from another which a logic proposes: I believe the underlying similarities represent some shared background assumptions and my notation represents that.

Dummett, *1977*, and also McCarty, *1983*, discuss the aptness of various other semantics for modeling the intuitionist's point of view, such as Beth trees and De Swart models (see also *De Swart, 1977*) for which comparable readings of **v** and **s** can be given. To the extent that any of these semantics capture the basis of intuitionism, so will ours. I will return to this point in discussing translations between logics in §X.B.7.

Dummett, *1973*, apparently rejects the kind of reading for the semantics of intuitionistic logic that I give. But his comments on theories of meaning sound much like the motivation for the general framework of Chapter IV.

> A theory of meaning, at least of the kind with which we are most familiar, seizes upon some one general feature of sentences ... as central: the notion of the content of an individual sentence is then to be explained in terms of this central feature. ... The justification for thus selecting some one single feature of sentences as central—as being that in which their individual meanings consist—is that it is hoped that every other feature of the use of sentences can be derived, in a uniform manner, from this central one.
>
> Dummett, *1973*, pp. 222–223

I agree with Dummett here: many logicians make the unreasonable claim that all features of use which are of significance to logic can be reduced to the one upon which their logic is based. If some feature, such as relevance, cannot be derived then it's argued that it must not be significant, or not logical.

But Dummett himself wishes us to believe that intuitionism manages to capture all features of the use of a sentence; "meaning is use" where

> The "use" of a sentence is not, in this sense, a *single* feature; the slogan simply restricts the *kind* of feature that may legitimately be appealed to as constituting or determining meaning. ... It is the multiplicity of the different features of the use of sentences, and the consequent legitimacy of the demand, given a molecular view of language, for harmony between them, that makes it possible to criticise existing practice, to call in question uses that are actually made of sentences of the language.
>
> Dummett, *1973*, p. 223

This argues for one overarching semantic theory that encompasses all or at least many of these features of sentences which are important to reasoning. Dummett argues that intuitionism does that, not only for mathematical statements but in general for natural language (*1977*, Chapter 7.1). Yet how can that be? The intuitionists' notion of content seems to me only one among many, hardly able to model such features of use as, say, relevance (axiom schema V is a standard fallacy of relevance according to some logicians). And even if it did model those in some general fashion it doesn't tell us how to reason in accord with any particular one. It seems to me that Dummett has fallen into the same error as the logicians he has criticized: he has taken one notion of content to be central, claiming that the numerous features of sentences which are significant to reasoning are thus taken into account.

3. Observations and refinements of the set-assignment semantics

a. If we wished to give a reading of $\mathsf{s}(A)$ as the constructive mathematical content of A, as I'd once hoped to do, dependent implication rather than dual dependent implication would be appropriate: $A \rightarrow B$ is true iff the constructive mathematical content of A contains that of B, and not both A is true and B is false. That has a nice sound to it, and it's not hard to modify the semantics here to give ones based on dependent implication. For instance, condition Int 1 would read $\mathsf{s}(A \wedge B) = \mathsf{s}(A) \cup \mathsf{s}(B)$: the constructive mathematical content of $A \wedge B$ is the constructive mathematical content of A plus the constructive mathematical content of B. But surprisingly I can find in the literature no explication of the constructive mathematical content of a proposition, nothing that would tell me how the content of $\forall x\,(x+2=2+x)$ differs from that of $\forall x \forall y\,(x+y=y+x)$ in such a way that I could see how the contents affect proofs or derivations: we can derive the former from the latter, but does that mean that the latter has *more* constructive content?

b. Is classical negation definable from the intuitionist connectives? That is, is there some schema N(A) built from $\neg, \rightarrow, \wedge, \vee$ which is evaluated by the following table in every **Int**-model $<\mathsf{v},\mathsf{s}>$?

A	N(A)
T	F
F	T

I do not know. If there is, then $\{\neg, \rightarrow, \wedge, \vee\}$ is functionally complete by an argument similar to the one given for **S** in Theorem III.5, noting that $<\mathsf{v},\mathsf{s}> \models (A \vee B) \rightarrow A$ iff $\mathsf{s}(A) \subseteq \mathsf{s}(B)$, and $<\mathsf{v},\mathsf{s}> \models \neg A$ iff $\mathsf{s}(A) = \varnothing$.

This question is closely related to whether there is a semantically faithful translation of **PC** into **Int** (see §X.B.4) and whether **Int** is narrower than **PC** (see the comment following Theorem 21, p.215). By each criteria of Chapter IV, Appendix 1.B (p.107), the semantics I have given for **Int** are incompatible with those for **PC**.

c. If for every every state of information and every A, either A or $\neg A$ is derivable, then we have classical logic (Corollary 22). In terms of the set-assignments, if we add to the conditions on an **Int**-model $\mathsf{s}(\neg A) = \overline{\mathsf{s}(A)}$ then we also have by Int 3 and Int 4, $\mathsf{s}(A \rightarrow B) = \overline{\mathsf{s}(A)} \cup \mathsf{s}(B)$. Thus the set-assignments would be restricted to Boolean algebras of sets, for which we have $\mathbf{PC} \vdash A$ iff $\mathsf{s}(A) = \mathsf{S}$

(see *Rasiowa, 1974*). Since $s(A) = \varnothing$ would imply $s(\neg A) = S$, for such models we would have $v(A) = \mathsf{T}$ iff $s(A) = S$, which is iff $\mathbf{PC} \vdash A$.

d. From Corollary 4 we can picture how set-assignments operate on negations, noting that in some models we may have $s(A) = s(\neg\neg A)$.

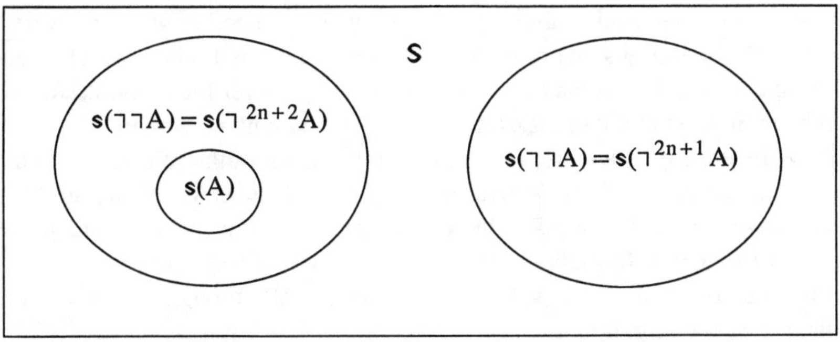

e. Do set-assignment semantics for **Int** allow an inductive definition of truth? Suppose we know $v(A)$, $s(A)$, and $v(B)$, $s(B)$. Then $v(A \wedge B)$, $v(A \vee B)$ can be calculated directly. We can also calculate $v(A \rightarrow B)$, and if it is true we know that we must assign $s(A \rightarrow B) = S$. But if it is false then the conditions that $s(A \rightarrow B)$ must satisfy are global, e.g., Int 5. Similarly, we may calculate $v(\neg A)$ and if it is true then we must assign $s(\neg A) = S$. If $v(A) = \mathsf{T}$ then by Int 10 we must assign $s(\neg A) = \varnothing$. However, if $v(A) = v(\neg A) = \mathsf{F}$, then the conditions which $s(\neg A)$ must satisfy are global. We need to know the content of these wffs in order to evaluate wffs of which they are constituent. So to the extent that we can assign $s(A \rightarrow B)$ and $s(\neg A)$ when these formulas are false we have an inductive definition of truth.

> The principal reason for suspecting these explanations [of the logical constants] of incoherence is their apparently highly impredicative character: if we know which constructions are proofs of the atomic statements of any first-order theory, then the explanations of the logical constants, taken together, determine which constructions are proofs of any of the statements of that theory; yet the explanations require us, in determining whether or not a construction is a proof of a conditional or of a negation, to consider its effect when applied to an arbitrary proof of the antecedent or of the negated statement, so that we must, in some sense, be able to survey or grasp some totality of constructions which will include all possible proofs of a given statement. The question is whether such a set of explanations can be acquitted of the charge of vicious circularity.
>
> Dummett, *1977*, p.390

f. The conditions on **Int**-models may be considerably weakened and still yield a class which is complete for **Int**. First, we can use the minimal intuitionist truth-table for ¬ from the next section: $v(\neg A) = T$ iff $v(A) = F$ and $s(A) \subseteq s(\neg A)$. In the presence of Int 10 this table is equivalent to the one already given. However, Int 10 is not needed and may be deleted: to verify that axioms X and XI are valid see the proof of Lemma 17 below. At least I don't believe that Int 10 is a consequence of Int 1–Int 9; Int 3 and Int 4 together yield only $s(A) \cap s(\neg A) \subseteq s(B)$ for all B. We could apparently have some subset $Q \subseteq S$ which is contained in $s(B)$ for every B: Q would be the conditions justifying the assertion of any contradiction. That reading is apt for the minimal calculus of the next section. However Int 10 would follow from the assumption that there is even one proposition with no content.

If we delete Int 10 and use the minimal intuitionist truth-table for ¬ then Int 9 can be weakened to: if $v(A) = T$ then for every B, $s(B) \subseteq s(A)$. However, Int 9 would follow from the assumption that every element of S is in some content set, that is, every state of information justifies the assertion of some proposition.

Int 4 can be changed to: $s(A \rightarrow B) \subseteq \overline{s(A)} \cup s(B)$, though it's not clear to me that this is a weaker condition.

Finally, in Int 7 and Int 8 '=' can be changed to '$\subseteq$'.

E. Johansson's Minimal Calculus J

1. The minimal calculus

In 1936 Johansson commented on Heyting's formal axioms for intuitionism.

> Among the logical axioms that Heyting set up for the derivation of the formal laws of intuitionistic logic there are two at which one starts:
>
> $\vdash B \rightarrow (A \rightarrow B)$
>
> $\vdash \neg A \rightarrow (A \rightarrow B)$
>
> The sense of these axioms is naturally only that the relation of implication in calculus has a different meaning than in ordinary speech. One can write $A \rightarrow B$ in the following three cases:
>
> 1. If B is recognized as a logical conclusion of A.
> 2. If B is recognized to be true.
> 3. If A is recognized to be false.
>
> One can easily become reconciled with the second case; however, the third case means an easily overlooked extension of the meaning of the conclusion. It is worth the effort to see if this can be avoided.
>
> Johansson, *1936*, p. 119

Johansson then proposed a reduced "minimal" calculus of intuitionism in which (3) need not be accepted. A similar analysis had been suggested by Kolmogoroff in *1925*.

J
in L(¬, →, ∧, ∨)
as for **Int** except delete axiom schema X. ¬A → (A→B)

The syntactic deductions for **Int** in §B.3 did not use Axiom X, so each is also correct for **J**. In particular, the Syntactic Deduction Theorem holds for **J**.

Johansson's axiomatization was in a different language. He notes that in line with an analysis given by Kolmogoroff, *1932*, one may introduce a propositional constant ⊥ and define ¬A as A→⊥. He explains this constant in the following manner.

> The interpretation of ⊥ as an undefined basic statement is related to the 'problem theoretical' meaning of the intuitionistic logic offered by Kolmogoroff. Namely, ¬A refers to (using Kolmogroff) the task 'assuming that the solution of A is given, to find a contradiction,' and that agrees with the definition of ¬A as A→⊥ if one interprets ⊥ as the task 'to obtain a contradiction.' This task is not defined; though it is an implicit assumption with Kolmogoroff that ⊥→B is valid, i.e., that the following problem has been solved: 'Assuming that a contradiction has been obtained, to solve an arbitrary problem.' If we leave out this assumption and thus obtain a sharper →, then we obtain a 'problem theoretical' meaning for the minimal calculus.
>
> Johansson, *1936*, p.131

Fitting, *1969*, gives Kripke-style semantics for **J** based on this reading by modifying those for **Int**. I will present those as modified to apply to L(¬, →, ∧, ∨). In §3 below I present an alternate axiomatization of **J** using ⊥ as primitive.

2. Kripke-style semantics

A *Kripke model for* **J** is ⟨W, R, Q, e⟩ where:

⟨W, R⟩ is transitive and reflexive
Q ⊆ W is R-closed (i.e., if $w \in Q$ and wRz, then $z \in Q$)
e : PV → Sub W

Q is to be thought of as those states of information which are inconsistent.

Validity in such a model is defined as for the Kripke semantics for **Int** with the exception of the evaluation of negations. We replace clause 4 by:

$w \models \neg A$ iff for all z such that wRz, $z \not\models A$ or $z \in Q$

Example (A ∧ ¬A) → B is not valid. Let wRz and wRy. Take $w, y, z \in e(p_1)$ and $w, y, z \in Q$. And take $y \notin e(p_2)$. So $w \models p_1$ and $w \models \neg p_1$. Yet $w \not\models p_2$. Hence $w \not\models (p_1 \wedge \neg p_1) \rightarrow p_2$.

Theorem 27 ***a.*** $\models_{\mathbf{J}} A$ iff A is validated by every Kripke model for **J**
b. $\models_{\mathbf{J}} A$ iff A is validated by every finite anti-symmetric Kripke model for **J**

A proof of this can be given along the lines of the one for **Int** in §B.4; see *Segerberg, 1968*. Strong completeness theorems can be derived by the same method, notably: for finite Γ, $\Gamma \models_{\mathbf{J}} A$ iff A is validated by every finite anti-symmetric Kripke model for **J** .

3. Translations and an alternate axiomatization

Classical logic can be translated into minimal logic by the double negation translation used to translate it into **Int**, as noted by Leivant, *1985*.

Theorem 28 $\Gamma \vdash_{\mathbf{PC}} A$ iff $\neg\neg\Gamma \vdash_{\mathbf{J}} \neg\neg A$

The proof is similar to the (syntactic) one for Theorem 18.

Here is an axiomatization of **J** in the language $L(\rightarrow, \wedge, \vee, \bot)$, where $\bot$ is a propositional constant. It is due to Segerberg, *1968*, p.30.

J
in $L(\rightarrow, \wedge, \vee, \bot)$
$\neg A \equiv_{\mathrm{Def}} A \rightarrow \bot$
axiom schemas

1. $(A \wedge B) \rightarrow A$
2. $(A \wedge B) \rightarrow B$
3. $A \rightarrow (A \vee B)$
4. $B \rightarrow (A \vee B)$
5. $(A \rightarrow C) \rightarrow ((B \rightarrow C) \rightarrow ((A \vee B) \rightarrow C))$
6. $(A \rightarrow B) \rightarrow ((A \rightarrow C) \rightarrow (A \rightarrow (B \wedge C))$
7. $(A \rightarrow (B \rightarrow C)) \rightarrow ((A \rightarrow B) \rightarrow (A \rightarrow C))$
8. $A \rightarrow (B \rightarrow A)$

rule $\dfrac{A, A \rightarrow B}{B}$

Segerberg shows that the closure of $\mathbf{J} \cup \{\bot \rightarrow A\}$ under *modus ponens* is strongly complete for the class of Kripke trees, where $\bot$ is evaluated as invalid at all elements of the tree. Hence we have the following axiomatization of **Int** .

Int
in $L(\rightarrow, \wedge, \vee, \bot)$
is the closure of $\mathbf{J} \cup \{\bot \rightarrow A\}$ under *modus ponens*

Leivant, *1985*, shows that in this language **Int** can be translated into **J**: let A^* be the result of replacing every nonatomic subformula B of A by $B \vee \bot$. Then $\Gamma \vdash_{\mathbf{Int}} A$ iff $\Gamma^* \vdash_{\mathbf{J}} A^*$.

4. Set-assignment semantics

We say that a set-assignment model $<v, s>$ uses *minimal intuitionist truth-conditions* if:

$\wedge$ and $\vee$ are evaluated classically

$\rightarrow$ is evaluated by the dual dependence table as for **Int**:

$v(A \rightarrow B) = T$ iff $s(A) \subseteq s(B)$ and not both $v(A) = T$ and $v(B) = F$

and $\neg$ is evaluated by the *minimal intuitionist negation* table:

A	$s(A) \subseteq s(\neg A)$	$\neg A$
any value	fails	F
T	holds	F
F		T

That is, $v(\neg A) = T$ iff $v(A) = F$ and $s(A) \subseteq s(\neg A)$.

We say that $<v, s>$ is a **J**-*model* if it uses the minimal intuitionist truth-conditions and satisfies Int 1, Int 2, Int 4–9 and

Int 3′. $s(B) \subseteq s(A \rightarrow B)$

Lemma 29 If $\mathbf{J} \vdash A$ then for every **J**-model $<v,s>$, $v(A) = T$.

Proof: The verification is just as in the proof of Lemma 23 except for axiom schemas V and XI. Using Int 3′, axiom schema V is easy to verify.

For axiom schema XI, suppose $v(A \rightarrow B) = T$ and $v(A \rightarrow \neg B) = T$. Then $s(A) \subseteq s(B) \cap s(\neg B)$. By Int 3′, $s(B) \subseteq s(A \rightarrow B)$ and $s(\neg B) \subseteq s(A \rightarrow \neg B)$. By Int 8 we then have $s(B) \cap s(\neg B) \subseteq s(\neg A)$. Hence $s(A) \subseteq s(\neg A)$. Also one of $B, \neg B$ must be false, hence $v(A) = F$. Thus by the minimal intuitionist table for negation, $v(\neg A) = T$. ■

Lemma 30 If $\mathbf{J} \nvdash E$ then for some **J**-model $<v,s>$, $v(E) = F$ and $s(A) = S$.

Proof: By Theorem 28, if $\mathbf{J} \nvdash \mathrm{E}$ then there is some Fitting model for **J**, $\langle W, R, Q, e \rangle$, and $w \in W$ such that $w \nvDash \mathrm{E}$. Pick such a model. Then define $<v, s>$ as in the proof of Lemma 24. It's not difficult to show that for this s we have:

$$w \vDash \mathrm{A} \text{ iff } s(\mathrm{A}) = S$$
$$s(\mathrm{B}) \subseteq s(\neg \mathrm{B}) \text{ iff } s(\neg \mathrm{B}) = S$$
$$s(\mathrm{B}) \subseteq s(\mathrm{C}) \text{ iff } s(\mathrm{B} \rightarrow \mathrm{C}) = S$$

The proof that $v(\mathrm{A}) = \mathsf{T}$ implies $s(\mathrm{A}) = S$ now follows as in Lemma 24. Hence $v(\mathrm{E}) = \mathsf{F}$. ■

Theorem 31 $\mathrm{J} \vdash \mathrm{A}$ iff for every **J**-model $<v, s>$, $v(\mathrm{A}) = \mathsf{T}$.

We can give almost the same reading to these semantics as to those for **Int**. The only difference is that now we admit that the information available to us at some particular time may be inconsistent, leading us to correctly assert, relative to that bad information, both some proposition A and its negation. But there may be nothing in that faulty information which would lead us to assert some other proposition B. So $\varnothing \neq s(\mathrm{A}) \cap s(\neg \mathrm{A})$ and $s(\mathrm{A}) \cap s(\neg \mathrm{A})$ may not be contained in $s(\mathrm{B})$; hence $(\mathrm{A} \wedge \neg \mathrm{A}) \rightarrow \mathrm{B}$ could fail. Note the similarity to the motivation for Dependence Logic (Chapter V).

But for Dependence Logic $\mathrm{B} \rightarrow (\mathrm{A} \rightarrow \mathrm{B})$ fails. Fifty years after Johansson wrote his paper one still starts at that axiom schema. In Kolmogoroff's terms it would read 'Given a solution to B, convert it to a method for *converting* a solution of A to one for B'. Perhaps this is acceptable if intuitionistic logic is to be applied only to mathematics. But Dummett argues that the justification of a semantics is in its extension to a theory of meaning for natural language. Outside the domain of mathematics it would seem unreasonable to assume that the "trivial" conversion of ignoring A and taking the proffered solution to B is not tantamount to "changing the subject," and hence unacceptable.

VIII Many-Valued Logics – $\mathbf{L_3}$, $\mathbf{L_n}$, $\mathbf{L_{\aleph}}$, $\mathbf{K_3}$, $\mathbf{G_3}$, $\mathbf{G_n}$, $\mathbf{G_{\aleph}}$, S5 –

In this chapter I will discuss the idea of introducing more than two truth-values into the semantics for a logic. First I will present the motivation for taking more than two truth-values and explain how it can be understood in terms of paying attention to different aspects of propositions. I will then give several examples of semantics which use truth-tables with more than two values, as well applications to the study of some logics presented in earlier chapters. I also include a general definition of many-valued logics (§B) for reference.

General references for many-valued logics are *Rescher, 1968* and *1969*, *Wójcicki, 1988*, and, for the philosophical issues surrounding them, *Haack, 1974* .

A. How Many Truth-Values?

1. History

The applicability of the classical dichotomy of true-false has been questioned since antiquity. In the beginning of the modern development of formal logic De Morgan considered dealing with more than two values for his calculus.

> But we should be led to extend our formal system if we considered propositions under three points of view, as true, false, or inapplicable. We may confine ourselves to single alternatives either by introducing not-true (including both false and inapplicable) as the recognized contrary of true; or else by confining our results to universes in which there is always applicability, so that true or false holds in every case. The latter hypothesis will best suit my present purpose.
>
> *De Morgan, 1847,* p.149

In the early twentieth century tables for the connectives using three or more values were used by several logicians to establish the independence of particular axioms in formal systems, an example of which I give in §G. Łukasiewicz (pronounced 'Woo-kah-sheay-vitch') and Tarski, *1930,* p.43 (footnote 5), give a short history of that early work.

The experience of working with those tables suggested to Łukasiewicz that a three-valued formal system would be appropriate to reason with future contingent propositions, such as 'There will be a sea battle tomorrow', which, he argued, are neither true nor false but rather possible. The first many-valued system proposed as a logic, that is as a formal system for reasoning, was set out by him in *Łukasiewicz, 1920,* and is presented here in §C.1. In that system, as in some others developed later, a third formal value is introduced into the truth-tables not as an additional truth-value but as a marker to indicate that a proposition to which it is assigned has no truth-value.

At the same time that Łukasiewicz presented his logic, Post, *1921*, formulated a class of many-valued systems as generalizations of the 2-valued classical calculus. These are called 'logics' only by reason of analogy with other logical systems, for to my knowledge no one has proposed any of them as a logic of propositions. Many-valued systems are often interesting to mathematicians for they constitute a clearly demarcated area of finite combinatorics which can be developed in analogy with formal logics. In §B I present the general definition of a many-valued system in order to have a uniform terminology and so that I can show in §E that most of the logics we've already studied cannot be characterized by any finite-valued system. But this chapter will be primarily concerned with many-valued systems which are either proposed as logics, that is as descriptive or prescriptive models of reasoning, or which were devised to reveal facets of other logics we have encountered in the previous chapters. An example of the latter are the systems of §F that Gödel devised to investigate intuitionistic logic.

In contrast to the view that some propositions can be reasoned with despite having no truth-value is the idea of Kleene, *1952*. He proposed a 3-valued system suited to reasoning with propositions such as undecidable arithmetical statements which are true or false, but which of these alternatives we do not or cannot know. I present his logic in §D. Other systems have been developed based on a similar motivation that there are degrees of truth or falsity roughly corresponding to degrees of certainty.

Another motive for many-valued systems has been to deal with paradoxical or inconsistent sentences. For instance, Moh Shah–Kwei, *1954*, proposed reserving the third value of Łukasiewicz's system for sentences such as 'This sentence is false'. He and later Kripke, *1975*, who instead used Kleene's logic, viewed such sentences as neither true nor false. On the other hand Yablo, *1985*, argues that we should think of them as both true *and* false and indicates how many-valued systems can be appropriate for reasoning on that basis. In the next chapter I'll present a many-valued logic which deals with inconsistencies in that manner.

For a more thorough treatment of the history of many-valued logics consult *Rescher, 1968*.

2. Hypothetical reasoning and aspects of propositions

In Chapter I I argued that in most applications of logic we reason *on the hypothesis* that this or that proposition is true or is false since we cannot make our communications precise enough to be unequivocal nor, in general, can we know with certainty the truth-value of the propositions we deal with. Of course we may have doubts about the validity of the hypothesis. It seems to me that many-valued logics factor those doubts into the logic by ascribing them to the content of the proposition.

$$\left.\begin{array}{c}\text{I can assent}\\ \text{or}\\ \text{I can not assent}\end{array}\right\}\text{ to a proposition.}$$

There is no third choice. I can be in doubt about whether I should assent or not. But that doubt is not a third choice; it is doubt about *whether* to make the one choice or the other. If the doubt predominates so that I don't assent then that is the choice: I do not assent. Or I may assent with doubt about the wisdom or propriety of assenting: but I have assented. Every many-valued logic recognizes this by partitioning the n values ascribed to propositions into two classes: designated and undesignated.

$$\left.\begin{matrix} 1 \\ \vdots \\ m \end{matrix}\right\} \text{ assent — the } \textit{designated values}$$

$$\left.\begin{matrix} m+1 \\ \vdots \\ n \end{matrix}\right\} \text{ do not assent}$$

A many-valued logic provides us with a calculus which enables us to know whether and how much to doubt a complex proposition from knowledge of the content of (doubt about) its constituents, and how to reason accordingly. But there is no need for a new notion of proposition: a proposition is a written or uttered declarative sentence with which we agree to proceed hypothetically as being either true or false, sometimes ascribing additional content to it.

Smiley has given a similar analysis.

> The way to defend [the method of designating truth-values] is to read 'true' for 'designated'. The method of defining logical consequence then needs no justification, for it now reads as saying that a proposition follows from others if and only if it is true whenever they are all true. What does need explaining is how there can be more than two truth-values. The answer is that propositions can be classified in other ways than as true or untrue, and by combining such a classification with the true/untrue one we in effect subdivide the true and untrue propositions into a larger number of types. For example, given any property φ of propositions, there are prima facie four possible types of proposition: true and φ, true and not φ, untrue and φ, untrue and not φ. If φ is unrelated to truth, like 'obscene' or 'having to do with geometry', all four types can exist and we get four truth-values, two being designated and two undesignated. If φ has any bearing on truth some of the types may be ruled out; e.g., if φ is (perhaps) 'about the future' or 'meaningless', the type 'true and φ' will be empty, leaving three truth-values of which just one is designated. One cannot foretell how the connectives will behave with respect to this or that classification of propositions, but to the extent that the types of compound propositions turn out to be functions of the types of their constituents, so we shall get a many-valued logic.
>
> *Smiley, 1976,* pp. 86–87

B. A General Definition of Many-Valued Semantics

The definitions of this section are useful for providing a uniform terminology for many-valued logics and are necessary later when we investigate whether the logics we have already studied have finite-valued semantics. These definitions are quite abstract and as such have been studied extensively, for instance by *Wójcicki, 1988*, and *Carnielli, 1987 B*. In the following sections I will repeat these definitions for specific logics; you may wish to look at those examples first, after which the definitions of this section will seem natural.

We take as our language $L(p_0, p_1, \ldots \neg, \rightarrow, \wedge, \vee)$. It will be obvious how to modify the definitions to apply to languages with other connectives.

M is a *matrix* if

$$\mathsf{M} = \langle \mathsf{U}, \mathsf{D}, \neg_{\mathsf{M}}, \rightarrow_{\mathsf{M}}, \wedge_{\mathsf{M}}, \vee_{\mathsf{M}} \rangle$$

where:

U is a set with $\mathsf{D} \subseteq \mathsf{U}$

$\rightarrow_{\mathsf{M}}, \wedge_{\mathsf{M}}, \vee_{\mathsf{M}}$ are binary operations on U

$\neg_{\mathsf{M}}$ is a unary operation on U

D is called the set of *designated*, or *distinguished elements* of M. The operations are called the *(truth-) tables for* $\neg, \rightarrow, \wedge, \vee$. M is called *finite-valued, infinite-valued,* or *n-valued* according to whether U is finite, infinite, or has n elements.

An *evaluation* e *with respect to* M is a function $\mathsf{e}: PV \rightarrow \mathsf{U}$ which is extended inductively to all wffs by:

$$\mathsf{e}(\neg A) = \neg_{\mathsf{M}} \mathsf{e}(A)$$
$$\mathsf{e}(A \rightarrow B) = \rightarrow_{\mathsf{M}} (\mathsf{e}(A), \mathsf{e}(B))$$
$$\mathsf{e}(A \wedge B) = \wedge_{\mathsf{M}} (\mathsf{e}(A), \mathsf{e}(B))$$
$$\mathsf{e}(A \vee B) = \vee_{\mathsf{M}} (\mathsf{e}(A), \mathsf{e}(B))$$

If we consider U to be a collection of truth-values then this definition says that every evaluation is truth-functional.

We say that e *validates* A, written $\mathsf{e} \models A$, if $\mathsf{e}(A) \in \mathsf{D}$. We write $\mathsf{e} \models \Gamma$ if for all $A \in \Gamma$, $\mathsf{e} \models A$. When more than one matrix is under consideration we write $\models_{\mathsf{M}}$ for $\models$.

A wff A is *valid* (*with respect to* M), written $\models_{\mathsf{M}} A$ or simply $\models A$, if for every evaluation e, $\mathsf{e} \models A$. That is, every evaluation of A takes a designated value. We say that M is a *characteristic matrix for a set of wffs* Γ if $\Gamma = \{A : \models_{\mathsf{M}} A\}$. In that case we say that Γ *has many-valued semantics*, or that Γ *is a many-valued logic*. The term *many-valued system* may sometimes be used to refer to either such a logic or the semantics of it.

Every many-valued matrix in this chapter will have only one designated value. In the next chapter we will see a logic characterized by a 3-valued matrix with two designated values.

We can define the notion of *semantic consequence* (*with respect to* M) in the usual way: $\Gamma \vDash A$ iff for every evaluation e, if $e \vDash \Gamma$ then $e \vDash A$. That is, a proposition follows from others if it takes a designated value whenever they do. For some of the logics below other notions of semantic consequence have been suggested: see, for example, *Smiley, 1976,* or *Wójcicki, 1988*.

It is useful to be able to give finite-valued semantics for a logic for then there is a simple decision method for validity like the one for the two-valued classical logic (§II.G.5). But some logics which are decidable, such as **Int** and **S5**, cannot be given finite-valued semantics, as I'll show below in §E.

In contrast, *any* set of sentences Γ which is closed under the rule of substitution can be seen to have infinite-valued semantics. Take the matrix M to be $\langle$ Wffs, Γ, $\neg$, $\rightarrow$, $\wedge$, $\vee \rangle$: the designated values are just the wffs of Γ and the operations are the connectives. An evaluation of A is then a substitution instance of it. But if Γ is first presented as a syntactic or semantic consequence relation, then the semantic consequence relation for this matrix may not be the same. Under certain restrictions, however, that, too, can be guaranteed: see *Dummet, 1977,* Chapter 5.1.

C. The Łukasiewicz Logics

Łukasiewicz argued that a two-valued logic is incompatible with the view that some propositions about the future are not predetermined.

> I can assume without contradiction that my presence in Warsaw at a certain moment of next year, e.g., at noon on 21 December, is at the present time determined neither positively nor negatively. Hence it is *possible,* but not *necessary*, that I shall be present in Warsaw at the given time. On this assumption the proposition 'I shall be in Warsaw at noon on 21 December of next year', can at the present time be neither true nor false. For if it were true now, my future presence in Warsaw would have to be necessary, which is contradictory to the assumption. If it were false now, on the other hand, my future presence in Warsaw would have to be impossible, which is also contradictory to the assumption. Therefore, the proposition considered is at the moment *neither true nor false* and must possess a third value, different from '0' or falsity and '1' or truth. This value we can designate by '$\frac{1}{2}$'. It represents 'the possible', and joins 'the true' and 'the false' as a third value.
>
> *Łukasiewicz, 1930,* pp.165–166

Łukasiewicz saw his work partly as a formalization of Aristotle's views concerning future contingent propositions and the notions of necessity and possibility: see in particular *Łukasiewicz, 1922* and *1930*; in the latter he has a history of the principle of *tertium non datur*. Whether his interpretation of Aristotle is correct, and if correct reasonable, has been discussed by a number of authors. Haack, *1974,* Chapter 3, argues that his whole program was misconceived on the basis of a modal fallacy, and in doing so reviews the literature on the subject. Prior, *1955,* pp. 230–250, gives a sustained defense and explication of Łukasiewicz's work, relating it to Aristotle's and Ockham's views. Without judging the matter beyond what I have already said in §A, I will try to give a sympathetic reading of Łukasiewicz's views by following in large part Prior's presentation.

1. The 3-valued logic L_3

a. The truth-tables and their interpretation

This explanation will, for the most part, follow *Prior, 1955* and *1967*.

A proposition may take one of three values under an interpretation: 1, $\frac{1}{2}$, or 0, where we interpret

$e(A) = 1$ to mean that A is determinately true
$e(A) = 0$ to mean that A is determinately false
$e(A) = \frac{1}{2}$ to mean that A is neither determinately true nor false; in this case we say that A is possible or "neuter"

The intent is that future contingent propositions are the ones which take value $\frac{1}{2}$, reflecting Łukasiewicz's rejection of determinism.

The matrix for Łukasiewicz's 3-valued logic is given by the following tables:

$A \wedge B$	B: 1	$\frac{1}{2}$	0
A 1	1	$\frac{1}{2}$	0
$\frac{1}{2}$	$\frac{1}{2}$	$\frac{1}{2}$	0
0	0	0	0

$A \vee B$	B: 1	$\frac{1}{2}$	0
A 1	1	1	1
$\frac{1}{2}$	1	$\frac{1}{2}$	$\frac{1}{2}$
0	1	$\frac{1}{2}$	0

A	$\neg A$
1	0
$\frac{1}{2}$	$\frac{1}{2}$
0	1

$A \to B$	B: 1	$\frac{1}{2}$	0
A 1	1	$\frac{1}{2}$	0
$\frac{1}{2}$	1	1	$\frac{1}{2}$
0	1	1	1

We define

$A \leftrightarrow B \equiv_{\mathrm{Def}} (A \to B) \wedge (B \to A)$

An $\mathbf{L_3}$-*evaluation* is a map $e: PV \to \{0, \frac{1}{2}, 1\}$ which is extended to all wffs of $L(p_0, p_1, \dots \neg, \to, \wedge, \vee)$ by these tables. The *sole designated value* is 1, so that $e \models A$ means $e(A) = 1$; then A *is valid*, written $\models A$, means that $e(A) = 1$ for all $\mathbf{L_3}$-evaluations. And $\Gamma \models_{L_3} A$ means that for every $\mathbf{L_3}$-evaluation e, if $e(B) = 1$ for every B in Γ, then $\mathbf{e}(A) = 1$. The collection of valid wffs we call $\mathbf{L_3}$, or the $\mathbf{L_3}$-*tautologies.*

Both $A \vee \neg A$ and $\neg(A \wedge \neg A)$ fail to be $\mathbf{L_3}$-tautologies in accord with Łukasiewicz's rejection of bivalence: they take the value $\frac{1}{2}$ if A has value $\frac{1}{2}$. We have instead a principle of trivalence. To express that, we first pick out a formula that identifies a proposition as having value $\frac{1}{2}$:

$$IA \equiv_{Def} A \leftrightarrow \neg A$$

which is meant to be read as ''A' is indeterminate'. The table for it is

A	IA
1	0
$\frac{1}{2}$	1
0	0

(One could perhaps give a strained reading of this as a connective, namely, 'It is indeterminate whether A'.)

With the aid of IA we can express what I call the *law of excluded fourth*: $A \vee \neg A \vee IA$. This is an $\mathbf{L_3}$-tautology, and I sometimes refer to it as a *principle of trivalence* for $\mathbf{L_3}$.

Moh Shaw–Kwei, *1954*, suggested that paradoxical sentences are those which are equivalent to their negation and that we interpret $\mathbf{L_3}$ as assigning $\frac{1}{2}$ to those. Then IA could be read as ''A' is paradoxical'. However, it is not enough to simply invoke 3 values for propositions in order to resolve the liar paradox, 'This sentence is false': Łukasiewicz's system spawns the equally problematic *strengthened liar paradox*: 'This sentence is false or paradoxical'.

Though not every classical tautology is valid in $\mathbf{L_3}$, we do have the converse, $\mathbf{L_3} \subset \mathbf{PC}$, because the $\mathbf{L_3}$-tables restricted to the values 0 and 1 are the **PC**-tables reading T for 1, and F for 0. We cannot retrieve **PC** by identifying 1 and $\frac{1}{2}$ with T, for the **PC**-tautology $\neg(A \rightarrow \neg A) \vee \neg(\neg A \rightarrow A)$ takes value 0 when A has value $\frac{1}{2}$.

We may reduce the primitives of the language of $\mathbf{L_3}$ to just $\neg$ and $\rightarrow$:

$$A \vee B \equiv_{Def} (A \rightarrow B) \rightarrow B$$

$$A \wedge B \equiv_{Def} \neg(\neg A \vee \neg B)$$

have the correct tables. However, $\{\neg, \rightarrow, \wedge, \vee\}$ is not functionally complete in the following sense. We may introduce into the language a 3-valued functional connective TA, called 'the Słupecki operator' which is evaluated as $\frac{1}{2}$ regardless of the value of A. Then TA is not definable from these connectives, for if it were then, since $\neg$TA and TA always have the same value, $\neg TA \rightarrow TA$ would be an $\mathbf{L_3}$-tautology, whereas no wff of the form $\neg B \rightarrow B$ is a **PC**-tautology and $\mathbf{L_3} \subset \mathbf{PC}$.

Łukasiewicz considered the following function to be appropriate to formalize possibility.

A	$\Diamond$A
1	1
$\frac{1}{2}$	1
0	0

A proposition is possible if it is definitely true or neuter, but not possible if it is false. Tarski noted that this is definable in $\mathbf{L_3}$ as

$$\Diamond A \equiv_{Def} \neg A \rightarrow A$$

(Note that this is the definition we used in Chapter VI for necessity!)

It is harmless to understand $\Diamond$A as an abbreviation useful in our metalogical investigations; but if we view $\Diamond$ as a connective we are under an obligation to explain why this is not a use-mention confusion, as discussed in §VI.B.2.

We may define

$$\Box A \equiv_{Def} \neg\Diamond\neg A$$

to correspond to a necessity operator. The table for it is:

A	$\Box$A
1	1
$\frac{1}{2}$	0
0	0

Prior, *1955*, p.421, argues that this connective serves to model Aristotle's notion of necessity. He quotes Aristotle as saying,

> Once it is, that which is is-necessarily, and once it is not, that which is not necessarily-is-not.

Correspondingly we have that $A \rightarrow (A \rightarrow \Box A)$ is an $\mathbf{L_3}$-tautology, though $A \rightarrow \Box A$ can fail for neuter A. Prior, *1955*, pp. 248–249, points out that $\Box$A should not be interpreted as 'It is logically necessary that A', as we did for **S5** in Chapter VI. On that reading we should have '$\Box$(Socrates is dead $\rightarrow$ Socrates is dead)' is true on the basis of its form, whereas '$\Box$(Socrates is dead)' would not be true. Yet on Łukasiewicz's interpretation of necessity they would both be true since 'Socrates is dead' is true.

> And should the view that there are "neuter" propositions be accepted, not only the law of excluded middle but the whole structure of two-valued logic can be preserved by the understanding that the "propositions" substitutable for its variables are only those referring to matters of present, past or otherwise [e.g. timeless] determinate fact.
>
> *Prior, 1955*, p. 250

For such propositions $\Box A$ and $\Diamond A$ are both equivalent to A.

As mentioned above, $\mathbf{L_3} \subsetneq \mathbf{PC}$. But we can interpret **PC** in $\mathbf{L_3}$ by means of a translation, based on the ideas of Tokarz (see *Wójcicki, 1988,* pp.71–72). First let's consider whether there is a Deduction Theorem for $\mathbf{L_3}$. The appropriate connective cannot be $\rightarrow$, because $A \wedge \neg A \vDash_{\mathbf{L_3}} \neg(A \rightarrow B)$ but $\nvDash_{\mathbf{L_3}} (A \wedge \neg A) \rightarrow \neg(A \rightarrow B)$. We define:

$$A \rightarrow_3 B \equiv_{\mathrm{Def}} A \rightarrow (A \rightarrow B)$$

Its table is

		B		
$A \rightarrow_3 B$		1	$\frac{1}{2}$	0
A	1	1	$\frac{1}{2}$	0
	$\frac{1}{2}$	1	1	1
	0	1	1	1

So $e(A \rightarrow B) \neq 1$ iff $e(A) = 1$ and $e(B) \neq 1$. Thus we have:

Theorem 1 ***(A Semantic Deduction Theorem for*** $\mathbf{L_3}$***)***

$$\Gamma \cup \{A\} \vDash_{\mathbf{L_3}} B \quad \text{iff} \quad \Gamma \vDash_{\mathbf{L_3}} A \rightarrow_3 B$$

The table for $\rightarrow_3$ was first given by Monteiro, *1967*, who defined the connective as $\Diamond\neg A \vee B$. He noted that we could take $\neg, \wedge, \rightarrow_3$ as primitives, defining $A \rightarrow B$ as $(A \rightarrow_3 B) \wedge (\neg B \rightarrow_3 \neg A)$. In the form given above, however, it easily generalizes to Łukasiewicz's n-valued logics (below) by iterating '$A \rightarrow$' in the definition. Wójcicki, *1988,* counting the number of arrows, calls this '$\rightarrow_2$'.

Now consider the map * from $L(\neg, \rightarrow)$ to itself given by:

$$(p)^* = p$$
$$(A \rightarrow B)^* = A^* \rightarrow_3 B^*$$
$$(\neg A)^* = A^* \rightarrow_3 \neg(A^* \rightarrow A^*)$$

and $\Gamma^* = \{A^* : A \in \Gamma\}$.

Theorem 2 The map * is a grammatical translation of **PC** into $\mathbf{L_3}$:

$$\Gamma \vDash_{\mathbf{PC}} A \quad \text{iff} \quad \Gamma^* \vDash_{\mathbf{L_3}} A^*$$

Proof: If e is an $\mathbf{L_3}$-evaluation and v a 2-valued model of **PC** such that $e(p) = 1$ iff $v(p) = T$, then for every B, $e(B^*) = 1$ iff $v(B) = T$, as you can check by induction on the length of B. Hence $\Gamma \vDash_{\mathbf{PC}} A$ iff $\Gamma^* \vDash_{\mathbf{L_3}} A^*$. ■

b. Wajsberg's axiomatization of $\mathbf{L_3}$

The following axiomatization of $\mathbf{L_3}$ is due to Wajsberg, *1931*.

$\mathbf{L_3}$
in $L(\neg, \rightarrow)$

axiom schemas

$L_3 1.$ $A \rightarrow (B \rightarrow A)$

$L_3 2.$ $(A \rightarrow B) \rightarrow ((B \rightarrow C) \rightarrow (A \rightarrow C))$

$L_3 3.$ $(\neg A \rightarrow \neg B) \rightarrow (B \rightarrow A)$

$L_3 4.$ $((A \rightarrow \neg A) \rightarrow A) \rightarrow A)$

rule $\dfrac{A, A \rightarrow B}{B}$

We write $\Gamma \vdash_{\mathbf{L_3}} A$ if there is a proof of A from Γ in this system.

Axiom $L_3 4$ is a weak form of the law of excluded middle. It's definitionally equivalent to $(A \rightarrow \neg A) \vee A$, which in **PC** is provably equivalent to $A \vee \neg A$.

Theorem 3 $\vdash_{\mathbf{L_3}} A$ iff A is an $\mathbf{L_3}$-tautology.

Wajsberg, *1931*, gives a constructive proof of this. His proof, as all others I've seen, is a complex combinatorial argument reducing tautologies to certain normal forms. It is not possible to prove the strong completeness of the axiomatization by such finitistic means (cf. §II.H.4 and §II.K.1). In the next section I will produce a finite axiomatization which is strongly complete for $\mathbf{L_3}$ by modifying the strong completeness proof for classical logic using the principle of trivalence above. The equivalence of the two systems ensures that Wajsberg's is strongly complete.

Nonetheless, we may use Theorem 3 to prove a *finite* strong completeness theorem for this axiomatization by first establishing a Syntactic Deduction Theorem.

Theorem 4 ***(A Syntactic Deduction Theorem for*** $\mathbf{L_3}$***)***

$\Gamma \cup \{A\} \vdash_{\mathbf{L_3}} B$ iff $\Gamma \vdash_{\mathbf{L_3}} A \rightarrow_3 B$

Proof: The proof follows as for classical logic, Theorem II.8 (p.47), because both $A \rightarrow_3 (B \rightarrow_3 A)$ and $(A \rightarrow_3 (B \rightarrow_3 C)) \rightarrow ((A \rightarrow_3 B) \rightarrow (A \rightarrow_3 C))$ are $\mathbf{L_3}$-tautologies and hence, by Theorem 2, are theorems of $\mathbf{L_3}$. ∎

Theorem 5 ***(Finite Strong Completeness for*** $\mathbf{L_3}$***)***

For finite Γ, $\Gamma \vdash_{\mathbf{L_3}} A$ iff $\Gamma \models_{\mathbf{L_3}} A$

Proof: Let $\Gamma = \{B_1, \dots, B_n\}$. Then

$$\begin{aligned} \Gamma \vdash_{\mathbf{L_3}} A \quad & \text{iff} \quad \{B_1, \dots, B_n\} \vdash_{\mathbf{L_3}} A \\ & \text{iff} \quad \vdash_{\mathbf{L_3}} B_1 \to_3 (B_2 \to_3 (\dots (B_n \to_3 A)) \dots) \\ & \text{iff} \quad \models_{\mathbf{L_3}} B_1 \to_3 (B_2 \to_3 (\dots (B_n \to_3 A)) \dots) \\ & \text{iff} \quad \{B_1, \dots, B_n\} \models_{\mathbf{L_3}} A \\ & \text{iff} \quad \Gamma \models_{\mathbf{L_3}} A \end{aligned}$$

■

There is a rule of substitution of equivalents in $\mathbf{L_3}$. For any evaluation, $\mathbf{e}$, we have that if $\mathbf{e} \models (A \to B) \wedge (IA \leftrightarrow IB)$, then $\mathbf{e}(A) = \mathbf{e}(B)$. So the rule

$$\frac{(A \leftrightarrow B) \wedge (IA \leftrightarrow IB)}{C(A) \leftrightarrow C(B)}$$

is valid in $\mathbf{L_3}$, where $C(B)$ is the result of replacing some but not necessarily all occurrences of A in C with B. Indeed, the stronger rule,

$$\frac{(A \leftrightarrow B) \wedge (IA \leftrightarrow IB)}{(C(A) \leftrightarrow C(B)) \wedge (IC(A) \leftrightarrow IC(B))}$$

is valid in $\mathbf{L_3}$, too.

c. A proof that $\mathbf{L_3}$ *is finitely axiomatizable*

The proof I give here is a modification of the one for classical logic, using the appropriate analogue to the notion of a classically complete and consistent set. The axioms arise naturally in the proof and I will note the first appearance of each.

$\mathbf{L_3}$
in $L(\neg, \to)$

$A \vee B \equiv_{\text{Def}} (A \to B) \to B$ $\qquad$ $A \wedge B \equiv_{\text{Def}} \neg(\neg A \vee \neg B)$

$A \to_3 B \equiv_{\text{Def}} A \to (A \to B)$ $\qquad$ $IA \equiv_{\text{Def}} A \leftrightarrow \neg A$

$A \leftrightarrow B \equiv_{\text{Def}} (A \to B) \wedge (B \to A)$

axiom schemas

1. $A \to (B \to (A \wedge B))$
2. a. $(A \wedge B) \to A$

   b. $(A \wedge B) \to B$
3. $A \leftrightarrow \neg\neg A$
4. a. $\neg(A \to B) \to (A \wedge \neg B)$

   b. $(A \wedge \neg B) \to_3 \neg(A \to B)$

5. $(IA \wedge \neg B) \rightarrow_3 I(A \rightarrow B)$
6. $\neg A \rightarrow (A \rightarrow B)$
7. $B \rightarrow (A \rightarrow B)$
8. $(IA \wedge IB) \rightarrow (A \rightarrow B)$
9. $(B \wedge \neg B) \rightarrow_3 A$
10. $(B \wedge IB) \rightarrow_3 A$
11. $(\neg B \wedge IB) \rightarrow_3 A$
12. $B \rightarrow_3 (A \rightarrow_3 B)$
13. $(A \rightarrow_3 (B \rightarrow_3 C)) \rightarrow ((A \rightarrow_3 B) \rightarrow (A \rightarrow_3 C))$
14. $((\neg A \rightarrow_3 A) \wedge (IA \rightarrow_3 A)) \rightarrow A$

rule $\dfrac{A,\, A \rightarrow B}{B}$

I will use $\vdash$ throughout this section for the consequence relation of this axiom system. I will leave to you to check that this system is sound for $\mathbf{L_3}$.

The definition of a complete consistent theory should correspond to the set of sentences true in a model. Given any $\mathbf{L_3}$-evaluation exactly one of $\mathbf{e}(A)$, $\mathbf{e}(\neg A)$, $\mathbf{e}(IA)$ has value 1, and which one does determines the value of $\mathbf{e}(A)$. So I define for a set of wffs Γ:

> Γ is *complete* (*relative to* $\mathbf{L_3}$) if for every A *at least one of* $A, \neg A, IA$ is in Γ
>
> Γ is *consistent* (*relative to* $\mathbf{L_3}$) if *at most one of* $A, \neg A, IA$ is a consequence of Γ

As usual, $\mathrm{Th}(\Gamma) \equiv_{\mathrm{Def}} \{A: \Gamma \vdash A\}$ and Γ is *theory* if $\Gamma = \mathrm{Th}(\Gamma)$.

It's easy to establish the following using the definition of $\rightarrow_3$ and axioms 1 and 2.

Lemma 6 *a.* $\{A,B\} \vdash A \wedge B$
b. $\{A \wedge B\} \vdash B \wedge A$
c. $\{A, A \rightarrow_3 B\} \vdash B$
d. $\{A \wedge B\} \vdash A$

Lemma 7 ***(A Syntactic Deduction Theorem)*** $A \vdash B$ iff $\vdash A \rightarrow_3 B$

Proof: The proof is as for classical logic, Theorem II.8, due to the presence of axioms 12 and 13. ■

Lemma 8 *a.* Γ is inconsistent iff for every B, $\Gamma \vdash B$.

b. If Γ is complete and consistent and $A \notin \Gamma$, then for every B $\Gamma \cup \{A\} \vdash B$.

c. If Γ is complete and consistent, then it is a theory.

d. If $\Gamma \nvdash A$ then either $\Gamma \cup \{\neg A\}$ or $\Gamma \cup \{IA\}$ is consistent.

e. If Γ is consistent, then one of $\Gamma \cup \{A\}$, $\Gamma \cup \{\neg A\}$, or $\Gamma \cup \{IA\}$ is consistent.

Proof: (a) and (b) follow from axioms 9, 10, and 11.

c. If Γ is complete and consistent suppose $\Gamma \vdash A$. Then $\text{Th}(\Gamma) = \text{Th}(\Gamma \cup \{A\})$. If $A \notin \Gamma$ then $\text{Th}(\Gamma)$ is inconsistent by part (b). So $A \in \Gamma$.

d. Suppose $\Gamma \nvdash A$ and both $\Gamma \cup \{\neg A\}$ and $\Gamma \cup \{IA\}$ are inconsistent. By part (a) we have $\Gamma \cup \{\neg A\} \vdash A$ and $\Gamma \cup \{IA\} \vdash A$, hence by our Deduction Theorem (Lemma 7), Lemma 6.c, and axiom 14, $\Gamma \vdash A$, which is a contradiction.

e. If Γ is consistent suppose both $\Gamma \cup \{\neg A\}$ and $\Gamma \cup \{IA\}$ are inconsistent. Then as in the proof of part (d), $\Gamma \vdash A$, so $\Gamma \cup \{A\}$ is consistent. ■

Lemma 9 The following are equivalent:

a. Γ is complete and consistent.

b. There is some $\mathbf{L_3}$-evaluation **e** such that $\Gamma = \{A : \mathbf{e} \vDash A\}$.

c. There is some $\mathbf{L_3}$-evaluation **e** such that

$$\mathbf{e}(A) = 1 \text{ iff } A \in \Gamma$$
$$\mathbf{e}(A) = \tfrac{1}{2} \text{ iff } IA \in \Gamma$$
$$\mathbf{e}(A) = 0 \text{ iff } \neg A \in \Gamma$$

Proof: The equivalence of the last two statements is easy to prove. I will show that the first and last are equivalent.

$\Leftarrow$ If there is such an **e**, then Γ must be complete for, as remarked above, exactly one of $A, \neg A, IA$ is assigned 1 by **e**. Consistency then follows as $\Gamma = \{A : \mathbf{e} \vDash A\}$ and hence is closed under deduction.

$\Rightarrow$ Suppose Γ is complete and consistent. Then by Lemma 8 it is a theory. If we set:

$$\mathbf{e}(A) = \begin{cases} 1 & \text{if } A \in \Gamma \\ \frac{1}{2} & \text{if } IA \in \Gamma \\ 0 & \text{if } \neg A \in \Gamma \end{cases}$$

then **e** is well-defined. It remains to show that **e** is an $\mathbf{L_3}$-evaluation. We begin with negation:

$$\mathbf{e}(\neg A) = 1 \text{ iff } \neg A \in \Gamma$$
$$\text{iff } \mathbf{e}(A) = 0$$

$$\begin{aligned} e(\neg A)=0 \ &\text{iff}\ \neg(\neg A)\in\Gamma \\ &\text{iff}\ A\in\Gamma \qquad \text{by axiom 3} \\ &\text{iff}\ e(A)=1 \end{aligned}$$

So by process of elimination, $e(\neg A)=\frac{1}{2}$ iff $e(A)=\frac{1}{2}$.

Next, for the evaluation of the conditional,

$$\begin{aligned} e(A\to B)=0 \ &\text{iff}\ \neg(A\to B)\in\Gamma \\ &\text{iff}\ (A\wedge\neg B)\in\Gamma \quad \text{by axioms 4a and 4b} \\ &\text{iff}\ A, \neg B\in\Gamma \qquad \text{by Lemma 6} \\ &\text{iff}\ e(A)=1 \text{ and } e(B)=0 \end{aligned}$$

Now suppose $e(A\to B)=1$. Then $A\to B\in\Gamma$. We must show that it's not the case that $e(A)>e(B)$. Suppose $e(A)=1$. Then $A\in\Gamma$, and since Γ is a theory we have $B\in\Gamma$, so $e(B)=1$. Suppose $e(A)=\frac{1}{2}$ and to the contrary that $e(B)=0$. Then $IA\in\Gamma$ and $\neg B\in\Gamma$, so by axiom 5 and the fact that Γ is a theory, $I(A\to B)\in\Gamma$, which contradicts the consistency of Γ. So $e(B)\neq 0$. Hence $e(A)\leq e(B)$.

Now suppose that $e(A)\leq e(B)$. We need to show that $A\to B\in\Gamma$ and hence that $e(A\to B)=1$. If $e(A)=0$ then $\neg A\in\Gamma$, so by axiom 6, $A\to B\in\Gamma$. If $e(A)=\frac{1}{2}$ and $e(B)=\frac{1}{2}$, then IA and IB are both in Γ and hence by axiom 8, $A\to B\in\Gamma$. Finally, if $e(B)=1$ then $B\in\Gamma$, so $A\to B\in\Gamma$ by axiom 7.

So by process of elimination, $e(A\to B)=\frac{1}{2}$ iff $(e(A)=1$ and $e(B)=\frac{1}{2})$ or $(e(A)=\frac{1}{2}$ and $e(B)=0)$. And hence e is an L_3-evaluation. ■

Lemma 10 If $\Gamma\nvdash A$ then there is a complete and consistent theory Σ such that $A\notin\Sigma$ and $\Gamma\subseteq\Sigma$.

Proof: Let $B_0, B_1, \ldots$ be a listing of all wffs. Define

$$\Sigma_0 = \begin{cases} \Gamma\cup\{\neg A\} & \text{if that is consistent} \\ \Gamma\cup\{IA\} & \text{otherwise} \end{cases}$$

and

$$\Sigma_{n+1} = \begin{cases} \Sigma_n\cup\{IB_n\} & \text{if that is consistent; if not, then} \\ \Sigma_n\cup\{\neg B_n\} & \text{if that is consistent; if not, then} \\ \Sigma_n\cup\{B_n\} & \end{cases}$$

By Lemma 8, each Σ_i is consistent. Hence $\Sigma=\bigcup_n \Sigma_n$ is consistent, and by the choice of Σ_0, $A\notin\Sigma$. By construction Σ is complete. ■

Theorem 11 **(*Strong Completeness for* L_3)**

a. $\Gamma\vdash A$ iff $\Gamma\models_{L_3} A$

b. $\Gamma\vdash_{L_3} A$ iff $\Gamma\models_{L_3} A$

Proof: a. I've already remarked that the axioms are sound. So the proof follows in the usual way using Lemmas 9 and 10.

b. $\Gamma \vdash A$ iff there are $B_1, \dots, B_n$ in Γ such that $\{B_1, \dots, B_n\} \vdash A$
iff there are $B_1, \dots, B_n$ in Γ such that $\vdash B_1 \rightarrow_3 (B_2 \rightarrow_3 (\dots (B_n \rightarrow_3 A)) \dots)$
iff there are $B_1, \dots, B_n$ in Γ such that $\models_{\mathbf{L_3}} B_1 \rightarrow_3 (B_2 \rightarrow_3 (\dots (B_n \rightarrow_3 A)) \dots)$
iff there are $B_1, \dots, B_n$ in Γ such that $\vdash_{\mathbf{L_3}} B_1 \rightarrow_3 (B_2 \rightarrow_3 (\dots (B_n \rightarrow_3 A)) \dots)$
iff there are $B_1, \dots, B_n$ in Γ such that $\{B_1, \dots, B_n\} \vdash_{\mathbf{L_3}} A$
iff $\Gamma \vdash_{\mathbf{L_3}} A$.

Therefore by part (a), $\Gamma \vdash_{\mathbf{L_3}} A$ iff $\Gamma \models_{\mathbf{L_3}} A$. ■

Carnielli, *1987 B*, presents a method for characterizing many-valued logics syntactically by analytic tableaux, based on a method similar to the proof above: in his paper read $A, \neg A, IA$ for $a_0(A), a_1(A), a_2(A)$ for the logic $\mathbf{L_3}$.

d. Set-assignment semantics for $\mathbf{L_3}$

I will first give set-assignment semantics for $\mathbf{L_3}$ which imitate the 3-valued tables by allowing only three choices for content sets. These use the intuitionist truth-tables which generalize to Łukasiewicz's infinite-valued logic in §C.2.c. Then I will present a richer semantics which allow greater variation in content, and will discuss how those could be interpreted to apply to paradoxical sentences.

$\langle v, s \rangle$ is an $\mathbf{L_3}$-*model* for $L(\neg, \rightarrow)$ if $\neg$ and $\rightarrow$ are evaluated by the intuitionist tables:

$$v(\neg A) = T \text{ iff } s(A) = \varnothing \text{ and } v(A) = F$$

$$v(A \rightarrow B) = T \text{ iff } s(A) \subseteq s(B) \text{ and (not both } v(A) = T \text{ and } v(B) = F)$$

and s satisfies

1. $s(\neg A) = \begin{cases} \overline{s(A)} & \text{if } s(A) = \varnothing \text{ or } s(A) = S \\ s(A) & \text{otherwise} \end{cases}$

2. $s(A \rightarrow B) = \begin{cases} S & \text{if } s(A) \subseteq s(B) \\ s(B) & \text{if } s(B) \subset s(A) \text{ and } s(A) = S \\ s(A) & \text{if } s(B) \subset s(A) \text{ and } s(A) \neq S \end{cases}$

3. $v(p) = T$ iff $s(p) = S$

4. If both $\varnothing \subset s(A) \subset S$ and $\varnothing \subset s(B) \subset S$, then $s(A) = s(B)$.

Note that condition 3 allows for an inductive definition of truth from the assignment of contents and truth-values to the atomic propositions. Condition 4 ensures that there are at most 3 possibilities for content sets: $\varnothing$, S, and some U such that $\varnothing \subset U \subset S$.

The proof of the following lemma is straightforward, though lengthy, and I will leave it to you.

Lemma 12

a. $\mathsf{v}(A) = \mathsf{T}$ iff $\mathsf{s}(A) = \mathsf{S}$

b. $\mathsf{v}(A \vee B) = \mathsf{T}$ iff $\mathsf{v}(A) = \mathsf{T}$ or $\mathsf{v}(B) = \mathsf{T}$

c. $\mathsf{s}(A \vee B) = \mathsf{s}(A) \cup \mathsf{s}(B)$

d. $\mathsf{v}(A \wedge B) = \mathsf{T}$ iff $\mathsf{v}(A) = \mathsf{T}$ and $\mathsf{v}(B) = \mathsf{T}$

e. $\mathsf{s}(A \wedge B) = \mathsf{s}(A) \cap \mathsf{s}(B)$

f. $\mathsf{s}(A \rightarrow B) = \mathsf{s}(\neg A) \cup \mathsf{s}(B)$

g. Given an $\mathbf{L_3}$-evaluation e: Wffs $\rightarrow \{0, \frac{1}{2}, 1\}$, define

$$\mathsf{v}(p) = \mathsf{T} \text{ iff } \mathsf{e}(p) = 1$$
$$\mathsf{s}(A) = \{x : x \in [0,1] \text{ and } x < \mathsf{e}(A)\}$$

where $[0,1]$ is the collection of real numbers between 0 and 1. Extend v to all wffs by the intuitionist tables for $\neg$ and $\rightarrow$. Then $<\mathsf{v}, \mathsf{s}>$ is an $\mathbf{L_3}$-model and $\mathsf{v}(A) = \mathsf{T}$ iff $\mathsf{e}(A) = 1$.

h. Given an $\mathbf{L_3}$-model $<\mathsf{v}, \mathsf{s}>$, define e: Wffs $\rightarrow \{0, \frac{1}{2}, 1\}$ via

$$\mathsf{e}(A) = \begin{cases} 1 & \text{if } \mathsf{s}(A) = \mathsf{S} \\ \frac{1}{2} & \text{if } \varnothing \subset \mathsf{s}(A) \subset \mathsf{S} \\ 0 & \text{if } \mathsf{s}(A) = \varnothing \end{cases}$$

Then e is an $\mathbf{L_3}$-evaluation and $\mathsf{e}(A) = 1$ iff $\mathsf{v}(A) = \mathsf{T}$.

Note that $\mathbf{L_3}$-models evaluate all four connectives $\{\neg, \rightarrow, \wedge, \vee\}$ by the intuitionist truth-conditions.

From parts (g) and (h) we have that the consequence relation for the set-assignment semantics is the same as for the $\mathbf{L_3}$ matrix, which by Theorem 4 coincides with the syntactic consequence relation.

Theorem 13 (Strong Completeness of the Set-Assignment Semantics)

$\Gamma \vDash_{\mathbf{L_3}} A$ iff every set-assignment $\mathbf{L_3}$-model which validates Γ also validates A iff $\Gamma \vdash_{\mathbf{L_3}} A$

We can give alternate strongly complete semantics for $\mathbf{L_3}$ which are not limited to using only three content sets in each model.

We define $<\mathsf{v}, \mathsf{s}>$ to be a *rich* $\mathbf{L_3}$*-model* if:

$\mathsf{v}(\neg A) = \mathsf{T}$ iff $\mathsf{s}(A) = \varnothing$ and $\mathsf{v}(A) = \mathsf{F}$ [as before]

$\mathsf{v}(A \rightarrow B) = \mathsf{T}$ iff $(\mathsf{s}(A) \subseteq \mathsf{s}(B)$ or both $\varnothing \subset \mathsf{s}(A) \subset \mathsf{S}$ and $\varnothing \subset \mathsf{s}(B) \subset \mathsf{S})$ and (not both $\mathsf{v}(A) = \mathsf{T}$ and $\mathsf{v}(B) = \mathsf{F}$)

and $\mathfrak{s}$ satisfies conditions (1) and (3) as before, as well as

$$2.\ \mathfrak{s}(A \rightarrow B) = \begin{cases} \mathsf{S} & \text{if } \mathfrak{s}(A) \subseteq \mathfrak{s}(B) \text{ or (both } \varnothing \subset \mathfrak{s}(A) \subset \mathsf{S} \text{ and } \varnothing \subset \mathfrak{s}(B) \subset \mathsf{S}) \\ \mathfrak{s}(B) & \text{if } \mathfrak{s}(B) \subset \mathfrak{s}(A) \text{ and } \mathfrak{s}(A) = \mathsf{S} \\ \mathfrak{s}(A) & \text{if } \mathfrak{s}(B) \subset \mathfrak{s}(A) \text{ and } \mathfrak{s}(A) \neq \mathsf{S} \end{cases}$$

Lemma 12.a, g, h can be proved for rich models, from which follows the strong completeness of these semantics for $\mathbf{L_3}$. Note that every $\mathbf{L_3}$-model is a rich $\mathbf{L_3}$-model.

In these semantics all content sets other than $\varnothing$ and S play the same role in determining the truth-value of complex wffs. Nonetheless the intermediate content sets need not all be the same. Dummett says of many-valued logics generally:

> On one intuitive interpretation of 'true', 'is true' can then be taken to mean 'has a designated value' and 'is false' to mean 'has an undesignated value'. The different individual designated values are then to be taken not as degrees of truth, but, rather, as corresponding to different ways in which a sentence might be true. We cannot determine the truth or falsity of a complex sentence just from the truth or falsity of its constituents; to do this we must know the particular ways in which they are true or false.
>
> *Dummett, 1977*, p.166

Consider then a proposition A such that $\varnothing \subset \mathfrak{s}(A) \subset \mathsf{S}$. It has the same content as its negation. So on Dummett's reading of content the ways in which A could be true are the same as those in which $\neg A$ could be true. This would be appropriate for paradoxical sentences not all of which need have the same content. Different circumstances might distinguish different paradoxical sentences though we may choose to ignore the distinctions in calculating truth-values.

2. The logics $\mathbf{L_n}$ and $\mathbf{L_{\aleph}}$

a. Generalizing the 3-valued tables

Łukasiewicz, in *Łukasiewicz and Tarski, 1930*, generalized the 3-valued matrix for $\mathbf{L_3}$ by allowing evaluations to take any value in $[0,1]$, the set of real numbers between 0 and 1.

An $\mathbf{L}$-*evaluation* is a map $\mathsf{e}: PV \rightarrow [0,1]$ which is extended to all wffs of $L(p_0, p_1, \ldots \neg, \rightarrow)$ by the following tables:

$$\mathsf{e}(\neg A) = 1 - \mathsf{e}(A)$$

$$\mathsf{e}(A \rightarrow B) = \begin{cases} 1 & \text{if } \mathsf{e}(A) \leq \mathsf{e}(B) \\ (1 - \mathsf{e}(A)) + \mathsf{e}(B) & \text{if } \mathsf{e}(B) < \mathsf{e}(A) \end{cases}$$

As before we define:

$A \vee B \equiv_{Def} (A \rightarrow B) \rightarrow B$

$A \wedge B \equiv_{Def} \neg(\neg A \wedge \neg B)$

$A \leftrightarrow B \equiv_{Def} (A \rightarrow B) \wedge (B \rightarrow A)$

These have the following tables:

$\mathbf{e}(A \vee B) = \max(\mathbf{e}(A), \mathbf{e}(B))$

$\mathbf{e}(A \wedge B) = \min(\mathbf{e}(A), \mathbf{e}(B))$

$$\mathbf{e}(A \leftrightarrow B) = \begin{cases} 1 & \text{if } \mathbf{e}(A) = \mathbf{e}(B) \\ (1 - \mathbf{e}(A)) + \mathbf{e}(B) & \text{if } \mathbf{e}(A) > \mathbf{e}(B) \\ (1 - \mathbf{e}(B)) + \mathbf{e}(A) & \text{if } \mathbf{e}(B) > \mathbf{e}(A) \end{cases}$$

Note that the function for disjunction is associative so that the way that disjuncts are associated does not affect the evaluation of an alternation.

We define for $n \geq 2$,

$\mathbf{L_n} = \{ A : \mathbf{e}(A) = 1$ for every **L**-evaluation $\mathbf{e} : PV \rightarrow \{ \frac{m}{n-1} : 0 \leq m \leq n-1 \} \}$

$\mathbf{L}_{\aleph_0} = \{ A : \mathbf{e}(A) = 1$ for every **L**-evaluation **e** which takes only rational values in $[0,1] \}$

$\mathbf{L}_{\aleph} = \{ A : \mathbf{e}(A) = 1$ for every **L**-evaluation $\mathbf{e} \}$

Note that when we restrict the values that **e** may take on PV in these definitions then the extension of **e** to all wffs obeys the same restriction.

Theorem 14 *a.* $\mathbf{L_2} = \mathbf{PC}$

b. $\mathbf{L_n} \supseteq \mathbf{L_{n+1}} \supseteq \mathbf{L}_{\aleph_0} \supseteq \mathbf{L}_{\aleph}$

c. $\mathbf{L_n} \neq \mathbf{L_{n+1}}$

d. $\mathbf{L_n} \neq \mathbf{L}_{\aleph_0}$

e. $\mathbf{L}_{\aleph_0} = \mathbf{L}_{\aleph}$

Proof: (a) is immediate.

For (b) note that every $\mathbf{L_n}$-evaluation is an $\mathbf{L_{n+1}}$- evaluation. Therefore, if A is $\mathbf{L_{n+1}}$-valid then it is $\mathbf{L_n}$-valid, too.

For (c) and (d) define the sequence of wffs

$$D_n = \bigvee_{1 \leq i \neq k < n+1} (p_i \leftrightarrow p_k)$$

where this means the disjunction of the indexed wffs, associating to the right. For example,

$$D_2 = (p_1 \leftrightarrow p_2) \vee ((p_1 \leftrightarrow p_3) \vee (p_2 \leftrightarrow p_3))$$

Then for $m \geq n$, $D_m \in \mathbf{L_n}$; for $m < n$, $D_m \notin \mathbf{L_n}$; and for all n, $D_n \notin \mathbf{L}_{\aleph_0}$.

Part (e) follows from Theorem 15 below. ■

In developing these systems Łukasiewicz said,

> It was clear to me from the outset that among all the many-valued systems only two can claim any philosophical significance: the three-valued and the infinite-valued ones. For if values other than '0' and '1' are interpreted as 'the possible', only two cases can reasonably be distinguished: either one assumes that there are no variations in degree of the possible and consequently one arrives at the three-valued system; or one assumes the opposite, in which case it would be most natural to suppose (as in the theory of probabilities) that there are infinitely many degrees of possibility, which leads to the infinite-valued propositional calculus. I believe the latter system is preferable to all others.
>
> *Łukasiewicz, 1930,* p. 173

In *1953* Łukasiewicz changed his mind and argued that $\mathbf{L}_4$ could be interpreted as a reconstruction of Aristotle's modal notions. But I will look only at the infinite-valued logic here. More information about these logics can be found in *Wójcicki, 1988,* Chapter 4.3.

b. An axiom system for $\mathbf{L}_{\aleph}$

$\mathbf{L}_{\aleph}$

in L($\neg$, $\rightarrow$)

axiom schemas

$$\left.\begin{array}{ll} \mathrm{L}_3 1. & A \rightarrow (B \rightarrow A) \\ \mathrm{L}_3 2. & (A \rightarrow B) \rightarrow ((B \rightarrow C) \rightarrow (A \rightarrow C)) \\ \mathrm{L}_3 3. & (\neg A \rightarrow \neg B) \rightarrow (B \rightarrow A) \end{array}\right\} \text{ as for } \mathbf{L}_3$$

$$\mathrm{L}_{\aleph}.\quad (A \rightarrow B) \vee (B \rightarrow A)$$

rule $\dfrac{A,\ A \rightarrow B}{B}$

I denote the consequence relation of this system as $\vdash_{\mathbf{L}_{\aleph}}$.

Theorem 15 $\vdash_{\mathbf{L}_{\aleph}} A$ iff $\models_{\mathbf{L}_{\aleph_0}} A$
iff $\models_{\mathbf{L}_{\aleph}} A$

Turquette, *1959*, gives the history and references for the proof of the first part of this theorem. The second equivalence follows because every axiom and hence every consequence of the system is an $\mathbf{L}_{\aleph}$-tautology, so $\mathbf{L}_{\aleph_0} \subseteq \mathbf{L}_{\aleph}$; and by Lemma 14.b, $\mathbf{L}_{\aleph} \subseteq \mathbf{L}_{\aleph_0}$.

c. Set-assignment semantics for $\mathbf{L}_{\aleph}$

It might seem obvious that set-assignment semantics for $\mathbf{L}_{\aleph}$ could be obtained by modifying those for $\mathbf{L}_3$ to require that the content sets be linearly ordered under inclusion. But those models do not validate $L_3 3$. It would seem that we need to explicitly postulate some measure or topological structure on the collection of content sets to mock the operations of $+$ and $-$ on $[0,1]$.

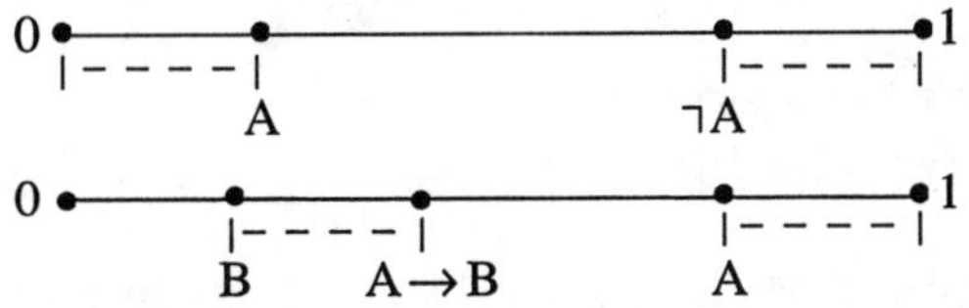

But the solution is simpler: just postulate enough structure on the collection $\{s(A) : \text{A is a wff}\}$ to validate $\mathbf{L}_{\aleph}$.

We say that $<v, s>$ is an $\mathbf{L}_{\aleph}$-*model* for $L(\neg, \rightarrow)$ if $\neg$ and $\rightarrow$ are evaluated by the intuitionist truth-conditions (as for $\mathbf{L}_3$):

$v(\neg A) = T$ iff $s(A) = \varnothing$ and $v(A) = F$

$v(A \rightarrow B) = T$ iff $s(A) \subseteq s(B)$ and (not both $v(A) = T$ and $v(B) = F$)

and s satisfies

L1. $s(A \rightarrow B) = S$ iff $s(A) \subseteq s(B)$

L2. $s(\neg A) = S$ iff $s(A) = \varnothing$

L3. $s(B) \subseteq s(A \rightarrow B)$

L4. $s(A \rightarrow B) \subseteq s((B \rightarrow C) \rightarrow (A \rightarrow C))$

L5. If $s(A) \subseteq s(B)$ then $s(B \rightarrow C) \subseteq s(A \rightarrow C)$.

L6. $s(\neg A \rightarrow \neg B) \subseteq s(B \rightarrow A)$

L7. $s(\neg B) \subseteq s(\neg A)$ iff $s(A) \subseteq s(B)$

L8. $s(A) \subseteq s(B)$ or $s(B) \subseteq s(A)$

L9. $v(p) = T$ iff $s(p) = S$

and for the defined connectives

L10. $s(A \vee B) = s(A) \cup s(B)$

L11. $s(A \wedge B) = s(A) \cap s(B)$

I suspect that it is possible to reduce the list of conditions on $\mathbf{L}_{\aleph}$-models either by deriving some from the others or by using different truth-conditions.

The following lemma is straightforward to prove.

Lemma 16 For every $\mathbf{L}_{\aleph}$-model $<v, s>$

a. $v(A) = T$ iff $s(A) = S$

b. $v(A \vee B) = T$ iff $v(A) = T$ or $v(B) = T$

c. $v(A \wedge B) = T$ iff $v(A) = T$ and $v(B) = T$

Thus $\mathbf{L}_{\aleph}$-models use the intuitionist truth-tables for all four connectives.

The algebra of sets $\{\mathsf{s}(A): A \text{ is a wff}\}$ inherits enough structure to prove that the semantics consequence relation of these set-assignment models is the same as the syntactic and semantic consequence relation of $\mathbf{L}_{\aleph}$.

Lemma 17 $\quad \vdash_{\mathbf{L}_{\aleph}} A$ iff for every $\mathbf{L}_{\aleph}$-model $<\upsilon,\mathsf{s}>$, $\upsilon(A) = \mathsf{T}$
iff for every $\mathbf{L}_{\aleph}$-model $<\upsilon,\mathsf{s}>$, $\mathsf{s}(A) = \mathsf{S}$

Proof: You can check that the semantics are sound.

If $\nvdash_{\mathbf{L}_{\aleph}} A$ then there is some **L**-evaluation e such that $\mathsf{e}(A) \neq 1$. Define $<\upsilon,\mathsf{s}>$ via $\mathsf{S} = (0,1]$, $\mathsf{s}(A) = (0,\mathsf{e}(A)]$, and $\upsilon(p) = \mathsf{T}$ iff $\mathsf{e}(p) = 1$, where $(0,0]$ is understood to mean $\varnothing$. Then $<\upsilon,\mathsf{s}>$ is an $\mathbf{L}_{\aleph}$-model, and by Lemma 16.a, $\upsilon(A) = \mathsf{F}$. ■

D. Kleene's 3-Valued Logic

1. The truth-tables

Kleene, *1952,* pp. 332–340, proposed a 3-valued logic as a way to reason with arithmetical propositions whose truth-value we either do not or cannot know.

The basic idea is that if we are given arithmetical predicates Q, R then predicates compounded from these, such as $Q \vee R$, should be computable (partial recursive) in Q and R. So, for example, if Q(n) and R(n) are both defined then $Q(n) \vee R(n)$ is defined and has a determinate truth-value. But what if one or both of Q(n), R(n) is not defined?

> From this standpoint the meaning of $Q \vee R$ is brought out clearly by the statement in words: $Q \vee R$ is true, if Q is true (here nothing is said about R) or if R is true (similarly); false if Q and R are both false; defined only in these cases (and hence undefined otherwise).
>
> *Kleene, 1952,* p. 336

Thus three "values" are postulated for propositions: T for 'true', F for 'false', and U for 'undefined' or 'unknown'. U is not considered to be a third truth-value. Rather in accord with Kleene's platonist view of propositions, it simply marks our ignorance of the actual truth-value of the proposition.

> Here 'unknown' is a category into which we can regard any proposition as falling whose value we either do not know or choose for the moment to disregard; and it does not then exclude the other two possibilities 'true' and 'false'. ...
>
> The strong 3-valued logic can be applied to completely defined predicates

Q(x) and R(x) from which composite predicates are formed using ¬, ∨, ∧, →, ↔ in the usual 2-valued meanings, thus. Suppose that there are fixed algorithms which decide the truth or falsity of Q(x) and of R(x), each on a subset of the natural numbers (as occurs, e.g., after completing the definition of any two partial recursive predicates classically). Let T, F, U mean 'decidable by the algorithms (i.e., by use of only such information about Q(x) and R(x) as can be obtained by the algorithms) to be true', 'decidable by the algorithms to be false', 'undecidable by the algorithms whether true or false'. [Or] assume a fixed state of knowledge about Q(x) and R(x) (as occurs, e.g., after pursuing algorithms for each of them up to a given stage). Let T, F, U mean 'known to be true', 'known to be false', 'unknown whether true or false'.

Kleene, 1952, pp. 335–336

Formally, we begin by taking as our language $L(p_0, p_1, \ldots \neg, \rightarrow, \wedge, \vee)$, defining $A \leftrightarrow B \equiv_{Def} (A \rightarrow B) \wedge (B \rightarrow A)$. Kleene gives the following tables for what he calls the *strong connectives*.

A ∧ B	B: T	U	F
A T	T	U	F
U	U	U	F
F	F	F	F

A ∨ B	B: T	U	F
A T	T	T	T
U	T	U	U
F	T	U	F

A → B	B: T	U	F
A T	T	U	F
U	T	U	U
F	T	T	T

A	¬A
T	F
U	U
F	T

A $\mathbf{K_3}$-*evaluation* is a map $e: PV \rightarrow \{T, F, U\}$ which is extended to all wffs by these tables. The *sole designated value* is T.

If we read 1 for T, 0 for F, and $\frac{1}{2}$ for U, then these tables agree with those for Łukasiewicz's 3-valued logic $\mathbf{L_3}$ with one exception: if $e(A) = e(B) = U$ then Łukasiewicz assigns $e(A \rightarrow B) = T$, whereas Kleene assigns value U. So with Kleene's tables we assign U to any complex proposition built from propositions all of which take value U. Kripke, *1975*, finds this an apt way to deal with paradoxical sentences such as 'What I am now saying is false', and uses $\mathbf{K_3}$-evaluations to give a theory of truth for a first-order language which contains its own truth predicate; see *Epstein, 1985* or *198?*, for a description and critique of Kripke's analysis. But now there are no tautologies.

Lemma 18 *a.* There is no wff A which takes value T for all $\mathbf{K_3}$-evaluations.

b. If both T and U are taken as designated values for these tables then the set of tautologies is **PC**. That is, $\models_{\mathbf{PC}} A$ iff for every $\mathbf{K_3}$-evaluation e, $e(A) = T$ or $e(A) = U$.

c. The consequence relation using both T and U as designated values does not coincide with $\models_{\mathbf{PC}}$.

Proof: a. If for all p, $e(p) = U$, then for all A, $e(A) = U$.

b. I'll show that $\not\models_{\mathbf{PC}} A$ iff there is some $\mathbf{K_3}$-evaluation e such that $e(A) = F$. Every 2-valued **PC**-model is a $\mathbf{K_3}$-evaluation, so if $\not\models_{\mathbf{PC}} A$ then there is a $\mathbf{K_3}$-evaluation e such that $e(A) = F$. If there is some $\mathbf{K_3}$-evaluation e such that $e(A) = F$ then define a **PC**-model v by $v(p) = T$ iff $e(A) = T$ or $e(A) = U$. It's then easy to prove by induction on the length of a wff that if $e(B) = T$ then $v(B) = T$, and if $e(B) = F$ then $v(B) = F$. So $v(A) = F$ and $\not\models_{\mathbf{PC}} A$.

c. If both T and U are designated then $\{A, A \to B\} \not\models B$, for we may have $e(A) = U$ and $e(B) = F$. ■

Since there are no tautologies for this logic we must understand the matrix as a semantic presentation of a logic in terms of a consequence relation only:

$$\Gamma \models_{\mathbf{K_3}} A \text{ iff for every } \mathbf{K_3}\text{-evaluation } e, \text{ if } e \models \Gamma \text{ then } e(A) = T$$

This relation is not empty: for instance, $A \to B \models_{\mathbf{K_3}} \neg B \to \neg A$, and $\neg(A \wedge \neg A) \models_{\mathbf{K_3}} A \vee \neg A$.

The absence of tautologies makes the usual Hilbert-style proof theory inappropriate. Cleave, *1974*, has given a natural deduction style definition of syntactic consequence which coincides with $\models_{\mathbf{K_3}}$.

2. Set-assignment semantics

The goal here is to give set-assignment semantics whose consequence relation coincides with $\models_{\mathbf{K_3}}$.

A pair $<v, s>$ is a $\mathbf{K_3}$-*model* for $L(p_0, p_1, \ldots \neg, \to, \wedge, \vee)$ if the truth-tables for $\wedge$ and $\vee$ are classical, and

$$v(A \to B) = T \text{ iff } v(\neg A) = T \text{ or } v(B) = T$$
$$v(\neg A) = T \text{ iff } v(A) = F \text{ and } s(A) = \varnothing$$

and s satisfies

K1. $s(A) \subseteq s(B)$ or $s(B) \subseteq s(A)$

K2. $s(A \to B) = s(\neg A) \cup s(B)$

K3. $s(\neg A) = \begin{cases} s(A) & \text{if } \varnothing \subset s(A) \subset S \\ \overline{s(A)} & \text{otherwise} \end{cases}$

K4. $s(A \wedge B) = s(A) \cap s(B)$

K5. $s(A \vee B) = s(A) \cup s(B)$

K6. $v(p) = T$ iff $s(p) = S$

The entire difference between these semantics and the classical ones lies in the table and set-assignments for negation.

Note that we require only that the content sets are linearly ordered under inclusion instead of restricting them to $\varnothing \subset U \subset S$ for some U as we originally did for $\mathbf{L_3}$.

Theorem 19

a. In every $\mathbf{K_3}$-model $<v,s>$, $v(A) = T$ iff $s(A) = S$

b. Given a $\mathbf{K_3}$-evaluation e, if we define $<v,s>$ by

$v(p) = T$ iff $e(p) = T$

$S = \{1,2\}$

$$s(A) = \begin{cases} S & \text{if } e(A) = T \\ \{1\} & \text{if } e(A) = U \\ \varnothing & \text{if } e(A) = F \end{cases}$$

then $<v,s>$ is a $\mathbf{K_3}$-model and $v(A) = T$ iff $e(A) = T$.

c. Given a $\mathbf{K_3}$-model $<v,s>$, if we set

$$e(A) = \begin{cases} T & \text{if } s(A) = S \\ U & \text{if } \varnothing \subset s(A) \subset S \\ F & \text{if } s(A) = \varnothing \end{cases}$$

then e is a $\mathbf{K_3}$-evaluation, and $e(A) = T$ iff $v(A) = T$.

d. $\Gamma \vDash_{\mathbf{K_3}} A$ iff for every $\mathbf{K_3}$-model $<v,s>$,
if for every $B \in \Gamma$, $v(B) = T$, then $v(A) = T$

The proofs are straightforward though long, and I will leave them to you.

Our $\mathbf{K_3}$-models reflect that all that's important about a wff is whether it takes an extreme value (T, F; or $s(A) = \varnothing, S$). We don't even need to require that the content sets be linearly ordered by inclusion; we can still prove Theorem 19 if we replace K1 and K2 by:

K7. $s(A \to B) = \begin{cases} S & \text{if } s(\neg A) = S \text{ or } s(B) = S \\ \varnothing & \text{if } s(A) = S \text{ and } s(B) = \varnothing \\ s(B) & \text{otherwise} \end{cases}$

E. Logics Having No Finite-Valued Semantics

1. General criteria

The following theorem about intuitionistic logics is due to Gödel, *1932*.

Theorem 20 There is no finite-valued matrix which characterizes either **Int** or **J**.

Proof: As in Lemma 14, for n ≥ 2 define

$$D_n = \bigvee_{1 \leq i \neq k < n+1} (p_i \leftrightarrow p_k)$$

where this means the disjunction of the indexed wffs, associating to the right. For example,

$$D_2 = (p_1 \leftrightarrow p_2) \vee ((p_1 \leftrightarrow p_3) \vee (p_2 \leftrightarrow p_3))$$

For each n, $\mathbf{Int} \nvdash D_n$, as you can check using the Kripke semantics for **Int** from §VII.B.2. Now assume to the contrary that M is an n-valued matrix characteristic for **Int**. I will show that $\mathsf{M} \models D_n$, which is a contradiction.

For any evaluation e with respect to M, there must be some i, k with $1 \leq i < k \leq n+1$, such that $e(p_i) = e(p_k)$. Choose such an i,k and then $e \models D_n$ iff $e \models D_n(p_i/p_k)$, where the latter formula is D_n with p_k substituted for p_i throughout. But then one of the disjunctions in the alternation $D_n(p_i/p_k)$ is $(p_k \leftrightarrow p_k)$, so that $\mathbf{Int} \vdash D_n(p_i/p_k)$. As M is characteristic for **Int** we must have that $e \models D_n(p_i/p_k)$, hence $e \models D_n$. Since e was an arbitrary evaluation, $\mathsf{M} \models D_n$.

The same proof applies to **J**. ■

In contrast to Theorem 20, *Dummett, 1959,* gives a countable infinite-valued matrix which characterizes **Int**.

The method of proof of Theorem 20 can easily be generalized to prove the following, as suggested by Walter Carnielli.

Theorem 21 Let **L** be a logic which is presented semantically. If there is a sequence of formulas E_n for n ≥ 2 in the language of **L** such that:

a. The propositional variables appearing in E_n are exactly $p_1, \ldots p_n, p_{n+1}$.

b. For all n, $\mathbf{L} \nvDash E_n$.

c. For all i, k ≤ n+1, if i ≠ k then $\mathbf{L} \models E_n(p_i/p_k)$.

then there is no finite-valued matrix which is characteristic for **L**.

Theorem 21 is also applicable to logics which are presented syntactically: simply replace ⊨ by ⊢ everywhere.

Corollary 22 None of the following logics has finite-valued semantics:

R, **S**, **D**, **Dual D**, **Eq** (Chapters III and V),

the modal logics **K**, **T**, **B**, **QT**, **S4**, **S5**, **S4Grz**, **MSI**, **G**, **G*** (Chapter VI)

Proof: Take E_n in Theorem 21 to be D_n from the proof of Theorem 20, using $\vee$ defined from $\neg$ and $\wedge$. ■

In essence, then, these logics recognize infinitely many possibilities for propositions.

2. Infinite-valued semantics for the modal logic S5

Though none of the classical modal logics of Chapter VI can be characterized by a finite-valued matrix, there is a simple and intuitive characterization of **S5** as an infinite-valued logic. My presentation of it will follow *Prior, 1967*.

Let's assume that we're given a Kripke model of **S5** which contains a countable number of possible worlds. Each proposition can be identified with the worlds in which it is true. So we can represent a proposition as an infinite sequence of 0's and 1's:

$$\mathbf{e}(A) = (a_1, \dots, a_n, \dots)$$

where for each i, a_i is 1 if A is true in the ith world, or 0 if A is false in the ith world. The evaluation of the connectives is:

$$\begin{aligned}\mathbf{e}(\neg A) &= (1, 1, 1, \dots, 1, \dots) - \mathbf{e}(A) \\ &= \text{the result of interchanging 0's and 1's in } \mathbf{e}(A)\end{aligned}$$

$$\begin{aligned}\mathbf{e}(A \wedge B) &= \mathbf{e}(A) \times \mathbf{e}(B) \quad \text{(the coordinatewise product)} \\ &= \text{the result of taking } \min(\mathbf{e}(A), \mathbf{e}(B)) \text{ at each coordinate}\end{aligned}$$

Recall that the necessity of **S5** models is logical necessity: every possible world is conceivable in every other. So the evaluation of the necessity operator is:

$$\mathbf{e}(\Box A) = \begin{cases} \text{all 1's} & \text{if } \mathbf{e}(A) \text{ is all 1's} \\ \text{all 0's} & \text{otherwise} \end{cases}$$

Recalling the definition $A \to B \equiv_{Def} \Box(\neg(A \wedge \neg B))$, the evaluation of conditionals is:

$$\mathbf{e}(A \to B) = \begin{cases} \text{all 1's if there is no index for which the entry in } \mathbf{e}(A) \\ \quad \text{is 1 and in } \mathbf{e}(B) \text{ is 0} \\ \text{all 0's otherwise} \end{cases}$$

The only designated value is the sequence consisting entirely of 1's.

We can use Theorem VI.49 to establish that this class of evaluations characterizes **S5** and that the semantic consequence relation of this matrix coincides with the consequence relation without necessitation for **S5** (Theorem VI.51). We only need to note that for any A, any finite Kripke model in which A is not valid can be extended to an infinite one in which A is not valid, which I'll leave to you.

By Theorem VI.49, **S5** is also characterized by the class of finite equivalence Kripke models, so we could use only arbitrarily long finite sequences instead of infinite ones.

F. The Systems $\mathbf{G_n}$ and $\mathbf{G_\aleph}$

In the proof of Theorem 20 we used Kripke semantics to show that $\mathbf{Int} \nvdash D_n$. In his paper Gödel, *1932*, devised a sequence of finite-valued matrices each of which validated **Int** but in which the D_n's successively failed. I'll present those here.

A **G**-*evaluation* is a map $e: PV \to [0,1]$ which is extended to all wffs of $L(p_0, p_1, \ldots \neg, \to, \wedge, \vee)$ by the following tables:

$$e(\neg A) = \begin{cases} 1 & \text{if } e(A) = 0 \\ 0 & \text{if } e(A) \neq 0 \end{cases}$$

$$e(A \to B) = \begin{cases} 1 & \text{if } e(A) \leq e(B) \\ e(B) & \text{otherwise} \end{cases}$$

$$e(A \wedge B) = \min(e(A), e(B))$$

$$e(A \vee B) = \max(e(A), e(B))$$

Defining $A \leftrightarrow B \equiv_{\text{Def}} (A \to B) \wedge (B \to A)$, we have:

$$e(A \leftrightarrow B) = \begin{cases} 1 & \text{if } e(A) = e(B) \\ \min(e(A), e(B)) & \text{otherwise} \end{cases}$$

For $n \geq 2$ define

$\mathbf{G_n} = \{ A : e(A) = 1$ for every **G**-evaluation $e: PV \to \{ \frac{m}{n-1} : 0 \leq m \leq n-1 \} \}$
$\mathbf{G_{\aleph_0}} = \{ A : e(A) = 1$ for every **G**-evaluation e which takes only rational values in $[0,1] \}$
$\mathbf{G_\aleph} = \{ A : e(A) = 1$ for every **G**-evaluation $e \}$

The tables for $\mathbf{G_3}$ are:

A	¬A
1	0
$\frac{1}{2}$	0
0	1

A → B		B = 1	B = $\frac{1}{2}$	B = 0
A	1	1	$\frac{1}{2}$	0
	$\frac{1}{2}$	1	1	0
	0	1	1	1

and (as for $\mathbf{L_3}$)

$A \wedge B$		B 1	$\frac{1}{2}$	0
A	1	1	$\frac{1}{2}$	0
	$\frac{1}{2}$	$\frac{1}{2}$	$\frac{1}{2}$	0
	0	0	0	0

$A \vee B$		B 1	$\frac{1}{2}$	0
A	1	1	1	1
	$\frac{1}{2}$	1	$\frac{1}{2}$	$\frac{1}{2}$
	0	1	$\frac{1}{2}$	0

These were originally devised by Heyting, *1930*, p.56, to show that $\mathbf{Int} \nvdash \neg\neg A \rightarrow A$.

The proof of the following is just as for Lemma 14. In particular, part (e) follows from Theorem 25 below.

Theorem 23 *a.* $\mathbf{G_2} = \mathbf{PC}$

b. $\mathbf{G_n} \supseteq \mathbf{G_{n+1}} \supseteq \mathbf{G_{\aleph_0}} \supseteq \mathbf{G_{\aleph}}$

c. $\mathbf{G_n} \neq \mathbf{G_{n+1}}$

d. $\mathbf{G_n} \neq \mathbf{G_{\aleph_0}}$

e. $\mathbf{G_{\aleph_0}} = \mathbf{G_{\aleph}}$

Theorem 24 $\mathbf{G_n} \supset \mathbf{G_{\aleph}} \supset \mathbf{Int} \supset \mathbf{J}$

Proof: From Lemma 23 we have that $\mathbf{G_n} \supset \mathbf{G_{\aleph}}$. We have that $\mathbf{Int} \supset \mathbf{J}$ because the axioms of the latter are among the former, whereas $\mathbf{J} \nvdash \neg A \rightarrow (A \rightarrow B)$ which is an axiom of **Int** (see §VII.B.1 and §VII.E.1,2). That $\mathbf{G_{\aleph}} \supseteq \mathbf{Int}$ follows by checking that every axiom of **Int** receives value 1 under every **G**-evaluation, and if $\mathbf{e}(A) = 1$ and $\mathbf{e}(A \rightarrow B) = 1$ then $\mathbf{e}(B) = 1$, and if $\mathbf{e}(A) = 1$ and $\mathbf{e}(B) = 1$ then $\mathbf{e}(A \wedge B) = 1$.

Finally, $\mathbf{G_{\aleph}} \neq \mathbf{Int}$ because $[(A \rightarrow B) \vee (B \rightarrow A)] \in \mathbf{G_{\aleph}}$, but that wff is not an **Int**-tautology as you can check using the Kripke semantics for **Int**. ■

Dummett, *1959*, has shown how to characterize $\mathbf{G_{\aleph_0}}$ syntactically.

$\mathbf{G_{\aleph_0}}$
in $L(\neg, \rightarrow, \wedge, \vee)$
as for **Int** (§VII.B.1, p.199) with the addition of $(A \rightarrow B) \vee (B \rightarrow A)$

Theorem 25 $\vdash_{\mathbf{G_{\aleph_0}}} A$ iff for every G-evaluation **e** which takes only rational values, $\mathbf{e}(A) = 1$.

This axiom system is also complete for $\mathbf{G_{\aleph}}$, as every A which is a theorem of this system is a $\mathbf{G_{\aleph}}$-tautology and $\mathbf{G_{\aleph}} \subseteq \mathbf{G_{\aleph_0}}$.

There are two ways in which we can give set-assignment semantics for $\mathbf{G_{\aleph}}$. In the first we say that $\langle v, s \rangle$ is a $\mathbf{G_{\aleph}}$-*model* if v uses the intuitionist truth-conditions, that is $\wedge$ and $\vee$ are classical, and

$$v(A \to B) = \mathsf{T} \text{ iff } s(A) \subseteq s(B) \text{ and (not both } v(A) = \mathsf{T} \text{ and } v(B) = \mathsf{F})$$
$$v(\neg A) = \mathsf{T} \text{ iff } v(A) = \mathsf{F} \text{ and } s(A) = \varnothing$$

and s satisfies

$$\text{G1.} \quad s(\neg A) = \begin{cases} \mathsf{S} & \text{if } s(A) = \varnothing \\ \varnothing & \text{otherwise} \end{cases}$$

$$\text{G2.} \quad s(A \to B) = \begin{cases} \mathsf{S} & \text{if } s(A) \subseteq s(B) \\ s(A) \cap s(B) & \text{otherwise} \end{cases}$$

$$\text{G3.} \quad s(A \wedge B) = s(A) \cap s(B)$$

$$\text{G4.} \quad s(A \vee B) = s(A) \cup s(B)$$

$$\text{G5.} \quad v(p) = \mathsf{T} \text{ iff } s(p) = \mathsf{S}$$

$$\text{G6.} \quad s(A) \subseteq s(B) \text{ or } s(B) \subseteq s(A)$$

The proof that the consequence relation for these semantics is the same as for the matrix for $\mathbf{G}_{\aleph}$ follows as in Lemma 12.

Observe that if we replace G6 by: there is some U such that $s : \text{Wffs} \to \{\varnothing, \mathsf{U}, \mathsf{S}\}$, then the consequence relation of that subclass of models coincides with the consequence relation for the matrix of $\mathbf{G}_3$.

Alternatively, we can add G6 to the list of conditions for an **Int**-model $<v,s>$ (see § VII.D.1, p. 216). We can then establish that the resulting models are complete for $\mathbf{G}_{\aleph}$ by using Theorem 25, first proving a version of Lemma 12. In those set-assignment models we also have

$$s(\neg A) = \begin{cases} \mathsf{S} & \text{if } s(A) = \varnothing \\ \varnothing & \text{otherwise} \end{cases}$$

I do not know if these latter set-assignment semantics for $\mathbf{G}_{\aleph}$ yield the same class of models as the first.

G. A Method for Proving Axiom Systems Independent

Given an axiom system we say that an *axiom is independent* of the others if it is not a consequence of them. If the axiomatization is by schema we say that an *axiom schema is independent* of the others if every instance of it is independent of all instances of the other schemas. An axiom (schema) which is a consequence of the others in a system is superfluous, though of course it may serve to shorten proofs.

One way to show that an axiom (schema) A is independent is to exhibit a finite matrix M which validates all the axioms (schemas) except for A, and validates the conclusion of a rule if it validates the hypotheses. In that case all consequences of the other axioms (schemas) will be validated by M. So A cannot be a consequence of the others. By using a finite matrix validity is (easily) decidable.

As an example, George Hughes has shown how to prove that the axiomatization of **PC** that I gave in Chapter II.J is independent.

PC
in $L(\neg, \rightarrow)$

axiom schemas

1. $\neg A \rightarrow (A \rightarrow B)$

2. $B \rightarrow (A \rightarrow B)$

3. $(A \rightarrow B) \rightarrow ((\neg A \rightarrow B) \rightarrow B)$

4. $(A \rightarrow (B \rightarrow C)) \rightarrow ((A \rightarrow B) \rightarrow (A \rightarrow C))$

rule $\dfrac{A,\ A \rightarrow B}{B}$

For each of the matrices below the only designated value is 1. And for each, if $e(A) = 1$ and $e(A \rightarrow B) = 1$, then $e(B) = 1$, so that deductions preserve the designated value. In each case we will use the following table for $\neg$.

A	¬A
1	4
2	3
3	2
4	1

To show schema 1 is independent we use the following table for $\rightarrow$:

	A → B	B 1	2	3	4
A	1	1	2	3	4
	2	1	1	3	4
	3	1	2	1	4
	4	1	1	1	1

This table and the table for $\neg$ above validate axiom schemas (2), (3), and (4): the proof is tedious but mechanical. However, they invalidate axiom schema 1: $\neg 2 \rightarrow (2 \rightarrow 4) = 4$.

To show schema (2) is independent we use the following table for $\rightarrow$, which together with the table for $\neg$ above validates schemas (1), (3) and (4).

$A \to B$		B: 1	2	3	4
A	1	1	3	3	3
	2	1	3	3	3
	3	1	1	1	1
	4	1	1	1	1

Axiom schema (2), however, is invalid: $2 \to (1 \to 2) = 3$. This table also invalidates $A \to A$, which shows that we can't deduce the law of identity from schemas (1), (3), and (4) only.

To show that schema (3) is independent we use the following table for $\to$, which together with the table for $\neg$ above validates schemas (1), (2), and (4).

$A \to B$		B: 1	2	3	4
A	1	1	2	3	4
	2	1	1	1	1
	3	1	2	1	2
	4	1	1	1	1

But axiom schema (3) is invalid: $(3 \to 3) \to ((\neg 3 \to 3) \to 3) = 3$.

To show schema (4) is independent we use the following table for $\to$, which together with the table for $\neg$ above validates schemas (1), (2), and (3).

$A \to B$		B: 1	2	3	4
A	1	1	2	3	4
	2	1	1	3	1
	3	1	2	1	2
	4	1	1	1	1

But axiom schema (4) is invalid : $(2 \to (4 \to 3)) \to ((2 \to 4) \to (2 \to 3)) = 3$.

It's tempting to conjecture that for any finite axiom system in which one of the axioms (schemas) is independent there is a finite-valued matrix which demonstrates the independence of that axiom (schema). But that's not so, as Gödel, *1933 C*, first showed. I'll present a counterexample based on the logic **Int** of Chapter VII, but any of the logics of Corollary 22 would do equally well.

In $L(p_0, p_1, \dots \neg, \to, \wedge, \vee, *)$ take the finite axiomatization of **Int** and add to it the schema :

$$*A \leftrightarrow **A$$

The Deduction Theorem for **Int** still holds, so adding this new schema is equivalent to adding the rules:

$$\frac{*A}{**A} \quad \text{and} \quad \frac{**A}{*A}$$

The connective $*$ is not definable because if it were then it would be unique, yet there are at least two definable connectives (in the set-assignment semantics) which satisfy the schema: $v(*A) = \mathsf{F}$ for all A, or $v(*A) = \mathsf{T}$ for all A, where for both of these we take $s(*A) = s(A)$.

The schema is independent since we can't derive a rule involving only $*$ using schemas and rules that do not involve $*$. Yet there is no finite-valued matrix which demonstrates this, since there is none which validates **Int**.

McKinsey, *1939,* has used finite matrices in a similar way to prove that the four connectives $\{\neg, \rightarrow, \wedge, \vee\}$ are independent in **Int**. As an example I'll show that $\neg$ cannot be defined from $\{\rightarrow, \wedge, \vee\}$.

Suppose it could. Then there is some schema S(A) in which $\neg$ does not appear such that $\mathbf{Int} \vdash S(A) \leftrightarrow \neg A$. Since the $\mathbf{G_3}$-tables validate **Int** we would therefore have that for every **G**-evaluation taking values in $\{0, \frac{1}{2}, 1\}$, $\mathsf{e}(S(A)) = \mathsf{e}(\neg A)$. But if $\mathsf{e}(p) = 1$ for all p, then $\mathsf{e}(S(A)) = 1$; whereas $\mathsf{e}(\neg A) = 0$. Therefore, there is no such S(A).

IX A Paraconsistent Logic: $\mathbf{J_3}$

in collaboration with **Itala M. L. D'Ottaviano**

A paraconsistent logic is one in which a nontrivial theory may include both a proposition and its negation. I will first introduce the general notion of a paraconsistent logic, and then analyze in detail the 3-valued paraconsistent logic $\mathbf{J_3}$. I will axiomatize $\mathbf{J_3}$, and in doing so will suggest that paraconsistent logics are inconsistent only with respect to classical semantics, not with respect to their own formal or informal semantic notions. An analysis of set-assignment semantics for $\mathbf{J_3}$ will highlight the way in which the general framework for semantics of Chapter IV uses falsity as a default truth-value.

A. Paraconsistent Logics

How does one proceed when faced with apparently contradictory sentences both of which seem equally plausible?

'It is raining' 'It is not raining'

For $\mathbf{z} = \{x\colon x \notin x\}$, '$\mathbf{z} \in \mathbf{z}$' '$\mathbf{z} \notin \mathbf{z}$'

The classical logician cannot incorporate both into a theory because from a proposition and its negation in classical logic one can deduce any other proposition of the semi-formal language. There is only one classically inconsistent formal theory, and that is the *trivial* one consisting of all wffs.

The classical logician resolves the matter by building separate theories based on first one and then the other proposition, comparing the consequences of each. Or he may say that the difficulty in the first pair is that the word 'raining' is vague, and he will strive to reach agreement on what that word means, making it sufficiently precise that one of the sentences is definitively true, the other false. But for the latter example such options won't work and the only course left is to exclude them as being incoherent, or place restrictions on what formulas define sets.

There is another tradition in logic, however, which embraces contradictions as either accurately representing reality, or as fruitful to study for the syntheses they may generate. In the words of the poet Whitman,

> Do I contradict myself?
> Very well then ... I contradict myself;
> I am large ... I contain multitudes.

Jaśkowski, *1948*, proposed constructing logical systems which would allow for nontrivial theories containing (apparent) contradictions. The motives for doing so, he said, were: to systematize theories which contain contradictions, particularly as they occur in dialectics; to study theories in which there are contradictions caused by vagueness; and to study empirical theories whose postulates or basic assumptions could be considered contradictory. He proposed the following problem.

> The task is to find a system of the sentential calculus which: 1) when applied to the contradictory systems would not always entail their [triviality], 2) would be rich enough to enable practical inference, 3) would have an intuitive justification.
>
> *Jaśkowski, 1948,* p.145

Jaśkowski himself devised a propositional calculus to satisfy these criteria which he called 'discursive'.

Jaśkowski's work first appeared in Polish in *1948* and was translated into English only in 1969. Quite independently of him da Costa in *1963* had developed a sequence of logics called $\mathbf{C_n}$ which allow for nontrivial theories based on (apparent) contradictions. His motives were similar to Jaśkowski's and are described in *da Costa, 1974,* where he summarizes the systems, their extensions to first-order logic, and his investigations of their use in resolving paradoxes of set theory. Those logics are presented entirely syntactically: no explanations of the connectives are given, though we can assume that they are formal versions of 'not', 'if ... then ...', 'and', 'or'.

Due primarily to da Costa's influence much work has been done on these and other systems which allow for nontrivial theories which may contain (apparent) contradictions, dubbed *paraconsistent logics* by F. Miró Quesada. Arruda, *1980* and *198?*, surveys this work and the history of the subject, while da Costa and Marconi, *198?*, also discuss the philosophical motivation.

We have already seen a paraconsistent logic in this volume: Johansson's minimal intuitionistic logic **J** (§VII.E). In this chapter we will study the paraconsistent logic $\mathbf{J_3}$ which was first proposed by D'Ottaviano and da Costa, *1970*, as a solution to Jaśkowski's problem, and which was later developed by D'Ottaviano, *1985 A*, *1985 B*, and *1987*.

B. The Semantics of $\mathbf{J_3}$

1. D'Ottaviano on the semantics basis of $\mathbf{J_3}$

The semantic intuitions behind $\mathbf{J_3}$ are described by D'Ottaviano and da Costa as follows.

> In the preliminary phase of the formulation of a theory (mathematical, physical, etc.) contradictions can appear which, in the definitive formulation, are eliminated; 0, 1, $\frac{1}{2}$ are the truth-values, where 0 represents the "false", 1 the "truth" and $\frac{1}{2}$ the provisional value of a proposition A, such that A and $\neg$A are theses of the theory under consideration in its provisional formulation; in the definitive form of the theory, the value $\frac{1}{2}$ will be reduced, at least in principle, to 0 or 1. ...
>
> The calculus $\mathbf{J_3}$ can also be used as a foundation for inconsistent and nontrivial systems ... In this case, $\frac{1}{2}$ represents the logical value of a formula which is, really, true and false at the same time. ...
>
> In the elaboration of a logic suitable to handle "exact concepts" and "inexact concepts" ... $\mathbf{J_3}$ also constitutes a solution.
>
> *D'Ottaviano and da Costa, 1970,* p.1351

Thus we will consider a 3-valued logic whose truth-values will be 0, $\frac{1}{2}$, 1. However, unlike the other many-valued logics studied in this volume, two of the truth-values will be designated: 1 and $\frac{1}{2}$. D'Ottaviano explains the idea behind this as follows:

> Łukasiewicz, in comparison, introduced the many valued logics $\mathbf{L_3}$, $\mathbf{L_4}$, ..., $\mathbf{L_{\aleph}}$, but he required that only the value 1 represents truth. In fact, he didn't open up the possibility of characterizing more truth, or degrees or levels of truth. He

characterized only different degrees of falsity. The idea of absolute truth (the value 1) was maintained and, in general, this might not be the case in nature.

I believe that not only absolute truth (value 1) and absolute falsity (value 0) but also different degrees, levels, or grades of truth and falsity must be assumed by the underlying logic for theories which represent reality.

$\mathbf{J_3}$ has only 3 truth-values. The aim is to work with these three values, trying to understand the mechanism underlying the existence of two designated truth-values. But a further motivation is to generalize $\mathbf{J_3}$ to logics with n designated truth-values and m undesignated.

If a sentence such as 'Chimpanzees can reason' is given value $\frac{1}{2}$ in a model it is because we wish to treat is as a provisional truth. Its negation, however, is no less likely, probable, or reasonable to assume true. Hence it, too, is assigned value $\frac{1}{2}$.

Differing from Łukasiewicz, a proposition which is possible is taken as suitable to proceed on as the basis of reasoning, to build theories with. But its negation is no less suitable.

From a classical perspective we would build one theory based on a proposition which is possible, and another on its negation, comparing the consequences of each. But here it is not a matter of knowing whether the proposition is true or false, or which of the proposition and its negation is most fruitful to be taken as the basis of a theory. Rather, as with paradoxical sentences, the proposition is neither absolutely true nor absolutely false, and it and its negation are inseparable. The appropriate methodology is to base one theory on both the proposition and its negation.

2. The truth-tables

The original presentation of $\mathbf{J_3}$ by D'Ottaviano and da Costa, *1970*, was in terms of three primitive connectives: negation, disjunction, and a possibility operator. In this section I will follow that approach in the main, differing only in taking conjunction rather than disjunction as primitive. In the next section I will give a presentation which reflects a radically different view of paraconsistency and the relation of $\mathbf{J_3}$ to classical logic, using quite different primitives. In that presentation the possibility operator will be derived as a metalogical abbreviation, avoiding the question of whether it involves a use-mention confusion if used as a connective.

We begin by introducing a new symbol, ~, for negation, the reasons for which I will explain in §C, §D, and particularly in §G. The table for this *weak negation* is:

A	~A
1	0
$\frac{1}{2}$	$\frac{1}{2}$
0	1

This is the same table as for $\neg$ in $\mathbf{L_3}$. But Łukasiewicz, taking $\frac{1}{2}$ as nondesignated, treats a proposition which is possible as provisionally false and its negation also as false. Here the import of the table is that both are treated as true.

The tables for conjunction and disjunction are:

$A \wedge B$		B: 1	$\frac{1}{2}$	0
A	1	1	$\frac{1}{2}$	0
	$\frac{1}{2}$	$\frac{1}{2}$	$\frac{1}{2}$	0
	0	0	0	0

$A \vee B$		B: 1	$\frac{1}{2}$	0
A	1	1	1	1
	$\frac{1}{2}$	1	$\frac{1}{2}$	$\frac{1}{2}$
	0	1	$\frac{1}{2}$	0

These two tables have their usual classical meanings in the sense that if the designated values 1 and $\frac{1}{2}$ are replaced by T, and 0 is replaced by F, then we have (with repetitions) the classical tables for $\wedge$ and $\vee$. We may take either as primitive. I will take conjunction, and then

$$A \vee B \equiv_{\text{Def}} \sim(\sim A \wedge \sim B)$$

We also have that $\sim(\sim A \vee \sim B)$ has the same table as $A \wedge B$.

Because of the significance of the notion of possibility in the semantic motivation, we symbolize it in the language with the operator $\Diamond$. Its table is:

A	$\Diamond A$
1	1
$\frac{1}{2}$	1
0	0

We then define the necessity operator as $\Box A \equiv_{\text{Def}} \sim(\Diamond \sim A)$ with table

A	$\Box A$
1	1
$\frac{1}{2}$	0
0	0

Though all these tables are the same as for $\mathbf{L_3}$, their interpretations are quite different due to both 1 and $\frac{1}{2}$ being designated. That difference makes the table for $\rightarrow$ of $\mathbf{L_3}$ inappropriate here. We take instead:

$A \rightarrow B$		B: 1	$\frac{1}{2}$	0
A	1	1	$\frac{1}{2}$	0
	$\frac{1}{2}$	1	$\frac{1}{2}$	0
	0	1	1	1

Here we acknowledge that if the antecedent is false then $A \rightarrow B$ is definitely true; if the antecedent is true, absolutely or provisionally, and the consequent is false, then $A \rightarrow B$ is false. Hence the table is again classical in the sense that if 1 and $\frac{1}{2}$ are replaced by T, and 0 is replaced by F we have the classical table. Therefore, the rule of *modus ponens* is valid. The remaining cases are when the antecedent is true, either definitely or provisionally, and the consequent is provisionally true (possible). In that case it is correct to ascribe only provisional truth (possibility) to $A \rightarrow B$.

It is not necessary to take $\rightarrow$ as primitive. We can define it:

$$A \rightarrow B \equiv_{\text{Def}} \sim(\Diamond A \wedge \sim B).$$

Several other connectives are important for $\mathbf{J_3}$. First,

$$A \leftrightarrow B \equiv_{\text{Def}} (A \rightarrow B) \wedge (B \rightarrow A)$$

Its table is

		B		
$A \leftrightarrow B$		1	$\frac{1}{2}$	0
A	1	1	$\frac{1}{2}$	0
	$\frac{1}{2}$	$\frac{1}{2}$	$\frac{1}{2}$	0
	0	0	0	1

If this seems puzzling, recall that a proposition which is provisionally true (possible) cannot be equivalent to a false one, while it can be provisionally equivalent to a true one.

In most paraconsistent logics, and in particular $\mathbf{J_3}$ and da Costa's *1974* systems $\mathbf{C_n}$, two negations are distinguished. The first we have already seen, $\sim$, which is called weak negation. The other is (in this development) a defined connective, $\neg$, and is called *strong* or *classical negation.* In $\mathbf{J_3}$ it is defined as

$$\neg A \equiv_{\text{Def}} \sim\Diamond A$$

with table

A	$\neg A$
1	0
$\frac{1}{2}$	0
0	1

This is classical in the sense that, unlike weak negation, the strong negation of a "true" proposition (one with designated value) is false, and of a false one is true.

This is not the first time in this volume we have seen two formalizations of a single English connective in one logic. The modal logics of Chapter VI used both

'$\rightarrow$' and '$\supset$'.

Finally, a metalogical abbreviation which is useful in axiomatizing $\mathbf{J_3}$ is

$©A \equiv_{Def} \neg(\Diamond A \wedge \Diamond{\sim}A)$

It has table:

A	©A
1	1
$\frac{1}{2}$	0
0	1

With this we can assert that A has a classical (absolute) truth-value. A strained reading of © as a connective might be 'It is classically (absolutely) true or false that ...'. An alternative definition for ©A is $\Box A \vee \Box{\sim}A$.

Note on notation: In *D'Ottaviano and da Costa, 1970,* and in D'Ottaviano's later work, $\Diamond A$ and $\Box A$ are written as ∇A and ΔA, and $\rightarrow$ is written as $\supset$. What is symbolized here as $\sim$ is written there as $\neg$, and what I write as $\neg$ they symbolize as $\neg^*$.

We now define a $\mathbf{J_3}$*-evaluation* for the language $L(p_0, p_1, \dots {\sim}, \wedge, \Diamond)$ to be a map $\mathbf{e}: PV \rightarrow \{0, \frac{1}{2}, 1\}$ which is extended to all wffs by the tables above for $\sim$, $\wedge$, $\Diamond$. The *designated values are* 1 *and* $\frac{1}{2}$, so that $\mathbf{e} \models A$ means $\mathbf{e}(A) = 1$ or $\frac{1}{2}$. And $\models A$ means that $\mathbf{e} \models A$ for all $\mathbf{J_3}$-evaluations $\mathbf{e}$. Finally, $\Gamma \models A$ means that for every $\mathbf{J_3}$-evaluation $\mathbf{e}$, if $\mathbf{e} \models B$ for all $B \in \Gamma$, then $\mathbf{e} \models A$. The collection of valid wffs is what we call $\mathbf{J_3}$, or the $\mathbf{J_3}$*-tautologies.* When these notions of validity and consequence are compared to others I'll write $\models_{\mathbf{J_3}}$.

Note that the Semantic Deduction Theorem holds in $\mathbf{J_3}$:

$\Gamma \cup \{A\} \models B$ iff $\Gamma \models A \rightarrow B$

3. Interdefinability of the connectives

The choice of which primitives are used in the development of $\mathbf{J_3}$ and how we symbolize them strongly reflects the way in which we understand paraconsistency, the relation of $\mathbf{J_3}$ to classical logic, and the adequacy of the general form of semantics of Chapter IV. Before we can understand why this is so, we need to know what choice of primitives we can use.

By 'definable' in the next theorem I mean the strong notion that for the connective in question there is a schema built from the other connectives which has the same 3-valued table. For instance, we'll see that a unary connective which is evaluated always as $\frac{1}{2}$ cannot be defined. However, there is a schema semantically equivalent to such a connective: $A \to A$. It always takes a designated value, and since both 1 and $\frac{1}{2}$ are designated the schema and connective are semantically equivalent.

Theorem 1 *a.* $\vee, \to, \neg, ©$ are definable from $\sim, \wedge, \Diamond$.

b. $\wedge, \to, \neg, ©$ are definable from $\sim, \vee, \Diamond$.

c. $\vee, \Diamond, ©$ are definable from $\neg, \to, \wedge, \sim$.

d. $\sim$ cannot be defined from $\wedge, \vee, \to, \Diamond$.

e. $\sim$ cannot be defined from $\wedge, \vee, \to, \Diamond, \neg$.

f. $\Diamond$ cannot be defined from $\sim, \wedge, \vee, \to$.

g. © cannot be defined from $\sim, \wedge, \vee, \to$.

h. $\neg$ cannot be defined from $\sim, \wedge, \vee, \to$.

j. No schema built from any of $\sim, \wedge, \Diamond, \vee, \to, ©, \neg$ takes only value $\frac{1}{2}$.

Proof: a. $A \vee B \equiv_{\text{Def}} \sim(\sim A \wedge \sim B)$ $\quad A \to B \equiv_{\text{Def}} \sim(\Diamond A \wedge \sim B)$

$\neg A \equiv_{\text{Def}} \sim\Diamond A$ $\quad \Box A \equiv_{\text{Def}} \sim\Diamond\sim A$

$©A \equiv_{\text{Def}} \Box A \vee \Box\sim A$

b. Observe that $A \wedge B$ has the same table as $\sim(\sim A \vee \sim B)$.

c. First note that for any B, $B \wedge \neg B$ takes value 0. Exploiting this we may define:

$\Diamond A \equiv_{\text{Def}} \neg(A \to (A \wedge \neg A))$

$©A \equiv_{\text{Def}} \neg[\neg(A \to (A \wedge \neg A)) \wedge \neg(\sim A \to (A \wedge \neg A)]$

Note that the latter is $\neg(\Diamond A \wedge \Diamond\sim A)$.

d. By a straightforward induction on the length of a schema S(A) built from $\wedge, \vee, \to, \Diamond$, we can show that if $\mathbf{e}(A) = 1$ or $\frac{1}{2}$, then $\mathbf{e}(S(A)) = 1$ or $\frac{1}{2}$. So $\sim$ cannot be defined.

e. Here we can prove by induction on the length of a schema S(A) built from $\wedge, \vee, \to, \Diamond, \neg$ that for every A if $\mathbf{e}(A) = 1$ or $\frac{1}{2}$, then $\mathbf{e}(S(A)) = 1$ or $\frac{1}{2}$ *or* if $\mathbf{e}(A) = 1$ or $\frac{1}{2}$, then $\mathbf{e}(S(A)) = 0$. That is, we cannot separate the values 1 and $\frac{1}{2}$. So $\sim$ cannot be defined.

f. For any schema S(A) built from $\sim, \wedge, \vee, \to$, if $\mathbf{e}(A) = \frac{1}{2}$ then $\mathbf{e}(S(A)) = \frac{1}{2}$.

g. Were © definable from $\sim, \wedge, \vee, \to$, then we could define $\Box A$ as $\sim(©A \to \sim A)$ and $\Diamond A$ as $\sim\Box\sim A$, contradicting (f).

h. Were $\neg$ definable from $\sim, \wedge, \vee, \to$, then by (c), $\Diamond$ would be, too, contradicting (f).

j. If $\mathbf{e}$ is any $\mathbf{J_3}$-evaluation such that $\mathbf{e}: \text{PV} \to \{0,1\}$, then its extension satisfies $\mathbf{e}$: Wffs $\to \{0,1\}$. So no connective taking only value $\frac{1}{2}$ can be defined. ■

There are two very different ways we may present $\mathbf{J_3}$. The first is to use $\sim$, $\wedge$, $\Diamond$ or $\sim$, $\vee$, $\Diamond$ as primitives. This is what we did above and is in accord with D'Ottaviano and da Costa's original motivation.

The alternative is to take $\neg$, $\rightarrow$, $\wedge$, $\sim$ as primitives. As we will see in §C and §D, to do so is to view $\mathbf{J_3}$ as an extension of classical logic. In what follows, *I will assume that the definitions of evaluation, validity, and semantic consequence for* $\mathbf{J_3}$ *are made with respect to either* $L(\sim, \wedge, \Diamond)$ or $L(\neg, \rightarrow, \wedge, \sim)$ as appropriate to the discussion at hand. Though we now have two different semantic consequence relations for two distinct languages, I will use the same symbol, $\vDash_{\mathbf{J_3}}$ or simply $\vDash$, for both. We can view them as "the same logic" formulated in two different languages (see the discussion in §II.G.4 and §X.B.6).

Note: We can also define a connective in $\mathbf{J_3}$ whose table is that of $\rightarrow$ in $\mathbf{L_3}$, namely $A \supset\!\rightarrow B \equiv_{\text{Def}} (\Diamond{\sim}A \vee B) \wedge (\Diamond B \vee {\sim}A)$. Because of the definability of the tables for $\vee, \Diamond$ from $\sim$ and $\supset\!\rightarrow$ which was shown in the discussion of $\mathbf{L_3}$ (§VIII.C.1.a), we could take $\sim$ and $\supset\!\rightarrow$ as primitives for $\mathbf{J_3}$. Besides being counterintuitive, we have that $\{A, A \supset\!\rightarrow B\} \vDash B$ is not a valid rule. Nonetheless, D'Ottaviano, *1985 A*, has exploited this close relation between $\mathbf{L_3}$ and $\mathbf{J_3}$ to give an axiomatization of $\mathbf{J_3}$ in $L(p_0, p_1, \ldots \sim, \supset\!\rightarrow)$. It does not, however, appear to be strongly complete. I do not know if there is a translation (consequence preserving map) of $\mathbf{J_3}$ into $\mathbf{L_3}$.

C. The Relationship Between $\mathbf{J_3}$ and Classical Logic

The way we see the relation between $\mathbf{J_3}$ and classical logic depends on how we understand the role of weak negation, $\sim$, in $\mathbf{J_3}$.

1. $\sim$ as standard negation

If we identify $\sim$ with negation in **PC**, then the fragment of $\mathbf{J_3}$ in the language of $\sim$, $\wedge$, $\vee$, $\rightarrow$ is contained in **PC**: the $\mathbf{J_3}$-tables for these connectives restricted to 0 and 1 are the classical tables, with 1 read as T, and 0 as F. However, this fragment does not equal **PC**. We have that $(A \wedge {\sim}A) \rightarrow B$ is not a $\mathbf{J_3}$-tautotlogy, for A may take value $\frac{1}{2}$ and B value 0. This accords with the design of $\mathbf{J_3}$ as a paraconsistent logic, ensuring that $\{A, {\sim}A\} \vDash_{\mathbf{J_3}} B$ is not a valid rule.

Other noteworthy classical tautologies which fail to be $\mathbf{J_3}$-tautologies when $\sim$ is identified with negation in **PC**, are:

$${\sim}A \rightarrow (A \rightarrow B)$$

$$(A \rightarrow B) \rightarrow ((A \rightarrow {\sim}B) \rightarrow {\sim}A)$$

$$(A \rightarrow B) \rightarrow ({\sim}B \rightarrow {\sim}A)$$

$$((A \vee B) \wedge \sim A) \to B$$

Perhaps surprisingly, however, both

$$\sim(A \wedge \sim A) \quad \text{and} \quad (A \vee \sim A)$$

are tautologies. Viewed individually each proposition apparently obeys these principles of bivalence, but not so in terms of its consequences. Instead of $(A \wedge \sim A) \to B$ we have that

$$(A \wedge \sim A \wedge ©A) \to B$$

is a $\mathbf{J_3}$-tautology; instead of $\sim(A \vee \sim A) \to B$, we have

$$\sim(A \vee \sim A \vee ©A) \to B$$

is a $\mathbf{J_3}$-tautology.

So long as we restrict our attention to propositions which are "classical" we can reason classically in $\mathbf{J_3}$. Define a map $*$ from the language $L(\neg, \to, \wedge, \vee)$ of **PC** to the language $L(\sim, \wedge, \Diamond)$ of $\mathbf{J_3}$ by first taking

$$A' = A \text{ with } \neg \text{ replaced by } \sim$$

with the understanding that $\vee$ and $\to$ are the defined connectives of $\mathbf{J_3}$. Then set

$$A^* = (\bigwedge\!\!\!\bigwedge_{\{p_i \text{ in } A\}} ©p_i) \to A'$$

where $\bigwedge\!\!\!\bigwedge$ means the conjunction of the indexed wffs, associating to the right.

Theorem 2 The map $*$ is a validity preserving map from **PC** to $\mathbf{J_3}$.
That is, $\models_{\mathbf{PC}} A$ iff $\models_{\mathbf{J_3}} A^*$

Proof: First note that for any evaluation e, if $\mathsf{e}(p) = \frac{1}{2}$ for some p in A, then the antecedent of A^* is evaluated to be 0. So we have:

$$\begin{aligned}\models_{\mathbf{PC}} A \quad & \text{iff} \quad \text{for every } \mathbf{PC}\text{-model } \upsilon,\ \upsilon(A) = \mathsf{T} \\ & \text{iff} \quad \text{for every } \mathbf{J_3}\text{-evaluation } \mathsf{e}: PV \to \{0, 1\},\ \mathsf{e}(A') = 1 \\ & \text{iff} \quad \text{for every } \mathbf{J_3}\text{-evaluation } \mathsf{e},\ \mathsf{e}(A^*) = 1 \\ & \text{iff} \quad \models_{\mathbf{J_3}} A^*\end{aligned}$$

■

However, $*$ does not preserve semantic consequence. We have $\{p, \neg p\} \models_{\mathbf{PC}} q$, but $\{©p \to p,\ ©p \to \sim p\} \not\models_{\mathbf{J_3}} ©q \to q$, for we may take $\mathsf{e}(p) = \frac{1}{2}$ and $\mathsf{e}(q) = 0$. The same remarks apply if we translate $\neg$ as the defined strong negation, $\neg$, of $\mathbf{J_3}$.

2. ¬ as standard negation

We may view $\mathbf{J_3}$ as an extension of **PC** if we present it in the language $L(\neg, \to, \wedge, \sim)$.

Theorem 3 The fragment of $\mathbf{J_3}$ in the language of $\neg$, $\rightarrow$, $\wedge$ is **PC**.
That is, for such wffs, $\Gamma \vDash_{\mathbf{PC}} A$ iff $\Gamma \vDash_{\mathbf{J_3}} A$.

Proof: This follows from the observation made in §B.2 that the tables for $\neg$, $\rightarrow$, $\wedge$ are the classical ones if we identify the designated values with T, and 0 with F. ∎

From this point of view $\mathbf{J_3}$ arises by adding an intensional connective, ~, to classical logic.

We also have that ~(~A ∧ ~B) is semantically equivalent in $\mathbf{J_3}$ to $\neg(\neg A \wedge \neg B)$, which you can check.

D. Consistency vs. Paraconsistency

1. Definitions of completeness and consistency for $\mathbf{J_3}$ theories

A theory is consistent if it contains no contradiction. A proposition is a contradiction if it is false due to its form only, or semantically, if the corresponding wff is false in all models.

A theory is complete if it is as full a description as possible of "the way the world is" relative to the atomic propositions we've assumed and the semantics, informal or formal, which we employ. It might be inconsistent, but if not it will contain as many complex propositions as possible relative to the atomic ones while still being consistent.

So a complete and consistent theory corresponds to a possible description of the world, relative to our semantic intuitions and choice of atomic propositions. Therefore, if we have formal semantics, a complete and consistent theory should correspond to the collection of propositions true in a model.

In classical logic we take the standard form for a contradiction to be $A \wedge \neg A$. A theory (collection of sentences closed under deduction) is then said to be consistent (with respect to classical logic) if for no A does it contain $A \wedge \neg A$; or, equivalently, for no A does it contain both A and $\neg A$. This reflects the classical semantic assumption that for no A can both A and $\neg A$ be true.

If we understand negation to be formalized by ~, then the classical notion of contradiction is clearly not applicable to $\mathbf{J_3}$. And the associated criterion of consistency is inappropriate for $\mathbf{J_3}$ in which we specifically assumed that it is acceptable to build a theory on the basis of both A and ~A. By the semantic assumptions of $\mathbf{J_3}$, a theory which contains both A and ~A for some A is not necessarily contradictory, for it can reflect a possible way the world could be. To formulate an appropriate criterion of consistency for $\mathbf{J_3}$ we first note the following.

Lemma 4 If e is a $\mathbf{J_3}$-evaluation, then

$$e \models A \quad \text{iff} \quad e(A) = 1 \text{ or } \tfrac{1}{2}$$
$$e \models {\sim}A \quad \text{iff} \quad e(A) = 0 \text{ or } \tfrac{1}{2}$$
$$e \models ©A \quad \text{iff} \quad e(A) = 0 \text{ or } 1$$

Thus if all three of A, ~A, ©A are assumed by a theory then we can have no model of it. Such a theory is inconsistent: it would contain the contradiction $A \wedge {\sim}A \wedge ©A$.

In classical logic we took a theory to be complete if for every A it contains at least one of A, ¬A. That is, it must decide between these and hence stipulate which is assumed to be true, or else embrace them both and be inconsistent. That criteria is inappropriate for $\mathbf{J_3}$ if we understand negation to be formalized by ~, for by choosing one of A, ~A we have not stipulated the truth-value of A. To do that we need to choose two of A, ~A, ©A. So we define:

Γ is *consistent* relative to $\mathbf{J_3}$ if for every A at most two of A, ~A, ©A are syntactic consequences of Γ.

Γ is *complete* relative to $\mathbf{J_3}$ if for every A at least two of A, ~A, ©A are in Γ.

The notion of syntactic consequence for $\mathbf{J_3}$ used here is either of those formalized in §E which, with hindsight, we know are strongly complete.

In Theorem 7 I'll demonstrate that these are the appropriate definitions for $\mathbf{J_3}$: Γ is complete and consistent relative to $\mathbf{J_3}$ iff there is a $\mathbf{J_3}$-evaluation which validates exactly Γ.

However, we may view $\mathbf{J_3}$ formulated in the language L(¬, →, ∧, ~) as an extension of classical logic (Theorem 3). In that case the standard way to formalize 'not' is with ¬, and a theory is *classically consistent* means that at most one of A, ¬A is a consequence of Γ; a theory is *classically complete* means that at least one of A, ¬A is in Γ. Because we have

$$\models_{\mathbf{J_3}} (A \wedge {\sim}A \wedge ©A) \leftrightarrow (A \wedge \neg A)$$

we can demonstrate in the next section that *for* $\mathbf{J_3}$*-theories,* Γ *is complete and consistent relative to* $\mathbf{J_3}$ iff Γ *is classically complete and consistent.*

2. The status of negation in $\mathbf{J_3}$

Relative to the connectives {∧, ∨, →}, both weak negation, ~, and strong negation, ¬, are primitive (Theorem 1).

The approach favored by paraconsistent logicians is to view weak negation as primitive. Then possibility is formalized as a connective, with appropriate deference made to the use-mention controversy surrounding that decision. Strong negation, ¬, is taken as merely a defined connective. Only under favorable circumstances where all the atomic propositions under discussion are classically (absolutely) true or false can we view negation, ~, as classical (Theorem 2). This is reflected by an alternative definition we can give of $\neg A$ as $\sim A \wedge ©A$. We may interpret 'not' as $\neg$ only for those propositions which satisfy $\Box A \vee \Box \sim A$.

From a more classical point of view we might argue that we never suggested that all logically significant uses of 'not' can be properly modeled by classical negation. The simple example of two assertions about a die: 'Three faces are even numbered', and 'Three faces are not even numbered', should convince us of that. The symbol $\neg$ should be reserved for formalizing 'not' in those cases where we agree that the proposition and the proposition with 'not' deleted cannot both be true (at the same time), and this is the analysis given in Chapter IV. We may choose to introduce a new connective, ~, to formalize other uses of 'not' where the truth of the proposition and the truth of the proposition with 'not' deleted are (apparently) inseparable. Such propositions might be ones using vague terms, such as 'It is not raining', or paradoxical sentences such as 'This sentence is not true'. The division of logical uses of 'not' is thus the same as the paraconsistent logician's, but comes from a very different perspective.

Paraconsistent logics contain "inconsistent nontrivial theories" only from the application of classical criteria to an admittedly nonclassical connective, ~. From their own semantic point of view such theories are consistent, corresponding to a possible description of the world. To call a theory inconsistent if it contains, for some A, both A and ~A is tantamount to understanding negation in its usual sense as assumed by all other logics in this volume: it cannot be that both A and ~A are true. That cannot be how we understand ~ in paraconsistent logics, for it would preclude building a theory based on both A and ~A. For $\mathbf{J_3}$ this strong view of negation is expressed by the table for $\neg$. Using that connective the classical criterion of consistency is apt, and we cannot have a nontrivial inconsistent theory: $(A \wedge \neg A) \models_{\mathbf{J_3}} B$.

I will return to this discussion of the status of negation in §F and §G.

E. Axiomatizations of $\mathbf{J_3}$

In this section I will give two axiomatizations. In the first I take the viewpoint of the paraconsistent logician that ~ is the standard interpretation of 'not' and treat $\mathbf{J_3}$ as a modal logic. Strong negation, $\neg$, does not appear in the axiomatization, and the only notions of completeness and consistency used are relative to $\mathbf{J_3}$.

In the second axiomatization I view $\mathbf{J_3}$ as an extension of classical logic in the language of $\neg$, $\rightarrow$, $\wedge$ with an intensional connective, ~. The notions of completeness and consistency will be the classical ones.

For both axiomatizations there are no inconsistent nontrivial theories.

1. As a modal logic

$\mathbf{J_3}$

in $L(\sim, \wedge, \Diamond)$

$A \vee B \equiv_{Def} \sim(\sim A \wedge \sim B)$ $\quad\quad$ $\Box A \equiv_{Def} \sim\Diamond\sim A$

$A \to B \equiv_{Def} \sim\Diamond A \vee B$ $\quad\quad$ $©A \equiv_{Def} \neg(\Diamond A \wedge \Diamond\sim A)$

$A \leftrightarrow B \equiv_{Def} (A \to B) \wedge (B \to A)$

axiom schemas

1. $B \to (A \to B)$
2. $(A \to (B \to C)) \to ((A \to B) \to (A \to C))$
3. $(B \to (A \to C)) \to ((A \wedge B) \to C)$
4. $A \to (B \to (A \wedge B))$
5. $(A \wedge \sim A \wedge ©A) \to B$
6. $((\sim A \wedge ©A) \to A) \to A$
7. $\sim\sim A \leftrightarrow A$
8. $©A \leftrightarrow ©\sim A$
9. $\sim\Diamond A \leftrightarrow (\sim A \wedge ©A)$
10. $©(\Diamond A)$
11. $[(\sim(A \wedge B) \wedge ©(A \wedge B)) \wedge B] \to (\sim A \wedge ©A)$
12. $(\sim A \wedge ©A) \to [\sim(A \wedge B) \wedge ©(A \wedge B)]$
13. $[(A \wedge B) \wedge ©(A \wedge B)] \leftrightarrow [(A \wedge ©A) \wedge (B \wedge ©B)]$

rule $\dfrac{A,\ A \to B}{B}$

I will denote by $\vdash_{\mathbf{J_3},\Diamond}$ the consequence relation of this axiom system. In this section only, §E.1, I will write $\vdash$ for $\vdash_{\mathbf{J_3},\Diamond}$.

Recall from the last section that Γ is *consistent relative to* $\mathbf{J_3}$ (*$\mathbf{J_3}$-consistent*) if for every A at most two of $A, \sim A, ©A$ are consequences of Γ. And Γ is *complete relative to* $\mathbf{J_3}$ (*$\mathbf{J_3}$-complete*) if for every A at least two of $A, \sim A, ©A$ are in Γ. As usual, Γ is a *theory* if Γ is closed under deduction.

Lemma 5 ***a.*** **(*The Syntactic Deduction Theorem*)** $\Gamma \cup \{A\} \vdash B$ iff $\Gamma \vdash A \to B$

b. $\vdash A \wedge B \to B$

c. $\vdash A \wedge B \to B \wedge A$

d. $\Gamma \cup \{A, B\} \vdash A \wedge B$

e. $\Gamma \cup \{A, B\} \vdash C$ iff $\Gamma \vdash (A \wedge B) \to C$

Proof: The proof of part (a) is as for **PC** (Theorem II.8, p.47) due to the presence of axioms 1 and 2.

Part (b) follows from axiom 3 taking B for C, and axiom 1. Part (c) uses axiom 3 taking $B \wedge A$ for C, and axiom 4. Part (d) uses axiom 4, and (e) uses axioms 3 and 4. ■

Lemma 6

a. Γ is $\mathbf{J_3}$-inconsistent iff for every B, $\Gamma \vdash B$.

b. If Γ is $\mathbf{J_3}$-complete and $\mathbf{J_3}$-consistent and $A \notin \Gamma$, then for every B, $\Gamma \cup \{A\} \vdash B$.

c. If Γ is $\mathbf{J_3}$-complete and $\mathbf{J_3}$-consistent, then Γ is a theory.

d. If $\Gamma \nvdash A$ then $\Gamma \cup \{\sim A, ©A\}$ is $\mathbf{J_3}$-consistent.

e. If Γ is $\mathbf{J_3}$-consistent, then one of $\Gamma \cup \{A, \sim A\}$, $\Gamma \cup \{A, ©A\}$, $\Gamma \cup \{\sim A, ©A\}$ is consistent.

Proof: a. From right to left is immediate.

In the other direction, if Γ is $\mathbf{J_3}$-inconsistent then for some A: $\Gamma \vdash A$, $\Gamma \vdash \sim A$, and $\Gamma \vdash ©A$. Hence by axiom 5, for every B, $\Gamma \vdash B$.

b. This is immediate from (a).

c. Suppose Γ is $\mathbf{J_3}$-complete and $\mathbf{J_3}$-consistent and $\Gamma \vdash A$. If $A \notin \Gamma$ then by the completeness of Γ, $\Gamma \cup \{A\}$ is $\mathbf{J_3}$-inconsistent. But $\text{Th}(\Gamma) = \text{Th}(\Gamma \cup \{A\})$, so Γ is $\mathbf{J_3}$-inconsistent, a contradiction. Hence $A \in \Gamma$.

d. Suppose $\Gamma \cup \{\sim A, ©A\}$ is $\mathbf{J_3}$-inconsistent. Then $\Gamma \cup \{\sim A, ©A\} \vdash A$. So by Lemma 5, $\Gamma \vdash (\sim A \wedge ©A) \rightarrow A$, and hence by axiom 6 and Lemma 5, $\Gamma \vdash A$.

e. Suppose $\Gamma \cup \{\sim A, ©A\}$ is $\mathbf{J_3}$-inconsistent. Then by part (a), $\Gamma \vdash (\sim A \wedge ©A) \rightarrow A$, hence by axiom 6, $\Gamma \vdash A$. Now suppose $\Gamma \cup \{A, ©A\}$ is also $\mathbf{J_3}$-inconsistent. Then by axiom 7, $\Gamma \cup \{\sim\sim A, ©A\}$ is $\mathbf{J_3}$-inconsistent. Hence $\Gamma \vdash (\sim\sim A \wedge ©A) \rightarrow \sim A$, so using axiom 6, $\Gamma \vdash \sim A$. Hence, $\Gamma \cup \{A, \sim A\}$ is $\mathbf{J_3}$-consistent. ■

Theorem 7 The following are equivalent:

a. Γ is $\mathbf{J_3}$-complete and $\mathbf{J_3}$-consistent.

b. There is some $\mathbf{J_3}$-evaluation **e** such that $\Gamma = \{A : \mathbf{e} \vDash A\}$.

c. There is some $\mathbf{J_3}$-evaluation **e** such that

$\mathbf{e}(A) = 1$ iff $A, ©A \in \Gamma$

$\mathbf{e}(A) = \frac{1}{2}$ iff $A, \sim A \in \Gamma$

$\mathbf{e}(A) = 0$ iff $\sim A, ©A \in \Gamma$

d. There is some $\mathbf{J_3}$-evaluation **e** such that

$\mathbf{e}(A) = 1$ iff $\sim A \notin \Gamma$

$\mathbf{e}(A) = \frac{1}{2}$ iff $©A \notin \Gamma$

$\mathbf{e}(A) = 0$ iff $A \notin \Gamma$

Proof: The equivalence of (b), (c), and (d) comes from Lemma 4. I will show that Γ is $\mathbf{J_3}$-complete and $\mathbf{J_3}$-consistent iff (c).

First suppose there is an $\mathbf{e}$ as in (c). Then for every A exactly two of A, ~A, ©A $\in \Gamma$. Hence Γ is $\mathbf{J_3}$-complete. By the equivalence of (b) and (c), the Semantic Deduction Theorem, and the fact that all the axioms are $\mathbf{J_3}$-tautologies, Γ is a theory. Hence Γ is $\mathbf{J_3}$-consistent.

Now suppose Γ is $\mathbf{J_3}$-complete and $\mathbf{J_3}$-consistent. Then by Lemma 6.c, Γ is a theory. Define $\mathbf{e}$ as in (c). It remains to show that $\mathbf{e}$ is a $\mathbf{J_3}$-evaluation. We begin with the evaluation of weak negation.

$\mathbf{e}(\sim A) = 1$ iff $\sim A$, ©$(\sim A) \in \Gamma$
iff $\sim A$, ©A $\in \Gamma$ by axiom 8
iff $\mathbf{e}(A) = 0$

$\mathbf{e}(\sim A) = \frac{1}{2}$ iff $\sim A, \sim\sim A \in \Gamma$
iff $\sim A, A \in \Gamma$ by axiom 7
iff $\mathbf{e}(A) = \frac{1}{2}$

Therefore, by process of elimination, $\mathbf{e}(\sim A) = 0$ iff $\mathbf{e}(A) = 1$, so $\mathbf{e}$ evaluates ~ correctly.

Turning now to the evaluation of the possibility operator, we have:

$\mathbf{e}(\Diamond A) = 0$ iff $\sim\Diamond A$, ©$(\Diamond A) \in \Gamma$
iff $\sim A$, ©A $\in \Gamma$ by axioms 9 and 10
iff $\mathbf{e}(A) = 1$

Since Γ is a theory, by axiom 10, ©$(\Diamond A) \in \Gamma$. So we cannot have $\mathbf{e}(\Diamond A) = \frac{1}{2}$. Thus $\mathbf{e}$ evaluates $\Diamond$ correctly.

For conjunction we begin by noting that

$\mathbf{e}(A \wedge B) = 0$ iff $\sim(A \wedge B)$, ©$(A \wedge B) \in \Gamma$

If $\mathbf{e}(A \wedge B) = 0$ and $\mathbf{e}(B) \neq 0$, then $B \in \Gamma$. So by axiom 11, $\sim A \wedge$ ©A $\in \Gamma$, and hence $\mathbf{e}(A) = 0$.

If $\mathbf{e}(A) = 0$, then $\sim A$, ©A $\in \Gamma$, so by axiom 12, $\sim(A \wedge B)$, ©$(A \wedge B) \in \Gamma$, so $\mathbf{e}(A \wedge B) = 0$. Using Lemma 5.c the same reasoning establishes that if $\mathbf{e}(B) = 0$ then $\mathbf{e}(A \wedge B) = 0$. Finally,

$\mathbf{e}(A \wedge B) = 1$ iff $A \wedge B$, ©$(A \wedge B) \in \Gamma$
iff A, ©A, B, ©B $\in \Gamma$ by axiom 13
iff $\mathbf{e}(A) = 1$ and $\mathbf{e}(B) = 1$

By process of elimination, $\mathbf{e}(A \wedge B) = \frac{1}{2}$ iff neither of $\mathbf{e}(A), \mathbf{e}(B) = 0$ and not both $\mathbf{e}(A), \mathbf{e}(B) = 1$. Hence $\mathbf{e}$ evaluates $\wedge$ correctly. ■

Lemma 8 If $\Gamma \nvdash A$ then there is some $\mathbf{J_3}$-complete and $\mathbf{J_3}$-consistent theory Σ such that $\Gamma \subseteq \Sigma$ and $A \notin \Sigma$.

Proof: Suppose $\Gamma \nvdash A$. Define $\Sigma_0 = \Gamma \cup \{\sim A, \copyright A\}$. This is $\mathbf{J_3}$-consistent by Lemma 6.d.

Let $B_0, B_1, \ldots$ be a listing of all wffs. Define:

$$\Sigma_{n+1} = \begin{cases} \Sigma_n \cup \{B_n, \sim B_n\} & \text{if that is } \mathbf{J_3}\text{-consistent; if not then} \\ \Sigma_n \cup \{B_n, \copyright B_n\} & \text{if that is } \mathbf{J_3}\text{-consistent; if not then} \\ \Sigma_n \cup \{\sim B_n, \copyright B_n\} \end{cases}$$

$$\Sigma = \bigcup_n \Sigma_n$$

By Lemma 6.e, for every n, Σ_{n+1} is $\mathbf{J_3}$-consistent. Hence Σ is $\mathbf{J_3}$-consistent, and by construction it is $\mathbf{J_3}$-complete. As $\Sigma_0 \subseteq \Sigma$, $A \notin \Sigma$. ■

Theorem 9 **(*Strong Completeness of*** $\vdash_{\mathbf{J_3},\diamond}$**)** $\Gamma \vdash_{\mathbf{J_3},\diamond} A$ iff $\Gamma \vDash_{\mathbf{J_3}} A$

The proof is standard using Theorem 7 and Lemma 8.

2. As an extension of classical logic

In §II.K.6 I gave an axiomatization of **PC** in the language $L(\neg, \rightarrow, \wedge, \vee)$. As mentioned there (p. 57), the first seven axioms and the rule of that system give a strongly complete axiomatization of **PC** in the language $L(\neg, \rightarrow, \wedge)$. If we allow any formula of the language $L(\neg, \rightarrow, \wedge, \sim)$ to be an instance of A, B, or C in those schema, then we have an *axiomatization of* **PC** *based on* $\neg, \rightarrow, \wedge$ *in the language of* $\mathbf{J_3}$.

$\mathbf{J_3}$

in $L(\neg, \rightarrow, \wedge, \sim)$

$\copyright A \equiv_{\text{Def}} \neg[\neg(A \rightarrow (A \wedge \neg A)) \wedge \neg(\sim A \rightarrow (A \wedge \neg A))]$

$\diamond A \equiv_{\text{Def}} \neg(A \rightarrow (A \wedge \neg A))$

axiom schemas

PC based on $\neg, \rightarrow, \wedge$, and

1. $(\sim A \wedge \copyright A) \leftrightarrow \neg A$
2. $\sim\sim A \leftrightarrow A$
3. $\copyright(\neg A)$
4. $[(A \wedge B) \wedge \copyright(A \wedge B)] \leftrightarrow [(A \wedge \copyright A) \wedge (B \wedge \copyright B)]$
5. $(\sim A \wedge \copyright A) \rightarrow \copyright(A \rightarrow B)$
6. $(B \wedge \copyright B) \rightarrow \copyright(A \rightarrow B)$

rule $\dfrac{A, A \rightarrow B}{B}$

I will denote by $\vdash_{\mathbf{J_3},\neg}$ the consequence relation of this axiom system. In this section only, §E.2, I will write $\vdash$ for $\vdash_{\mathbf{J_3},\neg}$.

Recall that a collection of wffs Γ is *classically consistent* if at most one of A, $\neg$A are consequences of Γ. And Γ is *classically complete and consistent* if for every A, exactly one of A, $\neg$A is in Γ.

Throughout the following I will liberally use results from **PC** as justified by this axiomatization. In particular, the Syntactic Deduction Theorem holds.

Lemma 10

a. Γ is classically consistent iff for every A at most two of A, ~A, ©A are in Γ

b. Γ is classically complete and consistent iff for every A exactly two of A, ~A, ©A are in Γ.

Proof: a. Γ is classically inconsistent iff (by **PC**) $\Gamma \vdash$B for every B. So if Γ is classically inconsistent, for every A, all three of A, ~A, ©A are consequences of Γ. If for some A all three of A, ~A, ©A are consequences of Γ, then by axiom 1 both A and $\neg$A are consequences of Γ, hence Γ is classically inconsistent.

b. Suppose Γ is classically complete and consistent. If A $\notin \Gamma$ then $\neg$A $\in \Gamma$, so by axiom 1, ~A, ©A $\in \Gamma$. If A $\in \Gamma$, then suppose ©A $\notin \Gamma$. In that case $\neg$©A $\in \Gamma$. Using **PC** and the definition of ©A, we have $\vdash$(A $\wedge \neg$©A) $\rightarrow$ ~A. So ~A $\in \Gamma$. ∎

Lemma 6 as above now follows if we replace '$\mathbf{J_3}$-consistent' by 'classically consistent' and '$\mathbf{J_3}$-complete and $\mathbf{J_3}$-consistent' by 'classically complete and consistent'. We only need to establish $\vdash$((~A $\wedge$ ©A) $\rightarrow$ A) $\rightarrow$ A. But by **PC**, $\vdash$($\neg$A $\rightarrow$ A) $\rightarrow$ A and $\vdash$[((B $\rightarrow$ C) $\wedge$ ((C $\rightarrow$ D) $\rightarrow$ D)] $\rightarrow$ ((B $\rightarrow$ D) $\rightarrow$ D), so using axiom 1 we have the desired theorem.

We now prove the analogue to Theorem 7.

Theorem 11 The following are equivalent:

a. Γ is classically complete and consistent.

b. There is some $\mathbf{J_3}$-evaluation **e** such that $\Gamma = \{A : \mathbf{e} \models A\}$

c. There is some $\mathbf{J_3}$-evaluation **e** such that

$$\mathbf{e}(A) = 1 \text{ iff } A, ©A \in \Gamma$$
$$\mathbf{e}(A) = \tfrac{1}{2} \text{ iff } A, \sim A \in \Gamma$$
$$\mathbf{e}(A) = 0 \text{ iff } \sim A, ©A \in \Gamma$$

d. There is some $\mathbf{J_3}$-evaluation **e** such that

$$\mathbf{e}(A) = 1 \text{ iff } \sim A \notin \Gamma$$
$$\mathbf{e}(A) = \tfrac{1}{2} \text{ iff } ©A \notin \Gamma$$
$$\mathbf{e}(A) = 0 \text{ iff } A \notin \Gamma$$

Proof: Using Lemma 10 the proof proceeds as for Theorem 7. The only new point is to establish that the connectives are evaluated correctly if **e** is defined as in (c).

I will leave to you to check that by **PC** and axiom 2 we have $\vdash ©A \leftrightarrow ©{\sim}A$. So the proof that $\sim$ is evaluated correctly is the same as for Theorem 7. Turning to the evaluation of $\neg$,

$$\begin{aligned} \mathbf{e}(A) = 0 \;\; &\text{iff } \; {\sim}A, ©A \in \Gamma \\ &\text{iff } \; \neg A \in \Gamma && \text{by axiom 1} \\ &\text{iff } \; \mathbf{e}(\neg A) = 1 && \text{by axiom 3} \end{aligned}$$

Now I will show that if $\mathbf{e}(\neg A) \neq 1$ then $\mathbf{e}(\neg A) = 0$. So suppose that $\mathbf{e}(\neg A) \neq 1$. Then one of $\neg A, ©\neg A \notin \Gamma$; so by axiom 3, $\neg A \notin \Gamma$, so $\mathbf{e}(\neg A) = 0$. Thus $\neg$ is evaluated correctly.

For conjunction,

$$\begin{aligned} \mathbf{e}(A \wedge B) = 0 \;\; &\text{iff } \; {\sim}(A \wedge B), ©(A \wedge B) \in \Gamma \\ &\text{iff } \; \neg(A \wedge B) \in \Gamma && \text{by axiom 1} \\ &\text{iff } \; A \notin \Gamma \text{ or } B \notin \Gamma && \text{by } \mathbf{PC} \\ &\text{iff } \; \mathbf{e}(A) = 0 \text{ or } \mathbf{e}(B) = 0 \end{aligned}$$

$$\begin{aligned} \mathbf{e}(A \wedge B) = 1 \;\; &\text{iff } \; A \wedge B, ©(A \wedge B) \in \Gamma \\ &\text{iff } \; A, ©A, B, ©B \in \Gamma && \text{by axiom 4} \\ &\text{iff } \; \mathbf{e}(A) = 1 \text{ and } \mathbf{e}(B) = 1 \end{aligned}$$

So $\wedge$ is evaluated correctly, the other case following by process of elimination.

Finally, we consider the conditional.

$$\begin{aligned} \mathbf{e}(A \to B) = 0 \;\; &\text{iff } \; {\sim}(A \to B), ©(A \to B) \in \Gamma \\ &\text{iff } \; \neg(A \to B) \in \Gamma && \text{by axiom 1} \\ &\text{iff } \; A \wedge \neg B \in \Gamma && \text{by } \mathbf{PC} \\ &\text{iff } \; A, \neg B \in \Gamma && \text{by } \mathbf{PC} \\ &\text{iff } \; \mathbf{e}(A) \neq 0 \text{ and } {\sim}B, ©B \in \Gamma && \text{by axiom 1} \\ &\text{iff } \; \mathbf{e}(A) \neq 0 \text{ and } \mathbf{e}(B) = 0 \end{aligned}$$

Suppose now that $\mathbf{e}(A \to B) = 1$. Then $A \to B, ©(A \to B) \in \Gamma$. Suppose that $\mathbf{e}(B) \neq 1$. Then $B \notin \Gamma$, so $\neg B \in \Gamma$. But by **PC**, $\vdash ((A \to B) \wedge \neg B) \to \neg A$, so $\neg A \in \Gamma$. And then by axiom 1, $\mathbf{e}(A) = 0$.

Suppose $\mathbf{e}(A) = 0$. Then $\neg A \in \Gamma$, so by **PC**, $A \to B \in \Gamma$. And by axiom 5, $©(A \to B) \in \Gamma$, so $\mathbf{e}(A \to B) = 1$.

If $\mathbf{e}(B) = 1$ then $B, ©B \in \Gamma$, so by **PC**, $A \to B \in \Gamma$. And by axiom 6, $©(A \to B) \in \Gamma$, so $\mathbf{e}(A \to B) = 1$.

Thus $\to$ is evaluated correctly, the other case following by process of elimination. ■

Lemma 12 If $\Gamma \nvdash A$ then there is a classically complete and consistent Σ such that $\Gamma \subseteq \Sigma$ and $A \notin \Sigma$.

The proof is as for **PC**.

Now it's routine to prove the following using Theorem 11 and Lemma 12.

Theorem 13 (Strong Completeness of $\vdash_{\mathbf{J_3},\neg}$***)*** $\Gamma \vdash_{\mathbf{J_3},\neg} A$ iff $\Gamma \vDash_{\mathbf{J_3}} A$

For both axiomatizations the classical rule of substitution fails:

$$\frac{A \leftrightarrow B}{C(A) \leftrightarrow C(B)}$$

We have $\vDash_{\mathbf{J_3}} (A \leftrightarrow A) \leftrightarrow (B \leftrightarrow B)$, whereas $\nvDash_{\mathbf{J_3}} \sim(A \leftrightarrow A) \leftrightarrow \sim(B \leftrightarrow B)$, for the latter may fail when $\mathbf{e}(A) = 1$ and $\mathbf{e}(B) = \frac{1}{2}$. D'Ottaviano, *1985 A*, defines the stronger equivalence

$$(A \equiv^* B) \equiv_{\mathrm{Def}} (A \leftrightarrow B) \wedge (\sim A \leftrightarrow \sim B)$$

Then for every evaluation $\mathbf{e}$, $\mathbf{e} \vDash A \equiv^* B$ iff $\mathbf{e}(A) = \mathbf{e}(B)$. That is, $A \equiv^* B$ is true in a model iff A and B are extensionally equivalent. And the relation

$$A \approx B \equiv_{\mathrm{Def}} \vDash_{\mathbf{J_3}} (A \equiv^* B)$$

defines a congruence relation on the set of wffs. The appropriate rule of substitution for $\mathbf{J_3}$ is then:

$$\frac{A \equiv^* B}{C(A) \equiv^* C(B)}$$

F. Set-Assignment Semantics for $\mathbf{J_3}$

In giving set-assignment semantics for $\mathbf{J_3}$ we must decide which negation, $\sim$ or $\neg$, should be modeled by the usual set-assignment table for negation. In this section I'll begin with $\neg$, which is most in keeping with the discussion in §D.2 and Chapter IV.

Accordingly, we first take $L(\neg, \rightarrow, \wedge, \sim)$ as our language for $\mathbf{J_3}$. Then $\langle v, s, S \rangle$ is a $\mathbf{J_3}$*-model* for $L(\neg, \rightarrow, \wedge, \sim)$ if

$\neg, \rightarrow, \wedge$ are evaluated classically

$v(\sim A) = T$ iff $s(A) \neq S$

and s, S satisfy:

1. $S \neq \varnothing$
2. $s(p) \neq \varnothing$ iff $v(p) = T$
3. $s(A) \subseteq s(B)$ or $s(B) \subseteq s(A)$

4. $s(A \wedge B) = s(A) \cap s(B)$

5. $s(A \to B) = \begin{cases} s(B) & \text{if } \varnothing \subset s(A) \subset S \\ \overline{s(A)} \cup s(B) & \text{otherwise} \end{cases}$

6. $s(\neg A) = \begin{cases} S & \text{if } s(A) = \varnothing \\ \varnothing & \text{otherwise} \end{cases}$

7. $s({\sim}A) = \begin{cases} s(A) & \text{if } \varnothing \subset s(A) \subset S \\ \overline{s(A)} & \text{otherwise} \end{cases}$

From this point of view $\mathbf{J_3}$ *is classical logic with the addition of one wholly intensional connective,* ~. To give a model we first assign each propositional variable (proposition) a truth-value (T or F), and then, choosing any collection of sets linearly ordered by inclusion, assign one set to each propositional variable in accord with condition (1), $s(p) \neq \varnothing$ iff $v(p) = T$. Conditions (3)–(6) then allow an inductive definition of the set-assignment on Wffs, and hence of the valuation on all wffs.

Which are the false propositions? Those with no content. If A has some content, but not the full content S, then both A and its weak negation, ~A, are true.

The proof of the following lemma is routine.

Lemma 14

a. $v(A) = T$ iff $s(A) \neq \varnothing$

b. $v(A \vee B) = T$ iff $v(A) = T$ or $v(B) = T$

c. If there are only three possible content sets, $\varnothing \subset U \subset S$, then

$$s(A \vee B) = s(A) \cup s(B)$$

Otherwise,

$$s(A \vee B) = \begin{cases} s(A) \cap s(B) & \text{if both } \varnothing \subset s(A) \subset S \text{ and } \varnothing \subset s(B) \subset S \\ s(A) \cup s(B) & \text{otherwise} \end{cases}$$

d. $v(\Diamond A) = T$ iff $v(A) = T$

e. $s(\Diamond A) = \begin{cases} S & \text{if } s(A) \neq \varnothing \\ \varnothing & \text{if } s(A) = \varnothing \end{cases}$

f. $s(\Diamond A) = \begin{cases} S & \text{if } v(A) = T \\ \varnothing & \text{if } v(A) = F \end{cases}$

g. Given a $\mathbf{J_3}$-evaluation $\mathbf{e}$: Wffs $\to \{0, \frac{1}{2}, 1\}$, define

$$s(A) = \{x : x < \mathbf{e}(A) \text{ and } x \in [0,1]\}$$
$$v(p) = T \text{ iff } \mathbf{e}(p) = 1 \text{ or } \tfrac{1}{2}$$

Then $<v,s>$ is a $\mathbf{J_3}$-model and $<v,s> \models A$ iff $\mathbf{e}(A) = 1$ or $\frac{1}{2}$.

g. Given a $\mathbf{J_3}$-model $<\mathsf{v},\mathsf{s}>$, define e: Wffs $\rightarrow \{0, \frac{1}{2}, 1\}$ by

$$\mathsf{e}(A) = \begin{cases} 1 & \text{if } \mathsf{s}(A) = \mathsf{S} \\ \frac{1}{2} & \text{if } \varnothing \subset \mathsf{s}(A) \subset \mathsf{S} \\ 0 & \text{if } \mathsf{s}(A) = \varnothing \end{cases}$$

Then e is a $\mathbf{J_3}$-evaluation and $\mathsf{e}(A) = 1$ or $\frac{1}{2}$ iff $<\mathsf{v},\mathsf{s}> \models A$.

Note that though $\Diamond A$ and A take the same truth-value in every model, they do not necessarily have the same content.

From parts (g) and (h), the consequence relation for these set-assignment semantics is the same as for the $\mathbf{J_3}$-matrix, which by Theorem 13 coincides with the syntactic consequence relation in this language.

Theorem 15 **(*Strong Completeness of the Set-Assignment Semantics*)**

$\Gamma \models_{\mathbf{J_3}} A$ iff every set-assignment $\mathbf{J_3}$-model which validates Γ also validates A iff $\Gamma \vdash_{\mathbf{J_3},\neg} A$

If we take $\mathbf{J_3}$ *to be formulated in the language* $L(\sim, \wedge, \Diamond)$, then we can define $<\mathsf{v},\mathsf{s},\mathsf{S}>$ to be a $\mathbf{J_3}$-*model* if

$\mathsf{v}(A \wedge B) = \mathsf{T}$ iff $\mathsf{v}(A) = \mathsf{T}$ and $\mathsf{v}(B) = \mathsf{T}$

$\mathsf{v}(\sim A) = \mathsf{T}$ iff $\mathsf{s}(A) \neq \mathsf{S}$

$\mathsf{v}(\Diamond A) = \mathsf{T}$ iff $\mathsf{v}(A) = \mathsf{T}$

and s, S satisfy:

1. $\mathsf{S} \neq \varnothing$
2. $\mathsf{s}(p) \neq \varnothing$ iff $\mathsf{v}(p) = \mathsf{T}$
3. $\mathsf{s}(A) \subseteq \mathsf{s}(B)$ or $\mathsf{s}(B) \subseteq \mathsf{s}(A)$
4. $\mathsf{s}(A \wedge B) = \mathsf{s}(A) \cap \mathsf{s}(B)$
5. $\mathsf{s}(\sim A) = \begin{cases} \mathsf{s}(A) & \text{if } \varnothing \subset \mathsf{s}(A) \subset \mathsf{S} \\ \overline{\mathsf{s}(A)} & \text{otherwise} \end{cases}$
6. $\mathsf{s}(\Diamond A) = \begin{cases} \mathsf{S} & \text{if } \mathsf{s}(A) \neq \varnothing \\ \varnothing & \text{if } \mathsf{s}(A) = \varnothing \end{cases}$

You can check that for these semantics we have:

$\mathsf{v}(\neg A) = \mathsf{T}$ iff $\mathsf{v}(A) = \mathsf{F}$

$\mathsf{v}(\sim A) = \mathsf{F}$ iff $\mathsf{v}(A) = \mathsf{T}$ and $\mathsf{s}(A) = \mathsf{S}$

$\mathsf{v}(A \rightarrow B) = \mathsf{T}$ iff $\mathsf{v}(A) = \mathsf{F}$ or $\mathsf{v}(B) = \mathsf{T}$

$\mathsf{v}(A \vee B) = \mathsf{T}$ iff $\mathsf{v}(A) = \mathsf{T}$ or $\mathsf{v}(B) = \mathsf{T}$

And as above we can establish the following theorem.

Theorem 16 $\Gamma \models_{J_3} A$ iff every set-assignment $\mathbf{J_3}$-model for $L(\sim, \wedge, \Diamond)$ which validates Γ also validates A
iff $\Gamma \vdash_{J_3, \Diamond} A$

G. Truth-Default Semantics

Let us take the paraconsistent logician's point of view that the English 'not' is to be formalized as $\sim$, and $\neg$ is simply a defined connective which happens to correspond to classical negation. In that case it is not possible to give set-assignment semantics within the general framework of Chapter IV in such a way that we can translate $\mathbf{J_3}$-evaluations to set-assignment models and vice versa while satisfying

$$e \models A \text{ iff } \langle v, s \rangle \models A$$

We do not have that if $e \models A$ then $e \not\models \sim A$. The following table cannot be realized:

A	N(A)	¬A
any value	fails	F
T	holds	F
F		T

The tables of the general framework of Chapter IV are based on the view that for a proposition to be true it must pass certain tests; if it fails any it is false. I have argued in Chapter IV and throughout this volume, as well as in *Epstein, 198?,* that this is correct: we analyze what it means for a proposition to be true, and every proposition which is not true is false.

The semantic intuitions behind $\mathbf{J_3}$ are a mirror image of this: we analyze what it means for a proposition to be false, and every proposition which is not false is true.

The general form of semantics of Chapter IV can be viewed as using falsity as the default truth-value. In $\mathbf{J_3}$ truth is taken as the default truth-value. Previously we have said: we cannot have both A and its negation false. For $\mathbf{J_3}$ we say: we cannot have both A and its negation true. It is precisely because these views are incompatible and yet are both represented in $\mathbf{J_3}$, albeit one of them derivatively, that I have used a different symbol, $\sim$, for this negation. The appropriate form of the set-assignment table for negation is the following:

(1)

A	N(A)	~A
any value	fails	T
T	holds	F
F		T

Because $\wedge$, $\vee$, and $\rightarrow$ are evaluated classically in $\mathbf{J_3}$ they can be presented by tables which take either truth or falsity as the default value. Truth-default tables for these connectives have the following form.

(2)

A	B	$\mathsf{B}(A,B)$	$A \rightarrow B$
any values		fails	T
T	T		T
T	F	holds	F
F	T		T
F	F		T

(3)

A	B	$\mathsf{C}(A,B)$	$A \wedge B$
any values		fails	T
T	T		T
T	F	holds	F
F	T		F
F	F		F

(4)

A	B	$\mathsf{A}(A,B)$	$A \vee B$
any values		fails	T
T	T		T
T	F	holds	T
F	T		T
F	F		F

Set-assignment semantics using tables (1)–(4) I call *truth-default set-assignment semantics*, though we would normally use the symbol $\neg$ for the connective of table (1). I will continue to call semantics of the general form of Chapter IV simply 'set-assignment semantics', but when there is a need to distinguish them from this alternate form I will refer to them as *falsity-default set-assignment semantics*. Similar definitions apply for relation based semantics. In §E and §F of Appendix 2 of Chapter IV (pp.112-114) I discuss further the general framework of truth-default semantics.

Consider now the set-assignment semantics for $\mathbf{J_3}$ in the language of $\sim, \wedge, \Diamond$ given in the last section. These are truth-default semantics if we interpret $\sim$ as the standard formalization of 'not' because the evaluation of the connectives can be expressed equivalently as:

$\upsilon(A \wedge B) = \mathsf{F}$ iff $\upsilon(A) = \mathsf{F}$ or $\upsilon(B) = \mathsf{F}$

$\upsilon(\sim A) = \mathsf{F}$ iff $\upsilon(A) = \mathsf{T}$ and $s(A) \neq S$

$\upsilon(\Diamond A) = \mathsf{F}$ iff $\upsilon(A) = \mathsf{F}$

and the defined connectives as:

$\upsilon(A \rightarrow B) = \mathsf{F}$ iff $\upsilon(A) = \mathsf{T}$ and $\upsilon(B) = \mathsf{F}$

$\upsilon(A \vee B) = \mathsf{F}$ iff $\upsilon(A) = \mathsf{F}$ and $\upsilon(B) = \mathsf{F}$

This points out that for all the set-assignment semantics for logics previously considered in this volume it is essential that the truth-value conditions "tag along" in the evaluation of the connectives even though the connective often could be evaluated as dependent only on the content of the constituent propositions.

We can also classify many-valued semantics as being either falsity-default or truth-default. We say that a table for $\neg$, $\wedge$, $\vee$, or $\rightarrow$ in a many-valued matrix is *standard* if by renaming the designated values as T and the undesignated values as F, then the table is (with repetitions) the classical table for that connective. We say the table is *falsity-weighted* if, renaming as above, any row of the classical table which takes value F also takes value F in this table, whereas a row which takes value T in the classical table may take value F in the renamed many-valued table. A *truth-weighted* table is one in which any row of the classical table which takes value T also takes value T in the renamed many-valued table.

Every standard table is both falsity-weighted and truth-weighted. Every many-valued logic of Chapter VIII uses standard or falsity-weighted tables for each of $\neg$, $\rightarrow$, $\wedge$, $\vee$. Only the table for $\sim$ in $\mathbf{J_3}$ is truth-weighted.

A many-valued logic which uses truth-weighted tables for $\neg$, $\rightarrow$, $\wedge$, $\vee$ at least one of which is not standard cannot be given the usual falsity-default set-assignment semantics in such a way that we can translate many-valued models to set-assignment ones and vice versa while preserving validity in a model. It seems likely to me, though, that every truth-weighted many-valued matrix can be presented in terms of truth-default semantics, and every falsity weighted many-valued matrix can be presented in terms of falsity-default semantics.

X Translations Between Logics

In the previous chapters we've seen many examples of interpretations of one logic in another. In §A of this chapter Stanisław Krajewski and I formalize the notion of a translation, review the translations we've already encountered, present further examples, and then raise some questions about them. Foremost will be how to judge whether a translation preserves meaning.

In §B I attempt to give criteria for when a translation preserves meaning. I first define what it means for a translation to be model preserving, and then define the stronger notion of a semantically faithful translation. I show that all the translations we have seen are semantically faithful with the exception of the translations of classical logic into intuitionist logics. I conclude with a discussion of whether semantically faithful translations can be said to preserve meaning.

Wójcicki, *1984*, §1.8.2, has a short history of other formalizations of the notion of translation.

A. Syntactic Translations

1. A formal notion of translation

Throughout this volume I have referred to various interpretations of one logic into another as translations. In ordinary speech 'a translation' means a changing of some text from one language to another. Within the study of logic, the text to be changed is the logic itself. But what is 'a logic'?

If logics are presented to us as collections of theorems, then any mapping which we would want to call a translation should preserve theoremhood, that is theorems of one should be mapped into theorems of the other. But if that is all we mean by a translation then the notion is trivial, for we can always enumerate the theorems of one logic and map them in order onto the theorems of another as we enumerate those. Similarly, the requirement that tautologies must be mapped to tautologies is trivial if we make the nonconstructive assumption that we can map any countable set into another. Nonetheless, particularly regular maps which preserve theoremhood or validity can be of interest and have sometimes been called 'translations' in the literature and I will discuss them below. But preservation of validity or theoremhood alone seems too weak a criteria for a mapping to be a translation.

Generally we have considered logics to be either semantic or syntactic consequence relations. In that case what should be preserved by a translation is the consequence relation. This accords closely to the use of the term in the literature, and any such mapping preserves the essential *syntactic* aspect of the logic, for a consequence relation, whether presented semantically or syntactically, is a relation on collections of wffs.

For ease of exposition, throughout this chapter I will assume that *every logic has strongly complete semantics*. By that I mean that either the logic is originally presented as a semantic consequence relation or else there is a class of models which determines a semantic consequence relation which coincides with the syntactic consequence relation. All the definitions of §A apply equally to logics which are presented as syntactic consequence relations by replacing $\models$ by $\vdash$ throughout, as I comment on that at the end of this section. I will use the bold face letters **L** and **M** to range over logics, and $L_{\mathbf{L}}$ and $L_{\mathbf{M}}$ for their respective languages.

Definition 1 A *validity mapping* of a propositional logic **L** into a propositional logic **M** is a map * from $L_{\mathbf{L}}$ to $L_{\mathbf{M}}$ such that for every A,

$$\models_{\mathbf{L}} A \ \text{ iff } \ \models_{\mathbf{M}} A^*$$

A mapping is a *translation* if for every Γ and A,

$$\Gamma \models_{\mathbf{L}} A \ \text{ iff } \ \Gamma^* \models_{\mathbf{M}} A^*$$

where $\Gamma^* = \{A^*: A \in \Gamma\}$. I write $\mathbf{L} \hookrightarrow \mathbf{M}$ if there is a translation of **L** into **M**.

This definition does not require a translation to preserve the structure of the language being translated. The reason is that the obvious requirement that the map be a homomorphism rules out maps which preserve the structure of the language in weaker but still regular ways and which seem to merit the name of translation, for example the double negation translation of classical logic into intuitionist logic (Example 7 in §A.2 below). Nonetheless, maps which are particularly regular are important to single out. To do that I need to refer to specific languages in the following definition; the definition generalizes to languages with different sets of connectives, fewer or more or even different propositional variables, or propositional constants (e.g., $\perp$), and I'll assume the appropriate generalizations below.

Definition 2 A map $*$ from the language $L(p_0, p_1, \ldots \neg, \rightarrow)$ to any formal language is called *grammatical* if there are schemas λ, φ, ψ of the latter language such that

$$p^* = \lambda(p)$$
$$(\neg A)^* = \varphi(A^*)$$
$$(A \rightarrow B)^* = \psi(A^*, B^*)$$

Propositional constants may appear in these schemas. But A^* may contain no variables other than those appearing in $\lambda(p)$ where p appears in A. That is, the mapping may not depend on any *parameters*. Thus a grammatical map is a homomorphism between languages.

A grammatical map is *homophonic* if it translates each connective to itself.

A *grammatical translation* is a grammatical map which is a translation. I write $\mathbf{L} \twoheadrightarrow \mathbf{M}$ if there is a grammatical translation of **L** into **M**.

The proof of the following lemma is straightforward.

Lemma 3 The composition of translations is a translation, and the composition of grammatical maps is grammatical.

It might seem that if two logics **L** and **M** both have a deduction theorem and $*$ is a grammatical map which preserves validity (theoremhood), then $*$ must be a translation. That is wrong, for the homophonic mapping of **PC** in $L(\neg, \wedge)$ to **Int** in $L(\neg, \rightarrow, \wedge, \vee)$ is grammatical but does not preserve consequences (Corollaries VII.19,20, and the remarks thereafter, p.213). That is,

Theorem 4 A grammatical validity mapping between logics each of which has a deduction theorem is not necessarily a translation.

The definitions given here apply equally to logics which are presented as syntactic consequences relations if $\models$ is replaced by $\vdash$ and 'validity' by 'theorem'. In that case it is enough that the mapping preserves finite consequences, as syntactic consequence relations are compact. For logics which have both a syntactic and semantic presentation we have the following theorem, the proof of which I leave to you.

Theorem 5 If **L** and **M** are logics with strongly complete axiomatizations, then a mapping * from $L_{\mathbf{L}}$ to $L_{\mathbf{M}}$ is a translation iff for every *finite* Γ, $\Gamma \vDash_{\mathbf{L}} A$ iff $\Gamma^* \vDash_{\mathbf{M}} A^*$.

In some cases it is useful to be able to classify a map as preserving finite consequences, and I call such a map a *finite consequence translation.*

2. Examples

Let's review the translations of the previous chapters in terms of our new terminology. In each case I will assume that the languages of both logics have the same stock of propositional variables $p_0, p_1, \ldots$. Unless noted otherwise, each propositional variable is translated to itself.

1. Classical logic to "itself"

$$\begin{array}{ccc} \mathbf{PC} & \twoheadrightarrow & \mathbf{PC} \\ L(\neg, \rightarrow, \wedge, \vee) & & L(\neg, \rightarrow) \end{array}$$

This was the first translation we saw (§II.G.4, pp.34–35). The question arose there what it meant for "the same logic" to be formulated in different languages. I'll return to that in §B.6 below.

2. Classical logic to subject matter relatedness logic

$$\begin{array}{ccc} \mathbf{PC} & \twoheadrightarrow & \mathbf{S} \\ L(\neg, \wedge) & & L(\neg, \rightarrow, \wedge) \end{array}$$

$$\begin{array}{ccc} \mathbf{PC} & \twoheadrightarrow & \mathbf{S} \\ L(\neg, \wedge) & & L(\neg, \rightarrow) \end{array}$$

When **S** is presented in $L(\neg, \rightarrow, \wedge)$ the homophonic map from $L(\neg, \wedge)$ is a translation (§III.G.1, pp.74–75). If only $\neg$ and $\rightarrow$ are used as primitives for **S**, then $\neg$ is translated homophonically and $\wedge$ is translated to the defined version of $\wedge$ in **S**:

$$(A \wedge B)^* = \neg(A^* \rightarrow (B^* \rightarrow \neg((A^* \rightarrow B^*) \rightarrow (A^* \rightarrow B^*))))$$

(see §III.H, p.78).

The same maps establish $\mathbf{PC} \twoheadrightarrow \mathbf{R}$ and $\mathbf{PC} \twoheadrightarrow$ **Dual D**. Similarly, $\mathbf{PC} \twoheadrightarrow \mathbf{D}$, except that for the second map the defined version of $\wedge$ within **D** is used:

$$(A \wedge B)^* = \neg(((A^* \rightarrow B^*) \rightarrow (A^* \rightarrow B^*) \rightarrow A^*) \rightarrow \neg B^*)$$

By composing maps, presentations of classical logic in other languages can also be translated grammatically into these logics (Lemma 3).

3. *Dependence Logic to Dual Dependence Logic*
Dual Dependence Logic to Dependence Logic

$$\begin{array}{ccc} \mathbf{D} & \twoheadrightarrow & \mathbf{Dual\ D} \\ L(\neg, \rightarrow, \wedge) & & L(\neg, \rightarrow, \wedge) \end{array}$$

$$\begin{array}{ccc} \mathbf{Dual\ D} & \twoheadrightarrow & \mathbf{D} \\ L(\neg, \rightarrow, \wedge) & & L(\neg, \rightarrow, \wedge) \end{array}$$

The map which translates $\neg$ and $\wedge$ homophonically and

$$(A \rightarrow B)^* = \neg B^* \rightarrow \neg A^*$$

is a translation *in both directions* (§V.C, pp.133–134). Note that by composing these we get translations of each logic to itself.

4. *The logic of equality of contents to Dependence Logic*

$$\begin{array}{ccc} \mathbf{Eq} & \twoheadrightarrow & \mathbf{D} \\ L(\neg, \rightarrow, \wedge) & & L(\neg, \rightarrow, \wedge) \end{array}$$

Both $\neg$ and $\wedge$ are translated homophonically, and

$$(A \rightarrow B)^* = (A^* \rightarrow B^*) \wedge (\neg B^* \rightarrow \neg A^*)$$

(see §V.D.2, p.136).

The same translation establishes $\mathbf{Eq} \twoheadrightarrow \mathbf{Dual\ D}$.

5. *Any classical modal logic* **L** *to "itself"*

$$\begin{array}{ccc} \mathbf{L} & \twoheadrightarrow & \mathbf{L} \\ L(\neg, \wedge, \Box) & & L(\neg, \rightarrow, \wedge) \end{array}$$

$$\begin{array}{ccc} \mathbf{L} & \twoheadrightarrow & \mathbf{L} \\ L(\neg, \rightarrow, \wedge) & & L(\neg, \wedge, \Box) \end{array}$$

In each case both $\neg$ and $\wedge$ are translated homophonically. For the first map,

$$(\Box A)^* = \neg A^* \rightarrow A^*$$

and for the second,

$$(A \rightarrow B)^* = \Box(A^* \supset B^*)$$

(see §VI.J.3, p.178)

6. *Heyting's intuitionist logic to the classical modal logic* **S4**

$$\begin{array}{ccc} \mathbf{Int} & \twoheadrightarrow & \mathbf{S4} \\ L(\neg, \rightarrow, \wedge, \vee) & & L(\neg, \wedge, \Box) \end{array}$$

Both $\wedge$ and $\vee$ are translated homophonically, and

$$p^* = \Box p$$
$$(A \to B)^* = \Box(A^* \supset B^*)$$
$$(\neg A)^* = \Box\neg(A^*)$$

(see Theorem VII.15, p. 210)

The same map is a finite consequence translation of **Int** into **S4Grz**.

7. *The double negation translation of classical logic to Heyting's intuitionist logic*

$$\begin{array}{ccc} \textbf{PC} & \hookrightarrow & \textbf{Int} \\ L(\neg, \to, \wedge, \vee) & & L(\neg, \to, \wedge, \vee) \end{array}$$

This translation (Theorem VII.18, pp. 211–212) takes every A to $\neg\neg A$. It preserves the structure of the language being translated, but *the smallest unit of the language whose structure is preserved is an entire sentence*. Grammatical translations require that there be a specific structure corresponding to each connective.

The same translation establishes **PC** $\hookrightarrow$ **J** (§VII.F.3, p. 226).

8. *Gentzen's translation of classical logic to Heyting's intuitionist logic*

$$\begin{array}{ccc} \textbf{PC} & \twoheadrightarrow & \textbf{Int} \\ L(\neg, \to, \wedge, \vee) & & L(\neg, \to, \wedge, \vee) \end{array}$$

In this translation (Theorem VII.21, p. 214) we have:

$$(p)^\circ = \neg\neg p$$
$$(A \vee B)^\circ = \neg(\neg A^\circ \wedge \neg B^\circ)$$

and each of $\neg$, $\to$, and $\wedge$ is translated homophonically.

9. *Heyting's intuitionist logic to Johansson's minimal calculus*

$$\begin{array}{ccc} \textbf{Int} & \hookrightarrow & \textbf{J} \\ L(\neg, \wedge, \vee, \bot) & & L(\neg, \wedge, \vee, \bot) \end{array}$$

This nongrammatical translation (p. 227) translates each wff by replacing every one of its nonatomic subformulas B by $B \vee \bot$.

10. *Classical logic to Łukasiewicz's 3-valued logic*

$$\begin{array}{ccc} \textbf{PC} & \twoheadrightarrow & \textbf{L}_3 \\ L(\neg, \to) & & L(\neg, \to) \end{array}$$

The translation (Theorem VIII.2, p. 238) takes

$$(A \to B)^* = A^* \to (A^* \to B^*)$$
$$(\neg A)^* = A^* \to (A^* \to \neg(A^* \to A^*))$$

11. Classical logic to the paraconsistent logic $\mathbf{J_3}$

$$\begin{array}{cc} \mathbf{PC} \twoheadrightarrow & \mathbf{J_3} \\ L(\neg, \rightarrow, \wedge) & L(\neg, \rightarrow, \wedge, \sim) \end{array}$$

$$\begin{array}{cc} \mathbf{PC} \twoheadrightarrow & \mathbf{J_3} \\ L(\neg, \rightarrow, \wedge, \vee) & L(\sim, \wedge, \Diamond) \end{array}$$

When $\neg$ is taken as a primitive for $\mathbf{J_3}$ the homophonic map is a translation (Theorem IX.3, p. 273). That translation can be extended to the language $L(\neg, \rightarrow, \wedge, \vee)$ of **PC** by first translating **PC** to itself.

With primitives $\sim$, $\wedge$, $\Diamond$ for $\mathbf{J_3}$ the definitions of $\neg$ and $\rightarrow$ within $\mathbf{J_3}$ are used (§IX.B.3, pp. 269–271).

The mapping of classical logic into the paraconsistent logic $\mathbf{J_3}$ given by assuming that each proposition is "classical",

$$A^* = (\bigwedge_{\{p_i \text{ in } A\}} \copyright p_i) \rightarrow A'$$

where A' is A with $\neg$ replaced by $\sim$, preserves validity but is not a translation (Theorem IX.2, p. 272). Though it is not grammatical, it does preserve the structure of the language of **PC** albeit only sentence by sentence.

3. Logics which cannot be translated grammatically into classical logic

It's noteworthy that in the list above we have no translation of a logic into classical logic, other than classical logic itself. I'll show that there can be no grammatical translation into **PC** of any logic we have studied. In doing so I'll outline a method due to Krajewski which can be applied to any logic which is presented in a language with $\neg$ and $\rightarrow$ among its primitives, though it easily generalizes to other languages.

Theorem 6 There is no grammatical translation of any of the following logics, presented in languages with $\neg$ and $\rightarrow$ among their primitives, into **PC**:

R and **S**	(relatedness logics of Chapter III)
D, **Dual D**, **Eq**	(dependence logics of Chapter V)
S4, **S5**, **S4Grz**, **T**, **B**, **K**, **QT**, **MSI**, **ML**, **G**, **G***	(the classical modal logics of Chapter VI)
Int and **J**	(intuitionist logics of Chapter VII)
$\mathbf{L_3}$	(a many-valued logic from Chapter VIII)
$\mathbf{J_3}$	(a paraconsistent logic from Chapter IX)

Indeed, there is not even a grammatical map which preserves validity.

Proof: I'll present the method for an arbitrary logic **L** and then instantiate it for each of the logics above.

I will assume that **PC** is given in the language of $\neg$ and $\wedge$ and will use $\supset$ and $\equiv$ as defined symbols; the proof generalizes to any other presentation of **PC**. For the purposes of this proof only I will write T for $\neg(p_1 \wedge \neg p_1)$, and $\bot$ for $(p_1 \wedge \neg p_1)$. I also adopt the convention of writing $\mathbf{L}\vDash$ for $\vDash_{\mathbf{L}}$ for a logic **L**.

To begin, note that for every wff A in the language of **PC** in which only one variable p appears, one of the following must hold:

$\mathbf{PC}\vDash A \equiv p$

$\mathbf{PC}\vDash A \equiv \neg p$

$\mathbf{PC}\vDash A \equiv T$

$\mathbf{PC}\vDash A \equiv \bot$

Similarly, there are exactly sixteen truth-functions of 2 variables, and every B in the language of **PC** in which exactly two propositional variables appears must be evaluated semantically in **PC** by one of those. Thus, up to semantic equivalence there are 256 possibilities for a triplet (λ, φ, ψ) which establishes a grammatical map into **PC** (Definition 2).

Now assume that we are given a grammatical map * from the language of one of these logics, **L**, to the language of **PC** that is purported to preserve validity. We start by narrowing the possibilities for the map φ which translates negation, where I will leave to you to exhibit the necessary examples.

We cannot have $\mathbf{PC}\vDash \varphi(A^*) \equiv \bot$.

For each logic **L** we can find a wff A such that $\mathbf{L}\vDash \neg A$. If * were a translation then we should have $\mathbf{PC}\vDash \varphi(A^*)$, but we have $\mathbf{PC}\nvDash \varphi(A^*)$.

We cannot have $\mathbf{PC}\vDash \varphi(A^*) \equiv T$.

For each logic we can find a wff A such that $\mathbf{L}\nvDash \neg A$, yet we would have $\mathbf{PC}\vDash \varphi(A^*)$.

We cannot have $\mathbf{PC}\vDash \varphi(A^*) \equiv A^*$.

For each logic we can find a wff A such that $\mathbf{L}\vDash A$ and $\mathbf{L}\nvDash \neg A$, yet we cannot have both $\mathbf{PC}\vDash A^*$ and $\mathbf{PC}\nvDash A^*$.

Therefore, we must have $\mathbf{PC}\vDash \varphi(A^*) \equiv \neg A^*$. So without loss of generality we can assume $(\neg A)^* = \neg(A^*)$.

Though inessential to this proof it is worth investigating λ, the map which translates the variables.

We cannot have $\mathbf{PC}\vDash \lambda(p) \equiv T$, since for each logic we have $\mathbf{L}\nvDash p$.

We cannot have $\mathbf{PC}\vDash \lambda(p) \equiv \bot$, for then, since $(\neg p)^* = \neg\lambda(p)$, we would have $\mathbf{PC}\vDash \neg\lambda(p)$, yet for each logic $\mathbf{L}\nvDash \neg p$.

Therefore, either $\mathbf{PC}\vDash \lambda(p) \equiv p$ or $\mathbf{PC}\vDash \lambda(p) \equiv \neg p$.

Consider now ψ, the map which translates the conditional. I will show a general form of counterexample to the various possibilities for ψ, and conclude the proof by instantiating the method for each of the logics. I will write, e.g.,

$\psi(\mathsf{F},\mathsf{T}) = \mathsf{F}$ to mean that the schema of ψ is evaluated so in **PC**.

a. We cannot have $\psi(\mathsf{T},\mathsf{T}) = \mathsf{F}$.
For each logic we can find wffs A and B such that $\mathbf{L} \vDash A$, $\mathbf{L} \vDash B$, and $\mathbf{L} \vDash A \to B$. So, since ψ is a translation, we would have $\mathbf{PC} \vDash A^*$, $\mathbf{PC} \vDash B^*$, and $\mathbf{PC} \vDash \psi(A^*, B^*)$. Yet this evaluation of ψ should yield $\mathbf{PC} \nvDash \psi(A^*, B^*)$.

b. We cannot have $\psi(\mathsf{T},\mathsf{F}) = \mathsf{T}$.
For each logic we can find wffs A and B such that $\mathbf{L} \vDash A$, $\mathbf{L} \vDash \neg B$, and $\mathbf{L} \nvDash A \to B$. In that case, as $(\neg B)^* = \neg(B^*)$, we would have $\mathbf{PC} \vDash A^*$, $\mathbf{PC} \vDash \neg B^*$, and $\mathbf{PC} \nvDash \psi(A^*, B^*)$. But if $\mathbf{PC} \vDash \neg B^*$ then B^* is false in every model, so this evaluation of ψ should yield $\mathbf{PC} \vDash \psi(A^*, B^*)$.

c. We cannot have $\psi(\mathsf{F},\mathsf{T}) = \mathsf{F}$.
For each logic we can find wffs A and B such that $\mathbf{L} \vDash \neg A$, $\mathbf{L} \vDash B$, and $\mathbf{L} \vDash A \to B$. Then as for (b), we would have $\mathbf{PC} \vDash \neg A^*$, $\mathbf{PC} \vDash B^*$, and $\mathbf{PC} \vDash \psi(A^*, B^*)$. And so A^* would have to be false in every **PC**-model and we should have $\mathbf{PC} \nvDash \psi(A^*, B^*)$.

d. We cannot have $\psi(\mathsf{F},\mathsf{F}) = \mathsf{F}$.
For each logic we can find wffs A and B such that $\mathbf{L} \vDash \neg A$, $\mathbf{L} \vDash \neg B$, yet $\mathbf{L} \vDash A \to B$. The argument then follows as in (c).

Hence we must have:

e. $\mathbf{PC} \vDash \psi(A, B) \equiv (A^* \supset B^*)$
For this possibility we will have to find a counterexample peculiar to each logic.

I'll now exhibit counterexamples for (a)–(e) for each of the logics of the theorem, under the assumption that $(\neg A)^* = \neg(A^*)$.

If **L** is **R**, **S**, **D**, **Dual D**, **Eq**, any of the modal logics of Chapter VI, or **Int**, the following serve as examples for (a)–(d):

a. $\mathbf{L} \vDash (p_1 \to p_1)$, and $\mathbf{L} \vDash (p_1 \to p_1) \to (p_1 \to p_1)$

b. $\mathbf{L} \vDash (p_1 \to p_1)$, $\mathbf{L} \vDash \neg(p_1 \wedge \neg p_1)$, and $\mathbf{L} \nvDash (p_1 \to p_1) \to (p_1 \wedge \neg p_1)$

c. $\mathbf{L} \vDash \neg(p_1 \wedge \neg p_1)$, $\mathbf{L} \vDash (p_1 \to p_1)$, and $\mathbf{L} \vDash (p_1 \wedge \neg p_1) \to (p_1 \to p_1)$

d. $\mathbf{L} \vDash \neg(p_1 \wedge \neg p_1)$, and $\mathbf{L} \vDash (p_1 \wedge \neg p_1) \to (p_1 \wedge \neg p_1)$

And, except for **Int**, a counterexample for (e) is:

e. $\mathbf{L} \nvDash p_1 \to (p_2 \to p_1)$ yet $\mathbf{PC} \vDash p_1^* \supset (p_2^* \supset p_1^*)$

For **Int**, to give a counterexample for (e) note that we would have $\mathbf{PC} \vDash (\neg\neg p_1 \to p_1)^* \equiv \neg\neg p_1^* \supset p_1^*$. And then

$\mathbf{Int} \nvDash \neg\neg p_1 \to p_1$, and $\mathbf{PC} \vDash \neg\neg p_1^* \supset p_1^*$

For **J** the example for (a) above will do, and a counterexample for (e) is the same as for **Int**. For (b)–(d) the examples above work if the **J**-anti-tautology $\neg(p_1 \rightarrow p_1)$ is substituted for $\neg(p_1 \wedge \neg p_1)$.

If **L** is $\mathbf{L_3}$ then the example for (a) above will do. For parts (b)–(d) simply substitute the $\mathbf{L_3}$-anti-tautology $(p_1 \wedge \neg p_1 \wedge \mathrm{I}\, p_1)$ for $(p_1 \wedge \neg p_1)$ in the examples above. For part (e), let A be $(p_1 \leftrightarrow \neg p_1)$. Then

$$\mathbf{L_3} \nvDash (\neg A \rightarrow A) \rightarrow A \quad \text{yet} \quad \mathbf{PC} \vDash (\neg A^* \supset A^*) \supset A^*$$

If **L** is $\mathbf{J_3}$ then the the example for (a) above will do, and for parts (b)–(d) simply substitute the $\mathbf{J_3}$-anti-tautology $(p_1 \wedge {\sim} p_1 \wedge ©p_1)$ for $(p_1 \wedge \neg p_1)$. Here we are using $\neg$ as a primitive of the language for $\mathbf{J_3}$. We could instead use ~, since the arguments concerning $\neg$ above apply equally to ~. So for (e) we have: $\mathbf{PC} \vDash ((p_1 \wedge {\sim} p_1) \rightarrow p_2)^* \equiv (p_1{}^* \wedge \neg p_1{}^*) \supset p_2{}^*$. And then

$$\mathbf{J_3} \nvDash (p_1 \wedge {\sim} p_1) \rightarrow p_2, \text{ but } \mathbf{PC} \vDash (p_1{}^* \wedge \neg p_1{}^*) \supset p_2{}^*. \qquad \blacksquare$$

Here is another example, the essentials of which can be found in §V.D.5, p.140.

Theorem 7 There is no grammatical translation of **D** into **Eq**.

4. Translations where there are no grammatical translations: R ↪ PC and S ↪ PC

From Theorem 5 we know that there is no grammatical translation of **R** into **PC**. Nonetheless, despite the apparently very much richer semantic basis of **R**, I will show that there is a translation. It is then easy to modify that translation to produce a one from **S** into **PC**. These preserve the structure of the language of **R** and **S**, though only in terms of sentences as wholes, and seem to reflect the underlying semantic assumptions of **R** and **S**.

The basic reason there is no grammatical translation of **R** into **PC** is that the only plausible candidate for $\psi(A,B)$ is $A^* \supset B^*$, but that does not allow us to take into account whether the two formulas are related. If we consider the fact of whether p_i is related to p_j *as a new proposition* then we can produce a translation. Note that in the semantics for **R** every relation governing the table for $\rightarrow$ is reflexive (though not necessarily symmetric), so the fact that p_i is related to p_i must be represented as a tautology.

In order to simplify the presentation of the translation I will translate from the language $L(p_0, p_1, \ldots \neg, \rightarrow, \wedge, \vee)$ of **R** to a language for **PC** with variables q_0, $q_1, \ldots$ as well as $\{d_{ij} : i,j \text{ are natural numbers}\}$, and primitives $\neg$, $\wedge$, i.e., $L(q_0, q_1, \ldots, \{d_{ij}: i,j \text{ are natural numbers}\}, \neg, \wedge)$. I take $\supset$ and $\vee$ to be defined for **PC** in the usual way. It may seem that by using this language for **PC** we are introducing two sorts of variables and that, therefore, we are no longer working with the usual classical logic. But a variable is still just a variable, no matter how we designate it. We can achieve the same effect by partitioning the usual list of variables $p_0, p_1, \ldots$ into two classes, taking one class to be those with even indices and the other with odd, but that is harder to read. In a remark below I'll set out the translation in that manner.

I will use the notation $\bigvee\!\!\!\!\bigvee_{A \text{ in } \Gamma} A$ to mean the disjunction of the finite number of wffs in Γ, reading from left to right in the usual ordering of those wffs and associating to the right. I write $\{p_i \text{ in } A\}$ for $\{ i\colon p_i \text{ appears in } A\}$.

The mapping $*$ is given by:

$(p_i)^* = q_i$

$(\neg A)^* = \neg(A^*)$

$(A \rightarrow B)^* =$

$$(A^* \supset B^*) \wedge [\bigvee\!\!\!\!\bigvee_{\{p_i \text{ in } A,\ p_j \text{ in } B\}} d_{ij} \vee \bigvee\!\!\!\!\bigvee_{\{p_i \text{ in } A \text{ and } p_i \text{ in } B\}} (d_{ii} \vee \neg d_{ii})]$$

That is, $A \rightarrow B$ is translated to the material implication conjoined with the disjunction of the propositions which are to be taken as asserting that the variables appearing in A are related to those of B.

Theorem 8 The map $*$ is a translation of **R** into **PC**.

Proof: First note that for the defined symbols $\wedge$, R of the language of **R** we can prove:

$\mathbf{PC} \vDash (A \wedge B)^* \equiv (A^* \wedge B^*)$

$\mathbf{PC} \vDash (R(A,B))^* \equiv \bigvee\!\!\!\!\bigvee_{\{p_i \text{ in } A,\ p_j \text{ in } B\}} d_{ij} \vee \bigvee\!\!\!\!\bigvee_{\{p_i \text{ in } A \text{ and } p_i \text{ in } B\}} (d_{ii} \vee \neg d_{ii})$

In particular, if $i \neq j$ then $\mathbf{PC} \vDash (R(p_i, p_j))^* \equiv d_{ij}$ and $\mathbf{PC} \vDash (R(p_i, p_i))^*$.

I will now reduce the problem to showing it is enough to prove that $*$ preserves validity. By Theorem 5 it is enough to show that for every finite Γ, $\Gamma \vDash_{\mathbf{R}} A$ iff $\Gamma^* \vDash_{\mathbf{PC}} A^*$. A Semantic Deduction Theorem for **R** holds in the form $\{A_1, \ldots, A_n\} \vDash_{\mathbf{R}} A$ iff $\vDash_{\mathbf{R}} \neg((A_1 \wedge \ldots \wedge A_n) \wedge \neg A)$. For **PC** we have it in the same form. Because $\neg$ is translated homophonically and the defined symbol $\wedge$ is translated to a wff semantically equivalent to its homophone, it thus suffices to prove $\mathbf{R} \vDash A$ iff $\mathbf{PC} \vDash A^*$.

To show that if $\mathbf{R} \vDash A$ then $\mathbf{PC} \vDash A^*$ we can use the complete axiomatization of **R** given in §III.J, p.81. It is not difficult to check that if A is an axiom of **R** then $\mathbf{PC} \vDash A^*$, since

$$\mathbf{R} \vDash R(A,B) \text{ iff } \mathbf{R} \vDash \bigvee\!\!\!\!\bigvee_{\{p_i \text{ in } A,\ p_j \text{ in } B\}} R(p,q)$$

and $\mathbf{R} \vDash R(p,p)$. Moreover, if $\mathbf{PC} \vDash (A \rightarrow B)^*$ and $\mathbf{PC} \vDash A^*$, then $\mathbf{PC} \vDash B^*$, so by induction on the length of a proof of A we have that if $\mathbf{R} \vdash A$ then $\mathbf{PC} \vDash A^*$.

Now suppose that $\mathbf{R} \not\models A$. Then there is a model $<v, R>$ such that $<v, R> \not\models A$. We want to define a **PC**-model w in which A^* is false, but I will do more. I will show that given any model $<v, R>$ of **R** there is a model w of **PC** such that for all B,

(†) $\quad v(B) = T$ iff $w(B^*) = T$

That proof can then be read as establishing a 1-1 correspondence between **PC**-models and **R**-models, since the same definition converts any model w of **PC** into one for **R** satisfying (†).

We define the model by:

$w(q_i) = T$ iff $v(p_i) = T$

$w(d_{ij}) = T$ iff $R(p_i, p_j)$ holds

We can then prove that $<v, R> \models B$ iff $w \models B^*$ by induction on the length of the formula B. The only interesting step is if B is $C \to D$. Then

$v(C \to D) = T$ iff $(v(C) = F$ or $v(D) = T)$ and $R(C, D)$

iff $(v(C) = F$ or $v(D) = T)$ and (some p_i in C, some p_j in D, $R(p_i, p_j)$ or some p_i appears in both C and D)

which by induction is

iff $(w(C^*) = F$ or $w(D^*) = T)$ and (some p_i in C, some p_j in D, $w(d_{ij}) = T$ or $(d_{ii} \vee \neg d_{ii})$ appears in $(C \to D)^*$)

iff $w((C^* \supset D^*) \wedge [\bigvee_{\{p_i \text{ in } C,\, p_j \text{ in } D\}} d_{ij} \vee \bigvee_{\{p_i \text{ in } C \text{ and } p_i \text{ in } D\}} (d_{ii} \vee \neg d_{ii})]) = T$

So in particular, for the model which does not validate A we have $w(A) = F$, and the proof is complete. ■

Theorem 9 $\quad \mathbf{S} \hookrightarrow \mathbf{PC}$

Proof: Recall (§III.J, p.81):

$\mathbf{S} = \mathrm{Th}(\mathbf{R} \cup \{R(A,B) \to R(B,A) \colon A, B \in \text{Wffs}\}$

Only a slight modification of the translation of **R** is needed: for $(A \to B)^*$ read $d_{ij} \wedge d_{ji}$ where d_{ij} appeared previously. You can check that **PC** validates the translation of $R(A,B) \to R(B,A)$, and the proof then is the same as for **R**. ■

Remarks on the translation:

a. Though the mapping of **R** into **PC** is not grammatical, it does preserve the structure of the language of **R**. If we first re-interpret $A \to B$ in **R** using the defined connectives $\wedge$, R, and (truth-functional inclusive) $\vee$ as:

$\gamma(A \to B) = \neg(A \wedge \neg B) \wedge$
$[\bigvee_{\{p_i \text{ in } A,\, p_j \text{ in } B\}} R(p_i, p_j) \vee (\bigvee_{\{p_i \text{ in } A \text{ and } p_i \text{ in } B\}} (R(p_i, p_i) \vee \neg R(p_i, p_i))]$

then * preserves the structure of these sentences, mapping $R(p_i, p_i)$ to d_{ij} (cf. the Normal Form Theorem for **S**, Corollary III.6, p.79).

Similar comments apply to **S**.

b. By the standards of subject matter relatedness logic many classically valid arguments are *enthymematic*. That is, one or more suppressed implicit premisses are needed to make them valid. For example, the following argument is valid in **PC**:

> If $1+1=2$, then Mary has two children.
> If Mary has two children, then John loves Mary.

Therefore:

> If $1+1=2$, then John loves Mary.

But it is enthymematic in **S**. We need the suppressed premisses:

> '$1+1=2$' has something in common with 'Mary has two children'.
> 'Mary has two children' has something in common with 'John loves Mary'.
> '$1+1=2$' has something in common with 'John loves Mary'.

The translation of **S** into **PC** described in (b) is the formal counterpart of this view.

c. Given a decision procedure for **PC** we can produce a decision procedure for **R** based on it which takes no more than a simple polynomial function of the time of the original procedure.

d. The promised translation of **R** into **PC** in $L(p_0, p_1, \ldots \neg, \rightarrow)$ is:

$(p_i)^* = p_{2i}$

$(\neg A)^* = \neg(A^*)$

$(A \rightarrow B)^* =$ as above reading $p_{3^i 5^j}$ for d_{ij}.

5. Some questions and comments

a. How ubiquitous are translations? Is there an example of two logics such that there is no translation from the first to the second? I suspect that there is no translation of **D** into **PC**, but have been unable to prove that even on the assumption that $(p)^* = p$, $(\neg A)^* = \neg(A^*)$, and $(A \wedge B)^* = (A^* \wedge B^*)$.

I also suspect that there is no translation of **Int** into **PC**.

b. Indeed, I suspect that there may be no nontrivial example of a logic which can be translated grammatically into **PC**. By 'nontrivial' I mean to exclude *ad hoc* examples and fragments of **PC** or of other logics which wouldn't normally be considered well-motivated logics in their own right.

c. In the proof of Theorem 8 we used the fact that the logics of the translation have semantic deduction theorems to reduce the question of whether a map is a translation to whether it preserves validity. The method is often applicable, though not universally so (Theorem 4). In summary the method is:

$\{A_1, \ldots, A_n\} \vDash_{\mathbf{L}} A$	iff	$\vDash_{\mathbf{L}} \beta_n(A_1, \ldots, A_n, A)$	by a Deduction Theorem for **L**
	iff	$\vDash_{\mathbf{M}} (\beta_n(A_1, \ldots, A_n, A))^*$	as * preserves validity
	iff	$\vDash_{\mathbf{M}} \delta_n(A_1{}^*, \ldots, A_n{}^*, A^*)$	because * is grammatical
	iff	$\vDash_{\mathbf{M}} \gamma_n(A_1{}^*, \ldots, A_n{}^*, A^*)$	by some internal translation within **M**
	iff	$\{A_1{}^*, \ldots, A_n{}^*\} \vDash_{\mathbf{M}} A^*$	by a Deduction Theorem for **M**

d. Do we get anything new by allowing parameters in translations? Call a map *grammatical with parameters* if parameters are allowed in the schemas of Definition 2. Is there a "natural" example of two logics **L** and **M** such that there is a grammatical translation with parameters from **L** into **M**, but none without parameters? I believe that there is no grammatical translation with parameters of any of the logics of Theorem 6 into **PC**.

e. In what ways are grammatical translations preferable to nongrammatical ones?

It can't be on the grounds of simplicity and regularity, as the double negation translation of **PC** into **Int** shows. But that is a global translation and does not reduce the connectives of the one logic to those of the other. Grammatical translations by preserving structure yield a *translation of the connectives.*

But then what about the translation of **R** into **PC**? The connectives are translated into specific structures, and once the semantic assumptions of these logics are understood the translation is natural. However, this translation involves metalogical machinery which seems not to be inherently propositional in nature: paying attention to the indices of variables or, in terms of comment (a) of the last section, translating a compound proposition into an atomic one.

Grammatical translations are those which translate connectives and all the structure of propositions in a manner consonant with the propositional nature of the logics.

B. Semantically Faithful Translations

What does it mean to say that a translation preserves meaning? The meanings of words and sentences in the one language should be reconstructed, understood, in terms of the meanings ascribed to words and sentences in the other language. To give a formal analysis of this idea within even the limited scope of translations between logics requires some uniform conception of how meanings are ascribed to languages. That is the role of the general framework for semantics of Chapter IV.

I will propose a definition here which I believe captures necessary conditions for a translation to preserve meaning. Whether the definition expresses sufficient conditions cannot easily be resolved, for that depends in part on whether particular semantics for a logic can reasonably be said to give meanings to the language of the logic. I will, however, show that the definition accords with our pre-formal intuition for the translations we have seen in this volume. In particular, the best known example of a translation between nonclassical logics which is generally thought to recapture the meaning of the one within the other, **Int** $\twoheadrightarrow$ **S4**, will be classified as preserving meaning, while the problematic, unintuitive translations of **PC** into **Int** do not preserve meaning with respect to their intended models.

The definitions I propose will only be for logics which have set-assignment semantics. I am not sure how to extend them to relation based logics.

1. A formal notion of semantically faithful translation

How do we justify that a mapping is a translation?

Given * a mapping from $L_{\mathbf{L}}$ to $L_{\mathbf{M}}$, the languages of logics **L** and **M** respectively, we need to show, for all Γ and A,

(a) If $\Gamma \vDash_{\mathbf{L}} A$ then $\Gamma^* \vDash_{\mathbf{M}} A^*$.

(b) If $\Gamma^* \vDash_{\mathbf{M}} A^*$ then $\Gamma \vDash_{\mathbf{L}} A$.

The first, (a), can sometimes be accomplished by using a strongly complete axiomatization of **L**: axioms of **L** are shown to translate into tautologies of **M**, and the rules of **L** correspond to derived rules on the set of translated wffs in $L_{\mathbf{M}}$. Deduction theorems for the logics are then invoked to complete the proof (cf. the proof of Theorem 8).

But that method, when available, does not generalize to (b); nor does it give any insight into how, or even whether the semantics of **L** can be reconstructed within those of **M**. We have seen, however, completely semantic justifications in this text, which proceed in the following way.

We prove (a) by showing that if $\Gamma^* \nvDash_{\mathbf{M}} A^*$ then $\Gamma \nvDash_{\mathbf{L}} A$. If $\Gamma^* \nvDash_{\mathbf{M}} A^*$ then there is a model M is of **M** such that $\mathsf{M} \vDash_{\mathbf{M}} \Gamma^*$ and $\mathsf{M} \nvDash_{\mathbf{M}} A^*$. So we find a model L of **L** such that for all B,

(†) $\mathsf{L} \vDash_{\mathbf{L}} B$ iff $\mathsf{M} \vDash_{\mathbf{M}} B^*$

Then $\mathsf{L} \vDash_{\mathbf{L}} \Gamma$ and $\mathsf{L} \nvDash_{\mathbf{L}} A$. A correspondence of models is set up; if every model of **L** can be derived from some M satisfying (†), then we also have (b).

What can we say about this correspondence of models? Consider $\text{Wffs}^* = \{A^* \in L_{\mathbf{M}} : A \in L_{\mathbf{L}}\}$. The model M restricted to Wffs* determines the model L of **L** via (†). Or at least it determines what wffs L validates, since there may be other models of **L** which validate the same wffs.

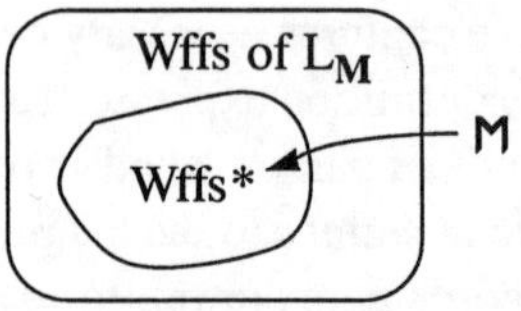

The model L, on the other hand, need not determine M. For example, for the translation **PC** $\twoheadrightarrow$ $\mathbf{L_3}$ (Theorem VIII.2, p.238) 2-valued models of **PC** are derived from 3-valued models of $\mathbf{L_3}$ via the condition

$$v(p) = T \text{ iff } e(p) = 1$$

The same 2-valued model arises whether the 3-valued evaluation assigns 0 or $\frac{1}{2}$ to variables to which it does not assign 1.

The correspondence, then, is from the class of models $\mathcal{M}$ of **M** to the class of models $\mathcal{L}$ of **L**. For simplicity let's denote it also by *.

$$L_L \xrightarrow{\ *\ } L_M$$

$$\mathcal{L} \xleftarrow{\ *\ } \mathcal{M}$$

Where this method was used in the text the model correspondences were established between the idiosyncratic models of each logic. How can we recast them within the general framework?

Given a model $<v, s, S>$ of **M**, how do we determine $<v^*, s^*, S^*>$? Condition (†) tells us what variables are accounted as true:

$$v^*(p) = v(p^*)$$

And the analogous condition to impose on content is:

$$s^*(A) = s(A^*)$$

This may be too strong, for it requires that $S^* \subseteq S$. Sufficient would be a map $\theta : S \to S^*$ such that

$$s^*(A) = \theta(s(A^*))$$

But such a θ would introduce serious complications into any discussion of whether a translation preserves meaning, and appears unnecessary if appropriate content sets are used for the semantics. So I make the following definitions.

Definition 10 Let **L**, **M** be logics, L_L, L_M their languages, and $\mathcal{L}, \mathcal{M}$ classes of set-assignment models for them.

a. Two models L and L′ of $\mathcal{L}$ are *elementarily equivalent* if for all B,

$$L \models_L B \text{ iff } L' \models_L B$$

b. A mapping $* : L_L \to L_M$ is *model preserving up to elementary equivalence* with respect to $\mathcal{L}$ and $\mathcal{M}$ if it induces a mapping $* : \mathcal{M} \to \mathcal{L}$,

$$\mathsf{M} = \langle \upsilon, s, S \rangle \mapsto \langle \upsilon^*, s^*, S^* \rangle = \mathsf{M}^*$$

via

$S^* \subseteq S$

$\upsilon^*(p) = \upsilon(p^*)$ for every propositional variable p of $L_{\mathbf{L}}$

$s^*(A) = s(A^*)$ for every wff $A \in L_{\mathbf{L}}$

and for every $A \in L_{\mathbf{L}}$,

$\upsilon^*(A) = \upsilon(A^*)$

And for every $\mathsf{L} \in \mathcal{L}$ there is some $\mathsf{M} \in \mathcal{M}$ such that M^* is elementary equivalent to L.

If the logic **L** is **PC** and $\mathcal{L}$ consists entirely of models which have no set-assignments, then all references to s and S in the definition are deleted.

c. A mapping $*: L_{\mathbf{L}} \to L_{\mathbf{M}}$ is *model preserving* with respect to $\mathcal{L}$ and $\mathcal{M}$ if it is model preserving up to elementary equivalence with respect to $\mathcal{L}$ and $\mathcal{M}$ and the mapping $*: L_{\mathbf{L}} \to L_{\mathbf{M}}$ is onto.

Theorem 11 If * is model preserving up to elementary equivalence with respect to strongly complete classes of models, then * is a translation.

Proof: Suppose that * is model preserving up to elementary equivalence with respect to the strongly complete classes of models $\mathcal{L}$ and $\mathcal{M}$ for the logics **L** and **M**. Then

$\Gamma^* \nvDash_{\mathbf{M}} A^*$ iff there is some $\mathsf{M} \in \mathcal{M}$ such that $\mathsf{M} \vDash_{\mathbf{M}} \Gamma^*$ and $\mathsf{M} \nvDash_{\mathbf{M}} A^*$

iff there is some $\mathsf{M} \in \mathcal{M}$ such that $\mathsf{M}^* \vDash_{\mathbf{L}} \Gamma$ and $\mathsf{M}^* \nvDash_{\mathbf{L}} A$

So if $\Gamma^* \nvDash_{\mathbf{M}} A^*$ then $\Gamma \nvDash_{\mathbf{L}} A$. And if $\Gamma \nvDash_{\mathbf{L}} A$ then there is some model $\mathsf{L} \in \mathcal{L}$ such that $\mathsf{L} \vDash_{\mathbf{L}} \Gamma$ and $\mathsf{L} \nvDash_{\mathbf{L}} A$. There is then some model $\mathsf{M} \in \mathcal{M}$ such that M^* is elementarily equivalent to L, so $\mathsf{M} \vDash_{\mathbf{M}} \Gamma^*$ and $\mathsf{M} \nvDash_{\mathbf{M}} A^*$. ■

Corollary 12 If * is model preserving with respect to strongly complete classes of models, then * is a translation.

Corollary 12 is much easier to prove directly, for in that case:

$\Gamma \vDash_{\mathbf{L}} A$ iff for every $\mathsf{L} \in \mathcal{L}$, if $\mathsf{L} \vDash_{\mathbf{L}} \Gamma$ then $\mathsf{L} \vDash_{\mathbf{L}} A$

iff for every $\mathsf{M} \in \mathcal{M}$, if $\mathsf{M}^* \vDash_{\mathbf{L}} \Gamma$ then $\mathsf{M}^* \vDash_{\mathbf{L}} A$

iff for every $\mathsf{M} \in \mathcal{M}$, if $\mathsf{M} \vDash_{\mathbf{M}} \Gamma^*$ then $\mathsf{M} \vDash_{\mathbf{M}} A^*$

iff $\Gamma^* \vDash_{\mathbf{M}} A^*$

We can define a mapping to be *content variant model preserving* (*up to elementary equivalence*) by requiring in the definitions above instead of $S^* \subseteq S$

that there be a map $\theta: \mathsf{S} \rightarrow \mathsf{S}^*$ such that $\mathsf{s}^*(A) = \theta(\mathsf{s}(A^*))$. I will not investigate such mappings here.

Earlier I said that for a translation to preserve meaning, the meanings of words and sentences in the one language should be reconstructed in terms of the meanings ascribed to words and sentences in the other language. Within the limited scope of propositional logic the only words which are of concern are those which are formalized by the connectives. To preserve the structure and meaning of those we know, from §A, that the mapping should be grammatical.

Definition 13 A translation is *semantically faithful* if it is grammatical and is model preserving with respect to strongly complete semantics for the logics.

I'll leave the proof of the following theorem to you.

Theorem 14 The composition of maps which are model preserving up to elementary equivalence is model preserving up to elementary equivalence.
The composition of model preserving maps is model preserving.
The composition of semantically faithful maps is semantically faithful.

Why not require semantically faithful translations to be simply model preserving up to elementary equivalence rather than model preserving? If we have a strongly complete class of models for a logic which contains elementary equivalent models, then either the distinctions between those models in terms of their content is significant, and hence should be respected by a mapping we wish to call semantically faithful. Or else the distinctions are inessential, and hence the class of models should be pared down to give a reduced class which is strongly complete for the logic and which still can be interpreted as giving meaning to the language.

With a semantically faithful translation we can completely reconstruct the semantics of the original logic within the semantics of the logic into which we are translating. In the correlation of models the translation of a connective has the same table as the connective in the original logic. Suppose that * is a semantically faithful translation from **L** to **M**, and L is a model of **L**. Then for some model of $\langle \upsilon, \mathsf{s} \rangle$ of **M**, $\mathsf{L} = \langle \upsilon^*, \mathsf{s}^* \rangle$. And, for example,

(#) $\upsilon^*(A \rightarrow B) = \mathsf{T}$ iff (not both $\upsilon^*(A) = \mathsf{T}$ and $\upsilon^*(B) = \mathsf{F}$) and $\mathsf{B}(\mathsf{s}^*(A), \mathsf{s}^*(B))$

But also, since * is grammatical,

$$\begin{aligned} \upsilon^*(A \rightarrow B) = \mathsf{T} \ &\text{iff} \ \upsilon((A \rightarrow B)^*) = \mathsf{T} \\ &\text{iff} \ \upsilon(\psi(A^*, B^*)) = \mathsf{T} \end{aligned}$$

Since ψ is schematically constructed from the connectives of $L_{\mathbf{M}}$, it is truth-and-content functional in the semantics of **M**. That is,

$\upsilon(\psi(A^*, B^*))$ depends only on $\upsilon(A^*)$, $\upsilon(B^*)$, $\mathsf{s}(A^*)$, and $\mathsf{s}(B^*)$

But then

$v(\psi(A^*, B^*))$ depends only on $v^*(A), v^*(B), s^*(A)$, and $s^*(B)$

And from (#) we know what that dependence must be:

$v(\psi(A^*, B^*)) = \mathsf{T}$ iff (not both $v^*(A) = \mathsf{T}$ and $v^*(B) = \mathsf{F}$) and $\mathsf{B}(s^*(A), s^*(B))$

The translation reconstructs the original semantics for each connective as well as for sentences as wholes.

2. Examples of semantically faithful translations

Let's review the translations we have seen in this volume, as summarized in §A.2. The semantic justifications which I quote below were established in the text relative to models idiosyncratic to the particular logics. They can be converted into model correspondences relative to set-assignment semantics in a straightforward way, and I will leave that to you.

a. Translations of a logic to "itself"
Every translation of a logic to itself which we have seen is semantically faithful. Each is designed to recreate models of the one language in terms of models of the other by defining the connectives in terms of others which are chosen to be primitive.

b. Translations of classical logic
Each grammatical translation of **PC** to another logic in the text is semantically faithful, as you can check. All, that is, except for the translation of **PC** into **Int**, which I discuss in §B.4 below.

c. Translations of **D**, **Dual D**, *and* **Eq**
The translations **D** $\twoheadrightarrow$ **Dual D** and **Dual D** $\twoheadrightarrow$ **D** are the simplest examples I have of semantically faithful translations between nonclassical logics. Recall that $\neg$ is translated homophonically and $(A \to B)^* = \neg B^* \to \neg A^*$. Thus the containment $\supseteq$ which governs the table for $\to$ in **D** is converted to the containment $\subseteq$ governing the table for $\to$ for **Dual D**, and vice versa. This is the only example I have of different logics which can be translated semantically faithfully into each other. I will discuss in §B.6 whether these translations should lead us to classify **D** and **Dual D** as being the "same" logic.

The translation **Eq** $\twoheadrightarrow$ **D** is semantically faithful, since the effect of $(A \to B)^* = (A^* \to B^*) \wedge (\neg B^* \to A^*)$ is to recreate the relation of equality of contents governing the table for $\to$ for **Eq** from the containment $\supseteq$ governing the table for $\to$ for **D**. Similarly, **Eq** $\twoheadrightarrow$ **Dual D** is semantically faithful.

By varying slightly the semantics for **D** the translation **Dual D** $\twoheadrightarrow$ **D** can provide an example of a grammatical translation which is *model preserving up to elementary equivalence but not model preserving* with respect to those semantics.

Define the class of models for **D** as in the text with the proviso that the content set **S** = the real numbers; define the class of models for **Dual D** as in the text with the proviso that the content set **S** = the natural numbers. These classes are strongly complete for the logic, but the translation **Dual D** $\twoheadrightarrow$ **D** cannot be model preserving with respect to them since the cardinality of the class for **D** is greater than that of the class for **Dual D**.

The remaining translations in §A.2 are those involving **Int** and those of relatedness logics to classical logic, which I discuss below.

3. The archetype of a semantically faithful translation: Int $\twoheadrightarrow$ S4

The best known translation between nonclassical logics which is generally thought to preserve meaning is that of **Int** into **S4**. I will show that it is semantically faithful.

Recall that the translation from the language $L(\neg, \rightarrow, \wedge, \vee)$ to $L(\neg, \wedge, \Box)$ is given by:

$$p^* = \Box p$$

$$(A \wedge B)^* = A^* \wedge B^* \qquad (\neg A)^* = \Box\neg(A^*)$$

$$(A \vee B)^* = A^* \vee B^* \qquad (A \rightarrow B)^* = \Box(A^* \supset B^*)$$

In this section I will use the symbol * for this map only. It is clearly grammatical.

We justified that * is a translation by reconstructing Kripke models of **Int** within Kripke models of **S4** (Theorem VII.15, p. 210). Since we have already established correlations of the Kripke models to set-assignment models of each logic, the proof that * is model preserving will be not be difficult.

The class of **S4** set-assignment models was defined in §VI.F, p. 166; it is strongly complete for **S4** (Theorem VI.52). The appropriate strongly complete class of set-assignment models for **Int** is $\mathcal{I}$, defined on p. 218 (Theorem VII.26).

Theorem 15 The translation * of **Int** to **S4** is semantically faithful with respect to $\mathcal{I}$ and the class of **S4** set-assignment models.

Proof: Consider the diagram

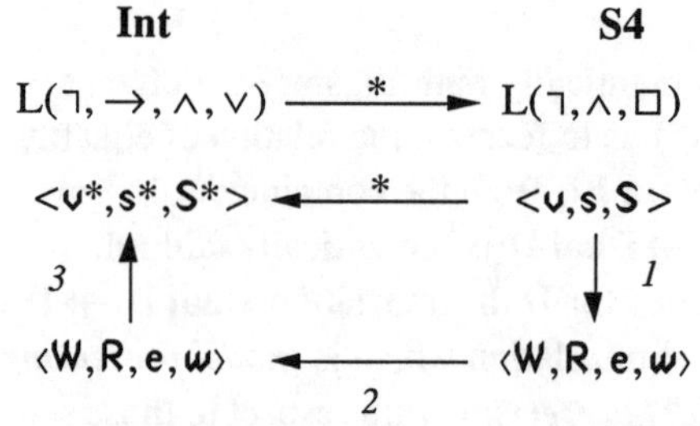

Map *1* from **S4** set-assignment models to Kripke models of **S4** is given in Theorem VI.9 (p.167). By Lemma VI.7 (p.166) it is onto.

Map *2* from Kripke models of **S4** to Kripke models of **Int** is simply the re-interpretation of the connectives, and is described in Theorem VII.15 (p.210). It is clearly onto.

Map *3* from Kripke models of **Int** to set-assignment models of **Int** which are in $\mathcal{I}$ is given in Theorem VII.26 (p.218) and Lemma VII.24, where it is shown to be onto.

The composition of these maps applied to a model $<v, s, S>$ of **S4** yields a model $<v^*, s^*, S^*>$ of **Int** given by:

$$S = S^*$$

$$v^*(p) = v(\Box p)$$

$$s^*(A) = s(A^*)$$

and $v^*(A) = v(A^*)$. Hence the composition of these maps, which is onto, is the induced map of set-assignment models, justifying that * is model preserving. Since * is grammatical, it is a semantically faithful translation. ■

Recall that * is also a finite consequence translation of **Int** into **S4Grz**. I suspect that it is model preserving with respect to the set-assignment models for **S4Grz**, and that those models are strongly complete for **S4Grz**.

4. The translations of PC into Int

Intuitionists say that classical logic is incoherent, it makes no sense to them. The classical logician points to the translations of **PC** into **Int** and argues that classical logic can be perfectly well understood by the intuitionist. But do the translations justify that? I will show that at least with respect to the standard models of **Int** and **PC**, the translations we have seen are not semantically faithful.

Lemma 16 Let $\mathcal{M}$ be a complete class of models for **Int** all of which use the intuitionist truth-conditions. Then for every propositional variable p there is some model $<v, s>$ in $\mathcal{M}$ such that $v(\neg\neg p) = F$ and $s(\neg\neg p) \neq \varnothing$.

Proof: Suppose to the contrary that for all $<v, s>$ in $\mathcal{M}$, $v(\neg\neg p) = T$, or $v(\neg\neg p) = F$ and $s(\neg\neg p) = \varnothing$. Then for each $<v, s>$, $v(\neg\neg p) = T$ or $v(\neg\neg\neg p) = T$, and hence $v(\neg\neg p \vee \neg\neg\neg p) = T$. So by completeness we would have $\vdash_{\mathbf{Int}} \neg\neg p \vee \neg\neg\neg p$, which is a contradiction, for we can have a Kripke model of **Int** satisfying:

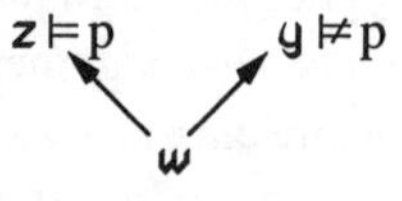

■

Again, recall that Gentzen's translation of **PC** into **Int** (p. 214) takes every variable to its double negation, translates $\neg$, $\rightarrow$, $\wedge$ homophonically, and takes $(A \vee B)^* = \neg(\neg A^* \wedge \neg B^*)$. The double negation translation of **PC** into **Int** simply takes every wff to its double negation (pp. 211–212). So, for example, the double negation translation takes $p \vee \neg p$ to $\neg\neg(p \vee \neg p)$, and Gentzen's translation takes $p \vee \neg p$ to $\neg(\neg\neg\neg p \wedge \neg\neg\neg\neg p)$.

Theorem 17 Let $\mathcal{M}$ be a complete class of models for **Int** all of which use the intuitionist truth-conditions. Then neither Gentzen's translation of **PC** into **Int** nor the double negation translation of **PC** into **Int** is model preserving with respect to $\mathcal{M}$ and the standard models for **PC**.

Proof: The proof is the same for both translations. Let $*$ be the translation and suppose it is model preserving. Let p be any propositional variable. By the previous lemma there is some $<v, s>$ in $\mathcal{M}$ such that $v(\neg\neg p) = \mathsf{F}$ and $s(\neg\neg p) \neq \varnothing$. So $v(\neg\neg\neg p) = \mathsf{F}$, too. Let v^* be the **PC**-evaluation which is induced by $*$ from v. Then

$$v^*(p) = v(p^*) = v(\neg\neg p) = \mathsf{F}$$

$$v^*(\neg p) = v((\neg p)^*) = v(\neg\neg\neg p) = \mathsf{F}$$

which contradicts that v^* is a **PC**-evaluation. Hence $*$ is not model preserving. ■

The proof actually shows more: if $\mathcal{M}$ uses just the intuitionist truth-table for negation and the classical table for disjunction then these translations cannot even be model preserving up to elementary equivalence with respect to $\mathcal{M}$.

Set-assignment semantics arising from any of the standard semantics for **Int** (Kripke, algebraic, topological, etc.) will, I believe, use the intuitionist truth-conditions, for in essence all these semantics are all "isomorphic". Thus at least with respect to the semantics currently proposed for **Int**, neither the double negation translation nor Gentzen's translation is model preserving.

The question remains whether there is some other class of models for **Int** with respect to which Gentzen's translation is semantically faithful, or whether there is some other semantically faithful translation of **PC** into **Int**. If there were one, then it would be grammatical, and hence we would have $(\neg A)^* = \varphi((A^*))$. So $v^*(\neg A) = \mathsf{T}$ iff $v(\varphi(A^*)) = \mathsf{T}$; and $v^*(A) = \mathsf{T}$ iff $v(A^*) = \mathsf{T}$. Thus $v(A^*) = \mathsf{T}$ iff $v(\varphi(A^*)) = \mathsf{F}$. That is, classical negation would be definable on the class of formulas Wffs* (cf. the query in §VII.D.3.b, p. 222).

The double negation mapping is also a translation of **PC** into the minimal intuitionist logic **J** (§VII.F.3), and the same discussion applies there. In particular, it is not model preserving with respect to any complete class of set-assignment models for **J** which use the minimal intuitionist table for negation and classical disjunction (§VII.F.4): simply replace $s(\neg\neg p) \neq \varnothing$ by $s(\neg p) \not\subseteq s(\neg\neg p)$ in Lemma 16 above.

5. The translation of S into PC

When I gave a translation of subject matter relatedness logic, **S**, into classical logic in §A.4 I remarked that the translation seems to reflect the underlying semantic assumptions. In this section I will show that it is model preserving.

Recall that the translation is

$(p_i)^* = q_i$

$(\neg A)^* = \neg(A^*)$

$(A \rightarrow B)^* =$

$$(A^* \supset B^*) \wedge [\bigvee_{\{p_i \text{ in } A,\ p_j \text{ in } B\}} d_{ij} \vee \bigvee_{\{p_i \text{ in } A \text{ and } p_i \text{ in } B\}} (d_{ii} \vee \neg d_{ii})]$$

This translation cannot be model preserving with respect to the standard models for **PC** because the set-assignments for the models for **S** must be derived from the set-assignments of the models for **PC**. So I first define a new class of models for **PC**.

Let ***P*** be the class of set-assignment models for the language

$$L(q_0, q_1, \ldots, \{d_{ij}: i,j \text{ are natural numbers}\}, \neg, \wedge)$$

which use the classical tables for the connectives, where S is countable, and for all i,j,

a. $s(q_i) \neq \varnothing$.

b. $v(d_{ij}) = T$ iff $s(q_i) \cap s(q_j) \neq \varnothing$

c. $s(A) = \bigcup\{ s(q_i) : q_i \text{ appears in } A\}$

To see that ***P*** is strongly complete for **PC** it's enough to note that given any standard **PC**-evaluation v for this language, we can define:

$S = \{(i,j): i \geq 0 \text{ and } j \geq 0\}$

$s(d_{ij}) = \varnothing$ for all i,j

$s(q_i) = \{(i,i)\} \cup \bigcup_j \{ (i,j), (j,i): v(d_{ij}) = T \}$

which is a set-assignment satisfying (a), (b) and (c).

Theorem 18 The translation * of **S** into **PC** is model preserving with respect to the standard models for **S** and the class of models ***P*** for **PC**.

Proof: Given a model $\langle w, t \rangle$ of **S**. I will show that there is a model $\langle v, s \rangle \in \boldsymbol{P}$ such that for all A, $v(A^*) = w(A)$ and $s(A^*) = t(A)$.

Define for all i,j:

$v(q_i) = T$ iff $w(p_i) = T$

$v(d_{ij}) = T$ iff $t(p_i) \cap t(p_j) \neq \varnothing$

$s(q_i) = t(p_i)$

$s(d_{ij}) = \varnothing$

$\mathsf{s}(B) = \bigcup\{\mathsf{s}(q_i) : q_i \text{ appears in } B\}$

Then $<\mathsf{v},\mathsf{s}> \in \mathbf{P}$, and for all A, $\mathsf{s}(A^*) = \mathsf{t}(A)$. I will prove by induction that for all A, $\mathsf{v}(A^*) = \mathsf{w}(A)$.

The only interesting case is if A is $B \rightarrow C$. Then

$$\begin{aligned} \mathsf{w}(B \rightarrow C) = \mathsf{T} \text{ iff } & (\text{not both } \mathsf{w}(B) = \mathsf{T} \text{ and } \mathsf{w}(C) = \mathsf{F}) \text{ and } \mathsf{t}(B) \cap \mathsf{t}(C) \neq \varnothing \\ \text{iff } & (\text{not both } \mathsf{v}(B^*) = \mathsf{T} \text{ and } \mathsf{v}(C^*) = \mathsf{F}) \text{ and } \mathsf{s}(B^*) \cap \mathsf{s}(C^*) \neq \varnothing \\ \text{iff } & \mathsf{v}(B^* \supset C^*) = \mathsf{T} \text{ and } \mathsf{s}(B^*) \cap \mathsf{s}(C^*) \neq \varnothing \end{aligned}$$

Now $\mathsf{s}(B^*) \cap \mathsf{s}(C^*) \neq \varnothing$ iff there is some i such that q_i appears in both B^* and C^*, or there are i, j such that q_i appears in B^*, q_j appears in C^*, and $\mathsf{s}(q_i) \cap \mathsf{s}(q_j) \neq \varnothing$ (see Lemma III.1). Hence $\mathsf{s}(B^*) \cap \mathsf{s}(C^*) \neq \varnothing$ iff for some i, $(d_{ii} \vee \neg d_{ii})$ appears in $(B \rightarrow C)^*$, or there are i, j such that $\mathsf{v}(d_{ij}) = \mathsf{T}$ and $\mathsf{v}(d_{ji}) = \mathsf{T}$ and $(d_{ij} \wedge d_{ji})$ appears in $(B \rightarrow C)^*$. So $\mathsf{w}(B \rightarrow C) = \mathsf{T}$ iff $\mathsf{v}((B \rightarrow C)^*) = \mathsf{T}$. ∎

6. Different presentations of the same logic and strong definability of connectives

What do we mean by saying that we have two different presentations of the same logic?

Model preserving intertranslatability cannot be enough, as the example of **S** and **PC** shows. Perhaps, then, the criteria should be that there are semantically faithful translations in both directions. That would classify **D** and **Dual D** as the same. Are they?

The examples we have seen in this book of cases where different presentations were called the same logic arose from choosing fewer or more connectives as primitive and defining others in terms of those. The notion of definability was, in each case, peculiar to the semantics under discussion, for example, the notion of definability with respect to the 3-valued semantics of $\mathbf{J_3}$ (§IV.B.3, pp. 269–270). Those various notions of definability can, I believe, be subsumed under one uniform notion, suggested by the discussion below Definition 13 that a semantically faithful translation recreates the table for each connective within the semantics of the other logic.

Definition 19 Let **L** be a logic with language $L_{\mathbf{L}}$, and $\mathcal{L}$ a strongly complete class of models for **L**. Let α be a connective of $L_{\mathbf{L}}$, and $L_{\mathbf{L}}/\alpha$ that language with α deleted. Denote by $\mathcal{L}/\alpha$ the class of models of $\mathcal{L}$ viewed as models for the language $L_{\mathbf{L}}/\alpha$. Then α is *strongly definable* with respect to $\mathcal{L}$ if there is a semantically faithful translation from $L_{\mathbf{L}}$ to $L_{\mathbf{L}}/\alpha$ with respect to $\mathcal{L}$ and $\mathcal{L}/\alpha$.

It is not clear to me how to define a notion of functional completeness for a collection of connectives in terms of strong definability.

Does this notion of strong definability lead to a more stringent criteria for two presentations to be classified as the same logic? Or is semantically faithful intertranslatability all we want? I do not know, though I suspect the resolution of these questions depends in part on what we mean when we say that a translation preserves meaning.

7. Do semantically faithful translations preserve meaning?

The translation $\mathbf{S} \hookrightarrow \mathbf{PC}$ is model preserving but does not, I believe, preserve meaning. It is not grammatical, and, as discussed in §A.5.e, it relies on metalogical machinery which is not inherently propositional. Moreover, the proof that it is model preserving suggests that any discussion of meanings being preserved should be reserved for translations which are semantically faithful with respect to classes of models which are agreed to ascribe meanings to the languages of the logics.

For **PC** the restriction would be clear: for a translation to preserve meaning it would have to be semantically faithful with respect to the standard 2-valued models of **PC**. For other logics it is not so evident how to restrict the choice: is there always a "standard" set of models? Indeed, even for **PC** the models $<\mathsf{v}, \mathsf{s}>$ where v evaluates the connectives classically and $\{\mathsf{s}(A): A \in \text{Wffs}\}$ is a Boolean algebra of sets (i.e., $\mathsf{s}(A \wedge B) = \mathsf{s}(A) \cap \mathsf{s}(B)$, etc.) could plausibly be said to ascribe meanings for **PC**.

Semantically faithful translations preserve all that is formally significant about a logic. If in addition the practitioners of the logic agree that the classes of models are accurate formalizations of how they understand the formal language, what more could be required for a translation to preserve meaning?

Most intuitionists would object to the idea that the interpretation of the intuitionist connectives in **S4** is a faithful explication of the logical notions of intuitionism. You do not understand intuitionism if you only understand it via modal logic, they would say. Yet for those of us who are not native speakers of intuitionism and feel that we shall never be fluent in a way that reflects the same understanding of logic and mathematics as an intuitionist, the translation affords us projective knowledge of intuitionism. That is, we can communicate with intuitionists and be certain that, confined to the language of propositional logic we will assert and deny the same propositions if we use the translation. That kind of projective knowledge is afforded by any translation. But here even the forms of how an intuitionist ascribes meaning to those sentences are recreated by us as we try to understand the language in the intuitionist's terms. Still, even those intuitionists who agree that the models of intuitionism accurately reflect their understanding of the logic continue to say,

> That is not it at all,
> That is not what I meant, at all.

Translations do not allow us to enter into another (logical) world view. But they can allow us projective knowledge of that world view. Whether we can ever do more, whether we can ever enter into another (logical) culture may be as problematic as whether we can ever fully enter into another person's world view who shares our own culture.

XI The Semantic Foundations of Logic

Concluding Philosophical Remarks

I believe that all logics can be understood in the same way: we start with our everyday language and abstract away certain aspects of linguistic units and take into account certain others by making idealizations of them. The aspects that we pay attention to determine our notion of truth.

I do not believe that there is a difference between logical and pragmatic aspects of what we call propositions, at least no difference which we can justify. We have no direct access to the world but only our uncertain perceptions of it. No two of us can share the same perceptions or thoughts so that ambiguity is essential to communication. Therefore, to call a sentence true is at best a hypothesis which we hope to share with others. This sharing, which in a sense amounts to objectivity, is brought about by common understandings which I call agreements. But this notion of truth is so basic to our experience and the fit of thought to the world that we can no longer allow ourselves to see that truth is in how we abstract, perceive, and agree, lest we have no language to talk.

The word 'agreement' is wrong, and 'convention' even more so. As Searle, *1983*, Chapter 5 § V, points out, we don't have any good nonintentional words with which to describe our backgrounds. Almost all our conventions, agreements, assumptions are implicit, tacit. They needn't be either conscious or voluntary. Many of them may be due to physiological, psychological, or, perhaps, metaphysical reasons: for the most part we shall never know. Agreements are manifested in lack of disagreement and the fact that people communicate. To be able to see that we have made (or been forced into, or simply have) a tacit agreement is to be challenged on it. If the assumption is sufficiently fundamental and widely held, we call the challenger 'mad'.

The explicit background is quite different from the implicit, though I use the same word 'agreement' for each. I think that's the best term, for it allows us to use the word 'disagreement' when our backgrounds clash. It points to the background as it affects our interactions with one another. And it is the drawing of the implicit background into explicitness by abstracting, idealizing, and simplifying that I am interested in, for that is the basis of formal logic and, as I see it, explicit objectivity.

It may be wrong to ascribe a uniformity to our backgrounds. As Peter Eggenberger has pointed out to me, from the comparatively uncontroversial 'For every act of communication there are some agreements on which it is founded', it does not follow that there are some agreements on which every act of communication is founded. But I have tried to show that this latter assumption is reasonable at least with respect to logical discourse.

In a conversation John Searle tried to convince me that truth and referring can't be a matter of agreement. He argued that when he says 'The moon is risen' and I say 'Which moon?' then he has successfully referred even if I keep saying 'Which moon?'

But I believe it is a matter of agreement that we say that the same object is in the sky each night, not 28 different ones, or one new one for each night of eternity. There may indeed be only one object "out there", but neither of us can ever know that with any certainty that transcends our perceptual framework, our background. Even relative to our background, anything that is beyond direct immediate experience must be a matter of agreement for referring; while much that is of direct immediate experience is so theory-laden as to be reasonably called a matter of agreement for referring, too.

But, Peter Eggenberger argues, suppose we're playing a game of chess. The rules are completely explicit: a rook moves in this fashion, pawns in that. The game ends with either a checkmate, a stalemate, or an agreed draw, and so on. Now suppose you move and announce, 'Checkmate'. Or as he puts it, suppose I'm in a position that "says" I'm in checkmate. Then I can't get out of it by saying that I disagree.

But of course I can, though it's extremely unlikely (it seems to us) that I could do so in good faith, that is, really not recognize it as checkmate by your "objective" standards. If I do say 'I don't agree', either through perversity or my actually not regarding such a position as checkmate, then pretty soon I'll find no one to play chess with me. Perhaps that's not such a great loss. But if I do not agree with the community's language agreements and assumptions then I cannot get anyone to talk with me. That is a loss. Objectivity arises because we'd all go mad (or be mad) if we didn't act in conformity with some implicit agreements and rules. We are all built roughly the same and we have to count on that in exchanging information about our experience. Given a particular shared background there will be plenty of room for experience and facts.

When we adopt a language we can't help but adopt the agreements on which it's based. For example, consider what happens when I teach real analysis, the theoretical foundations of calculus. I tend towards a constructivist view of mathematics and do not believe that there are any infinite entities, or at least none corresponding to what we call 'the real numbers'. But when I teach real analysis I have to adopt the language of the classical mathematician. I could preface each remark of mine with 'Of course, this is assuming that infinite totalities exist, which I don't really believe.' But I don't; I make that comment at the beginning of the course and then, slowly, forget it myself. I have to in order to be able to talk in the language of real analysis, a language which has grown out of human experience and is therefore accessible to me. I am only able to talk coherently in that language if I accept it's background assumptions. The more I talk the language, the more likely I am to forget that my acceptance was hypothetical. I convince myself that I have not betrayed my beliefs by saying that the theory I am teaching is an idealization of experience. But it could be said that I have not betrayed my beliefs in the same way when I learned Polish and began to speak it in Poland, and in doing so adopted its assumptions and conventions. I have not forgotten the categories of the world of my language, I have just put them aside in order to communicate with people who do not have or use them.

But I mustn't be fooled into thinking that a logic or language can be justified by its utility. An example comes from Michael Wrigley's paper, *1980*, and a discussion I had with him.

We use Peano Arithmetic, **PA**, because we believe it's "right". Now suppose we encountered someone who used a deviant arithmetic, **DA**, which we see is inconsistent. Turing said that we could distinguish the two arithmetics because if we built bridges using **DA** they'd fall down (more often). Well, in a sense he's right and in a sense he's wrong.

We ought to suppose that the practitioners of **DA** have their own background assumptions which surely must differ from ours, for with our background if we used **DA** to build bridges they would fall down more often. But why shouldn't we assume that relative to their background assumptions the practitioners of **DA** would build perfectly fine bridges? After all, **DA** would reflect their background assumptions well or they wouldn't use it. There is a temptation here to say that we do not have to account for the deviant arithmetic because it's no arithmetic at all: they must not be talking about the natural numbers. But we were assuming that we had some good reason for calling it 'arithmetic' in the first place. Disagreements do not disappear by saying that the subject has been changed. Moreover, how do we know that even among ourselves we all understand **PA** in the same way, that there is a common subject? We are back again at the not unreasonable hypothesis that we have a shared background so that we can communicate. And only relative to a particular background do criteria of utility have force. Turing was trying to fill in the dotted line in the following diagram.

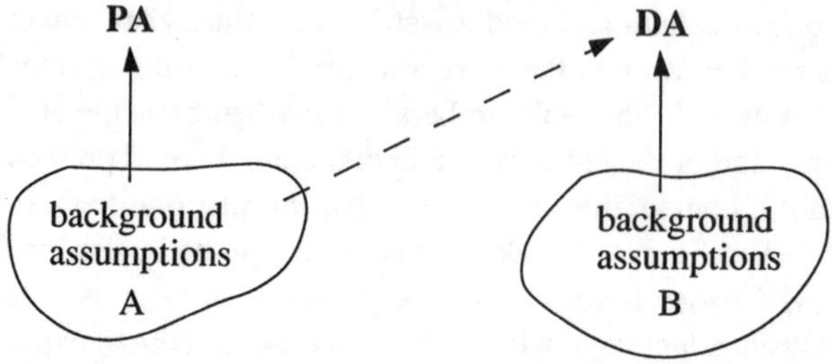

As Benson Mates pointed out to me, you don't believe it because it's useful: it's useful because you believe it. There's pragmatic value in believing that it's not a matter of pragmatics. Even to think that our fundamental background assumptions could be challenged is paralyzing to most users of language.

Well then, why should we question them? Because we have disagreements. There *are* many logics. It isn't a "let's pretend" situation as with the example of deviant arithmetic. The objectivity of our backgrounds really has been questioned.

Yet the classical logician, the intuitionist, the modal logician, the practitioner of relevance, or quantum, or a many-valued logic all do communicate and work together. They write journal articles and read ones written by the others, they collaborate however uneasily, they discuss. One possible explanation of this is that the usual daily languages of all these people share so many background assumptions that the practitioners of, say, intuitionism, cannot free themselves of the classical way of thinking any more than I can free myself of the classical real analysis way of thinking when I am helping my mathematics students. Perhaps intuitionism cannot be the challenge the intuitionists intended it to be because they only adopted a new way of mathematically talking (and thinking) while retaining the language and assumptions of the culture in which they live.

The relevance of this to doing logic and seeing the unity of logics is that because the practitioners of the various logics share the same background assumptions and agreements that come with the use of Western languages and culture (or perhaps that come with any human culture), those assumptions must be evident in their work. Those more fundamental assumptions are what I explicitly use in setting up the structural overview of propositional logic that I've presented in Chapter IV: there is a common notion of a smallest linguistic unit, called 'a proposition', which can be called true, and common notions of the connectives as portrayed in the truth-tables I give. Those tables reflect that some notion of meaning or content is ascribed to propositions and a connection of meanings or contents must be made for a compound proposition to be accepted. If the connection is there, then only the truth-values of the constituents remain to be considered and we evaluate the connectives according to the standard classical tables. Always present is the Yes-No, Accept-Reject dichotomy that we impose on (or is imposed on us by) experience.

The differences between the logics must be superficial relative to these assumptions, and the differences are, as I see it, in the choice of which, if any, less fundamental aspect (or possibly aspects) of propositions other than truth-values are to be taken into account in reasoning. Even then some aspects are so close to fundamental, so near the heart of the background, that a challenge to them seems incomprehensible to most of us. Such aspects or notions are the ones we tend to label 'logical'. To me whether a notion is called logical is a measure of how fundamental it is to our reasoning, our communication, our view of reality, not how close it *is* to reality. The following diagram pictures this view.

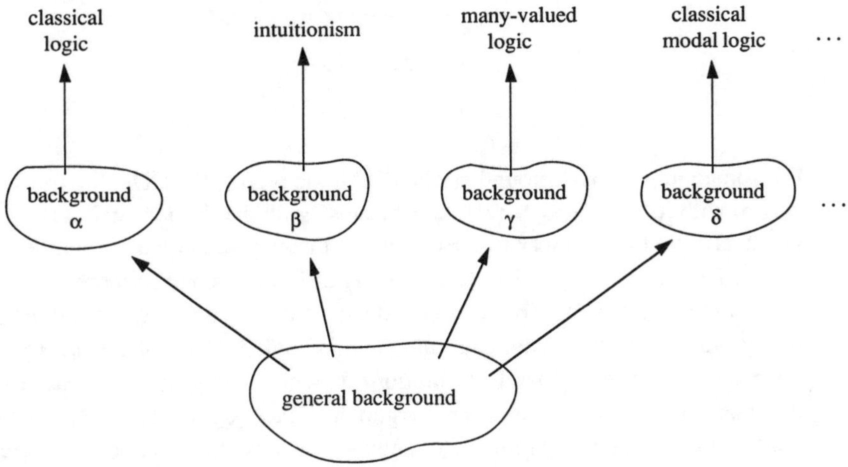

The backgrounds α, β, γ, δ, ... are a spectrum of assumptions one can have relative to our more general background; some of these overlap, others are apparently incompatible.

I believe that the structural analysis I give formalizes the general background. In the various chapters on modal, intuitionistic, many-valued, and paraconsistent logics I've tried to show that the fundamental assumptions on which I build that analysis are implicit in the reasoning of the practitioners of all these logics. I believe that similar analyses can be made for other logics which I have not included here. I may be wrong in my belief that I have uncovered an implicit background, yet if my view is compelling enough it may come to be seen as the general background.

Where is objectivity, where is 'must' in this picture? It used to be thought that it lay within one of the backgrounds α, β, γ, δ, ... , and that's been the basis of philosophical debates between practitioners of the various logics. I now believe that it lies between the backgrounds α, β, γ, δ, ... and their respective logics as depicted in the following diagram.

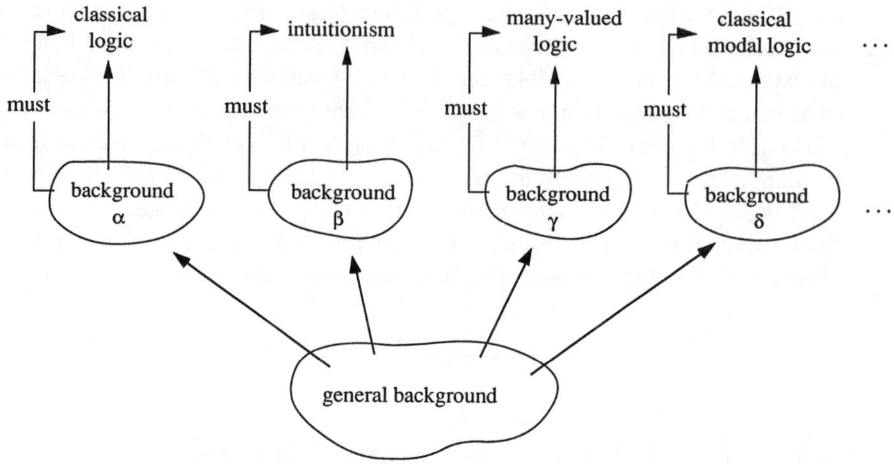

For someone with background α classical logic seems objective; we must reason in accord with it, he argues. And that is because with that background it is inconceivable to act otherwise. But it is not inconceivable to have another background, as I've argued, only extremely difficult to enter into one.

If there is a nonrelative sense of 'must' in this picture it would be in the general background. Nothing I have presented in this book depends on the answer to the question whether the general assumptions on which these logics are based are necessary truths. The metaphysical arguments may begin again, only relocated. We don't get rid of the background, for language must be anchored to the world. We only show that more or less of it as represented in language and logic can reasonably be considered universal.

Yet all our formalizations, all I've done in this book, are false if taken to be exact representations of our implicit backgrounds. What is thought remains thought, what is the external world remains the external world, and what we say is and can only be what we say, either about, because of, or through the world and thought. It is false that what is said can be directly of these other realms, or a perfect representation of them. The explicit agreements we codify are abstractions, idealizations, simplifications, and hence a distortion of our real background. These explicit agreements may make it easier for us to communicate, and we may henceforth act in accord with them; that can sometimes feed back into our backgrounds becoming imperceptibly, in a new nonexplicit way, our unconscious, implicit background.

It seems hard for me to conceive of someone who reasons who doesn't have a smallest unit of language to which a Yes-No, True-False dichotomy is applied, and hence to conceive of someone who uses connectives which are not expressible by the general tables of Chapter IV. So perhaps the story of propositional logics which I give is universal. For it to be useful for us, for us to develop it, we will probably have to believe that.

In Volume 2 I will present a general framework for predicate logics in which the aspects of predicates and names are taken as primitive, deriving from those the aspects of propositions which determine a logic. That framework seems much less likely to be universal. The object-predicate analysis of propositions is based on the assumption that the world is made up of things and that facts about the world are facts about things. An equally compelling and quite foreign assumption is that the world is made up of processes. That kind of assumption just won't mesh with the object-predicate distinction. To try to force, e.g., category theory or reasoning about actions into a predicate logic only distorts them. So perhaps the picture that will develop will be the following.

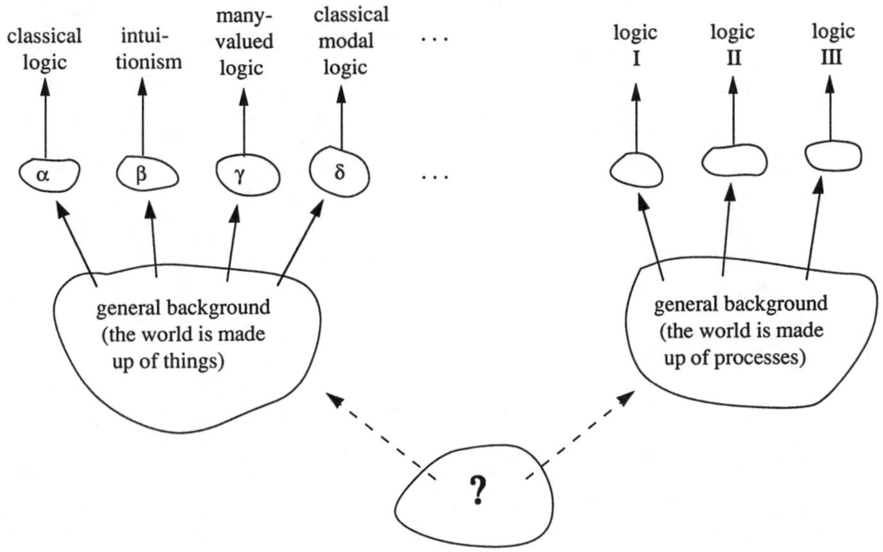

If so I am sure that efforts will be made to fill in the dotted lines, to find a still more general background, to relocate our notion of necessity so that we can live with a relativism that threatens to undermine our confidence in the way we talk and reason.

Summary of Logics

Unless noted otherwise, the axiomatization or semantics listed below is strongly complete for the logic. That is, $\Gamma \vdash A$ iff $\Gamma \vDash A$.

Axiomatizations are given in terms of schemas even in those cases where the rule of substitution was used by the originators.

For set-assignment semantics, unless noted otherwise S is countable.

A. Classical Logic, PC

A model for classical logic is a function $v: PV \to \{T, F\}$ which is extended to all wffs of $L(\neg, \to, \wedge, \vee)$ by the following tables:

A	¬A
T	F
F	T

A	B	A → B
T	T	T
T	F	F
F	T	T
F	F	T

A	B	A ∧ B
T	T	T
T	F	F
F	T	F
F	F	F

A	B	A ∨ B
T	T	T
T	F	T
F	T	T
F	F	F

Axiomatizations

PC *in* $L(\neg, \rightarrow)$ (§II.J.1)

1. $\neg A \rightarrow (A \rightarrow B)$
2. $B \rightarrow (A \rightarrow B)$
3. $(A \rightarrow B) \rightarrow ((\neg A \rightarrow B) \rightarrow B)$
4. $(A \rightarrow (B \rightarrow C)) \rightarrow ((A \rightarrow B) \rightarrow (A \rightarrow C))$

rule $\dfrac{A, A \rightarrow B}{B}$ (*modus ponens*)

PC *in* $L(\neg, \rightarrow, \wedge, \vee)$ (§II.K.6)

1. $\neg A \rightarrow (A \rightarrow B)$
2. $B \rightarrow (A \rightarrow B)$
3. $(A \rightarrow B) \rightarrow ((\neg A \rightarrow B) \rightarrow B)$
4. $(A \rightarrow (B \rightarrow C)) \rightarrow ((A \rightarrow B) \rightarrow (A \rightarrow C))$
5. $A \rightarrow (B \rightarrow (A \wedge B))$
6. $(A \wedge B) \rightarrow A$
7. $(A \wedge B) \rightarrow B$
8. $A \rightarrow (A \vee B)$
9. $B \rightarrow (A \vee B)$
10. $((A \vee B) \wedge \neg A) \rightarrow B$

rule $\dfrac{A, A \rightarrow B}{B}$

PC *in* $L(\neg, \rightarrow, \wedge)$ (§II.K.6)

Delete from the axiomatization of **PC** in $L(\neg, \rightarrow, \wedge, \vee)$ axiom schemas 8, 9, and 10.

PC *in* $L(\neg, \wedge)$ (*Rosser, 1953*)

$A \supset B \equiv_{Def} \neg(A \wedge \neg B)$

$A \supset (A \wedge A)$

$(A \wedge B) \supset A$

$(A \supset B) \supset (\neg(B \wedge C) \supset \neg(C \wedge A))$

rule $\dfrac{A, A \supset B}{B}$ (material detachment)

PC *in* $L(\neg, \rightarrow)$ (*Łukasiewicz and Tarski, 1930*)

$$(p_1 \rightarrow p_2) \rightarrow ((p_2 \rightarrow p_3) \rightarrow (p_1 \rightarrow p_3))$$

$$(\neg p_1 \rightarrow p_1) \rightarrow p_1$$

$$p_1 \rightarrow (\neg p_1 \rightarrow p_2)$$

rules $\dfrac{A, A \rightarrow B}{B}$ $\dfrac{\vdash A(p)}{\vdash A(B)}$ (substitution)

Axiomatizations of **PC** *in* $L(\neg, \rightarrow, \wedge, \vee)$ *relative to* **Int**

For axiomatizations of **Int** see pp. 335–336 below.

PC is the closure under *modus ponens* of **Int** plus any one of the following schemas (Corollary VII.6 and Theorem VII.22):

$$(A \rightarrow B) \rightarrow ((\neg A \rightarrow B) \rightarrow B)$$

$$\neg\neg A \rightarrow A$$

$$A \vee \neg A$$

B. Relatedness and Dependence Logics S, R, D, Dual D, Eq, DPC

General Form of Semantics

Except for the logic **R**, a model is a triplet $<v, s, S>$ where $v: PV \rightarrow \{T, F\}$, S is countable, and s: Wffs $\rightarrow$ Subsets of S. To extend v to all wffs, these logics use the classical tables for $\neg$ and $\wedge$, and evaluate $\rightarrow$ by:

$v(A \rightarrow B) = T$ iff $B(A, B)$ and (not both $v(A) = T$ and $v(B) = F$)

The relations B governing the tables for $\rightarrow$ for these logics are:

S	$s(A) \cap s(B) \neq \emptyset$, where for every A, $s(A) \neq \emptyset$
D	$s(A) \supseteq s(B)$
Dual D	$s(A) \subseteq s(B)$
Eq	$s(A) = s(B)$
DPC	$s(A) \supseteq s(B$

The class of *union set-assignments* are used for all these logics except **DPC**. They are defined as those satisfying:

$s(A) = \bigcup \{ s(p) :$ p appears in A$\}$

The union set-assignments can be characterized as those satisfying:

U1. $s(\neg A) = s(A)$

U2. $s(A \wedge B) = s(A) \cup s(B)$

U3. $s(A \rightarrow B) = s(A) \cup s(B)$

For each of the logics except **DPC** alternate *relation based semantics* are defined by a class of models $<v, B>$ where $v: PV \rightarrow \{T, F\}$ and $B \subseteq$ Wffs $\times$ Wffs; the class of relations for the logic is characterized below. The extension of v to all wffs uses the truth-conditions above.

S, Subject Matter Relatedness Logic

Set-Assignment Semantics (§III.D, §III.F.2)

The set-assignments are the union set-assignments which satisfy $s(A) \neq \varnothing$. The relation governing the truth-table for $\rightarrow$ is $s(A) \cap s(B) \neq \varnothing$, and $\neg$ and $\wedge$ are classical.

$R(A, B) \equiv_{Def} A \rightarrow (B \rightarrow B)$ and $<v, s> \models R(A, B)$ iff $s(A) \cap s(B) \neq \varnothing$.

$A \wedge B$ is equivalent to $\neg(A \rightarrow (B \rightarrow \neg[(A \rightarrow B) \rightarrow (A \rightarrow B)]))$.

Relation based semantics (§III.C, §III.F.1)

The class of relations governing the table for $\rightarrow$, called *subject matter relatedness relations*, are those satisfying

R1. $R(A, B)$ iff $R(\neg A, B)$

R2. $R(A, B \wedge C)$ iff $R(A, B \rightarrow C)$

R3. $R(A, B)$ iff $R(B, A)$

R4. $R(A, A)$

R5. $R(A, B \wedge C)$ iff $R(A, B)$ or $R(A, C)$

Alternatively, R may be taken as any symmetric, reflexive relation on PV $\times$ PV which is extended to all wffs by R1–R5.

Axiomatization (§III.J)

S *in* $L(\neg, \rightarrow, \wedge)$

$R(A, B) \equiv_{Def} A \rightarrow (B \rightarrow B)$ $\quad A \leftrightarrow B \equiv_{Def} (A \rightarrow B) \wedge (B \rightarrow A)$

In axiom schema 4, $C \vee D$ abbreviates $\neg(\neg C \wedge \neg D)$; that can be replaced by $\neg C \rightarrow D$.

1. $R(A, A)$
2. $R(B, A) \rightarrow R(A, B)$
3. $R(A, \neg B) \leftrightarrow R(A, B)$

4. $R(A,B\rightarrow C) \leftrightarrow (R(A,B)\vee R(A,C))$
5. $R(A,B\wedge C) \leftrightarrow R(A,B\rightarrow C)$
6. $(A\wedge B)\rightarrow A$
7. $A\rightarrow(B\rightarrow(A\wedge B))$
8. $(A\wedge B)\rightarrow(B\wedge A)$
9. $A\leftrightarrow\neg\neg A$
10. $(A\rightarrow B)\leftrightarrow(\neg(A\wedge\neg B)\wedge R(A,B))$
11. $A\rightarrow(\neg(B\wedge A)\rightarrow\neg B)$
12. $\neg(A\wedge B) \rightarrow (\neg(C\wedge\neg B)\rightarrow\neg(A\wedge C))$
13. $\neg((C\rightarrow D)\wedge(C\wedge\neg D))$

rule $\dfrac{A,\ A\rightarrow B}{B}$

R, Non-Symmetric Relatedness Logic

Set-Assignment Semantics

No set-assignment semantics are known.

Relation based semantics (§III.F.3)

The conditions defining the relations are R1, R2, R4, and R5 as for **S**, plus:

R6. $\mathsf{R}(B,A)$ iff $\mathsf{R}(B,\neg A)$

R7. $\mathsf{R}(B\wedge C,A)$ iff $\mathsf{R}(B\rightarrow C,A)$

R8. $\mathsf{R}(B\wedge C,A)$ iff $\mathsf{R}(B,A)$ or $\mathsf{R}(C,A)$

Alternatively, R may be taken to be any reflexive relation on PV×PV which is extended to all wffs by R1, R2, R4–R8.

$R(A,B) \equiv_{Def} A\rightarrow(B\rightarrow B)$ and $\langle v,\mathsf{R}\rangle\models R(A,B)$ iff $\mathsf{R}(A,B)$

$A\wedge B$ is equivalent to $\neg(A\rightarrow(B\rightarrow\neg[(A\rightarrow B)\rightarrow(A\rightarrow B)]))$.

Axiomatization (§III.J)

R *in* $L(\neg,\rightarrow,\wedge)$

Delete axiom schema 2 from the axiom system for **S** and add:

$R(B,A) \leftrightarrow R(\neg B,A)$

$R(B\rightarrow C,A) \leftrightarrow (R(B,A)\vee R(C,A))$

$R(B\wedge C,A) \leftrightarrow R(B\rightarrow C,A)$

D, Dependence Logic

Set-Assignment Semantics (§V.A.3)
Union set-assignments are used. The relation governing the truth-table for $\rightarrow$ is $s(A) \supseteq s(B)$, and $\neg$ and $\wedge$ are classical.

$D(A,B) \equiv_{Def} A \rightarrow (B \rightarrow B)$ and $<v,s> \vDash D(A,B)$ iff $s(A) \supseteq s(B)$.

$A \wedge B$ is equivalent to $\neg([(\{A \rightarrow B] \rightarrow [A \rightarrow B]) \rightarrow A] \rightarrow \neg B)$.

Relation based semantics (§V.A.4)
The class of relations governing the table for $\rightarrow$, called *dependence relations*, are those satisfying:

D is reflexive.

D is transitive.

$D(A,B)$ iff $D(A,p)$ for every p in B

Alternatively, the class of dependence relations can be characterized as those satisfying:

1. D is reflexive
2. D is transitive
3. $D(A,B \wedge C)$ iff $D(A,B)$ and $D(A,C)$
4. $D(A,B \rightarrow C)$ iff $D(A,B)$ and $D(A,C)$
5. $D(\neg A,A)$
6. $D(A,\neg A)$

Axiomatization (§V.A.7)

D *in* $L(\neg, \rightarrow, \wedge)$

$D(A,B) \equiv_{Def} A \rightarrow (B \rightarrow B)$ $\quad A \leftrightarrow B \equiv_{Def} (A \rightarrow B) \wedge (B \rightarrow A)$

1. $(D(A,B) \wedge D(B,C)) \rightarrow D(A,C)$
2. a. $D(A,\neg A)$
   b. $D(\neg A,A)$
3. $D(A,B \rightarrow C) \leftrightarrow D(A,B) \wedge D(A,C)$
4. $D(A,B \wedge C) \leftrightarrow D(A,B) \wedge D(A,C)$
5. a. $(A \wedge B) \rightarrow A$
   b. $(A \wedge B) \rightarrow (B \wedge A)$
6. $(D(A,B) \wedge \neg(A \wedge \neg B)) \leftrightarrow (A \rightarrow B)$
7. $\neg\neg A \leftrightarrow A$
8. $(\neg(A \wedge B) \wedge B) \rightarrow \neg A$
9. $(\neg(A \wedge \neg B) \wedge \neg(B \wedge C)) \rightarrow \neg(A \wedge C)$
10. $(A \leftrightarrow B) \leftrightarrow (\neg A \leftrightarrow \neg B)$

rules $\quad \dfrac{A,\ A \rightarrow B}{B} \qquad \dfrac{A,\ B}{A \wedge B}$ (adjunction)

Dual D, Dual Dependence Logic (§V.C)

Set-Assignment Semantics

Union set-assignments are used. The relation governing the truth-table for $\rightarrow$ is $s(A) \subseteq s(B)$, and $\neg$ and $\wedge$ are classical.

$D(A,B) \equiv_{Def} A \rightarrow (B \rightarrow B)$ and $<v,s> \models D(A,B)$ iff $s(A) \subseteq s(B)$.

$A \wedge B$ is equivalent to $\neg(A \rightarrow (B \rightarrow \neg[(A \rightarrow B) \rightarrow (A \rightarrow B)]))$.

Relation based semantics

Dual dependence relations are characterized as those satisfying the same laws as for dependence relations with the entries in reverse order, e.g., for (3) above, $\mathsf{D}(B \wedge C, A)$ iff $\mathsf{D}(B,A)$ and $\mathsf{D}(C,A)$.

Axiomatization

Dual D *in* $L(\neg, \rightarrow, \wedge)$

$D(A,B) \equiv_{Def} A \rightarrow (B \rightarrow B)$ $\quad A \leftrightarrow B \equiv_{Def} (A \rightarrow B) \wedge (B \rightarrow A)$

1. $D(A,B) \rightarrow \neg(D(B,C) \wedge \neg D(A,C))$
2. a. $D(A, \neg A)$
   b. $D(\neg A, A)$
3. $D(A \rightarrow B, C) \leftrightarrow (D(A,C) \wedge D(B,C))$
4. $D(A \wedge B, C) \leftrightarrow (D(A,C) \wedge D(B,C))$
5. a. $A \rightarrow (B \rightarrow (A \wedge B))$
   b. $(A \wedge B) \rightarrow (B \wedge A)$
6. a. $D(A,B) \rightarrow (\neg(A \wedge \neg B) \rightarrow (A \rightarrow B))$
   b. $(A \rightarrow B) \rightarrow D(A,B)$
   c. $(A \rightarrow B) \rightarrow \neg(A \wedge \neg B)$
7. $\neg\neg A \leftrightarrow A$
8. $(\neg(A \wedge B) \wedge B) \rightarrow (\neg A \wedge B)$
9. $(\neg(A \wedge \neg B) \wedge \neg(B \wedge C)) \rightarrow (\neg(A \wedge \neg B) \wedge \neg(A \wedge C))$
10. $(A \leftrightarrow B) \leftrightarrow (\neg A \leftrightarrow \neg B)$
11. $\neg A \rightarrow \neg(A \wedge B)$

rules $\dfrac{A, A \rightarrow B}{B}$ $\quad \dfrac{A \wedge B}{B}$

Eq, a Logic of Equality of Contents (§V.D)

Set-Assignment Semantics

Union set-assignments are used. The relation governing the truth-table for $\rightarrow$ is $s(A) = s(B)$, and $\neg$ and $\wedge$ are classical.

$E(A,B) \equiv_{Def} A \rightarrow (B \rightarrow B)$ and $<v,s> \models E(A,B)$ iff $s(A) = s(B)$.

Relation based semantics

The *Eq-relations* are those satisfying:

1. E is an equivalence relation.
2. If $\mathsf{E}(A,B)$, then $\mathsf{E}(A\wedge C, B\wedge C)$.
3. $\mathsf{E}(A,\neg A)$
4. $\mathsf{E}(A\wedge B, B\wedge A)$
5. $\mathsf{E}(A\rightarrow B, A\wedge B)$
6. $\mathsf{E}(A\wedge(B\wedge C),(A\wedge B)\wedge C)$
7. If $\mathsf{E}(A, B\wedge C)$ and $\mathsf{E}(B, A\wedge D)$, then $\mathsf{E}(A,B)$.
8. $\mathsf{E}(A, A\wedge A)$

Axiomatization

Eq *in* $L(\neg, \rightarrow, \wedge)$

$E(A,B) \equiv_{Def} A\rightarrow(B\rightarrow B) \qquad A\leftrightarrow B \equiv_{Def} (A\rightarrow B)\wedge(B\rightarrow A)$

1. $E(A,A)$
2. $E(A,B) \rightarrow E(B,A)$
3. $(E(A,B) \wedge E(B,C)) \rightarrow (E(A,C) \wedge E(A,B))$
4. $(E(A,B) \wedge E(C,C)) \rightarrow E(A\wedge C, B\wedge C)$
5. $E(A,\neg A)$
6. $E(A\wedge B, B\wedge A)$
7. $E(A\rightarrow B, A\wedge B)$
8. $E(A\wedge(B\wedge C), (A\wedge B)\wedge C)$
9. $(E(A,B\wedge C) \wedge E(B,A\wedge D)) \rightarrow ((E(A,B) \wedge E(C,C)) \wedge E(D,D)$
10. $E(A, A\wedge A)$
11. $(A\wedge B) \rightarrow (B\wedge A)$
12. $(E(A,B) \wedge \neg(A\wedge\neg B)) \leftrightarrow (A\rightarrow B)$
13. $A\leftrightarrow \neg\neg A$
14. $(\neg(A\wedge B)\wedge B) \rightarrow (\neg A\wedge B)$
15. $(\neg(A\wedge\neg B)\wedge\neg(C\wedge B)) \rightarrow (\neg(A\wedge C)\wedge E(B,B))$
16. $\neg(\neg A\wedge(A\wedge B))$

rules $\dfrac{A, A\rightarrow B}{B} \qquad \dfrac{A, B}{A\wedge B} \qquad \dfrac{A\wedge B}{B}$

DPC, Classically-Dependent Logic (§V.F)

Set-Assignment Semantics

There is only one set-assignment, $\mathsf{s}(A) = \{B : A\vdash_{PC} B\}$. The relation governing the truth-table for $\rightarrow$ is $\mathsf{s}(A)\supseteq\mathsf{s}(B)$, and $\neg$ and $\wedge$ are classical.

C. Classical Modal Logics

S4, S5, S4Grz, T, B, K, Q, T, MSI, ML, G, G*

Axiomatizations in L(¬,∧,□)

$A \to B \equiv_{Def} \Box(A \supset B) \qquad A \supset B \equiv_{Def} \neg(A \wedge \neg B)$

PC is the axiomatization of classical logic in L(¬,∧,□) (see p. 162).

logic	*axioms*	*rules*
K	**PC**	material detachment $\dfrac{A, \neg(A \wedge \neg B)}{B}$
	$\Box(A \supset B) \supset (\Box A \supset \Box B)$	necessitation $\dfrac{A}{\Box A}$
QT	**K**	material detachment
	$\Box A \supset A$	
T	**K**	material detachment
	$\Box A \supset A$	necessitation
B	**K**	material detachment
	$\Box A \supset A$	necessitation
	$A \supset \Box\Diamond A$	
S4	**K**	material detachment
	$\Box A \supset A$	necessitation
	$\Box A \supset \Box\Box A$	
S5	**S4**	material detachment
	$\Diamond A \supset \Box\Diamond A$	necessitation
S4Grz	**S4**	material detachment
	$(\Box(\Box(A \supset \Box A) \supset A)) \supset A$	necessitation
G	**K**	material detachment
	$\Box(\Box A \supset A) \supset \Box A$	necessitation
G*	**G**	material detachment
	$\Box A \supset A$	
ML	**PC** closed under necessitation	material detachment
	$\Box(A \supset B) \supset (\Box A \supset \Box B)$	
	$\Box A \supset A$	

Axiomatizations in $L(\neg, \rightarrow, \wedge)$

$\Box A \equiv_{Def} \neg A \rightarrow A \qquad A \equiv B \equiv_{Def} (A \supset B) \wedge (B \supset A)$

The axiom schema

$\Box(\neg(A \wedge \neg B)) \equiv (A \rightarrow B)$

is added to the axiomatization of each logic above, except **ML**.

ML	*axioms*	*rule*
	PC closed under necessitation	material detachment
	$\Box(A \supset B) \supset (\Box A \supset \Box B)$	
	$(A \wedge (A \rightarrow B)) \supset B$	
	$\Box(A \supset B) \leftrightarrow (A \rightarrow B)$	

Kripke Semantics (§VI.B.1)

A *model* is $\langle W, R, e \rangle$ where W is a nonempty set whose elements are called *possible worlds*, e is an *evaluation*, $e: W \rightarrow \text{Sub PV}$, and R is a binary relation on W called the *accessibility relation between possible worlds*. For $w \in W$ define $w \vDash A$ inductively, where $w \nvDash A$ means 'not $w \vDash A$':

$w \vDash p$ iff $p \in e(w)$

$w \vDash A \wedge B$ iff $w \vDash A$ and $w \vDash B$

$w \vDash \neg A$ iff $w \nvDash A$

$w \vDash A \rightarrow B$ iff for all z such that wRz, not both $z \vDash A$ and $z \nvDash B$

The derived condition for $\rightarrow$ is:

$w \vDash A \rightarrow B$ iff for all z such that wRz, not both $z \vDash A$ and $z \nvDash B$

This is taken as definition if $\rightarrow$ is a primitive.

$\langle W, R, e, w \rangle$ is called a *model with designated world* w.
$\langle W, R \rangle$ is called a *frame*.

Validity is defined:

$\langle W, R, e, w \rangle \vDash A$ iff $w \vDash A$

$\langle W, R, e \rangle \vDash A$ iff for all $w \in W$, $\langle W, R, e, w \rangle \vDash A$

$\langle W, R \rangle \vDash A$ iff for all e, for all $w \in W$, $\langle W, R, e, w \rangle \vDash A$

The following classes of frames are *complete* for the respective logics. That is, $\vdash A$ iff A is valid in every frame in the class (§VI Appendix.A, except for **S4Grz** and **G**, for which see *Boolos, 1979*).

K all frames

T all reflexive frames

B all reflexive and symmetric frames

S4	all reflexive and transitive frames
S5	all equivalence frames
S4Grz	all finite weak partial order frames (i.e., reflexive, transitive, and anti-symmetric)
QT	all frames with designated world w such that wRw
G	all finite strict partial order frames (i.e., anti-reflexive, transitive, and anti-symmetric)

For each of these logics the subclass of finite frames satisfying the conditions above is also complete (§VI Appendix.B).

For definitions of syntactic consequence relations and strong completeness theorems see §VI Appendix 1.C.

For **G*** a different notion of validity is used: see §VI.K.3.

Set-Assignment Semantics For Modal Logics (§VI.C.1)

A *modal semantics for implication* in $L(\neg, \rightarrow, \wedge)$ is a set-assignment model or class of models $<v, s, S>$ where $v: PV \rightarrow \{T, F\}$, s satisfies M1–M3 below, and v is extended to all wffs by evaluating $\neg$ and $\wedge$ classically, and $\rightarrow$ by:

(1) $v(A \rightarrow B) = T$ iff $s(A) \subseteq s(B)$ and (not both $v(A) = T$ and $v(B) = F$)

(the dual dependence truth-conditions). Every modal semantics for implication satisfies:

(2) $v(\Box A) = T$ iff $v(A) = T$ and $s(A) = S$

(3) $v(\Diamond A) = T$ iff $v(A) = T$ or $s(A) \neq \varnothing$

To define *modal semantics for implication* in $L(\neg, \wedge, \Box)$ replace (1) by (2). Then (1) and (3) are derivable.

MSI formulated in either of these languages is the collection of tautologies of all modal semantics of implication (§VI.J.3).

To define *weak modal semantics for implication*

in $L(\neg, \rightarrow, \wedge)$ replace (1) by: $v(A \rightarrow B) = T$ iff $s(A) \subseteq s(B)$
(the weak table for the conditional)

in $L(\neg, \wedge, \Box)$ replace (2) by: $v(\Box A) = T$ iff $s(A) = S$

In both languages it follows that $v(\Diamond A) = T$ iff $s(A) \neq \varnothing$.

Conditions on set-assignment models

M1. $s(A \wedge B) = s(A) \cap s(B)$

M2. $s(\neg A) = \overline{s(A)}$

M3. $s(A \rightarrow B) = s(\Box(A \supset B))$

M4. $s(\Box(A \supset B)) \subseteq s(\Box A \supset \Box B)$

M5. If $s(A) = S$ then $s(\Box A) = S$.

M6. $s(\Box A) \subseteq s(A)$

M7. $s(\Box A) \subseteq s(\Box\Box A)$

M8. $s(\Diamond A) = s(\Box\Diamond A)$

M9. If $v(A) = T$ or $s(A) \neq \varnothing$, then $s(\Diamond A) = S$.

M10. $s(\Box(\Box(A \supset \Box A) \supset A)) \subseteq s(A)$

M11. If $v(A) = F$ then $s(\Box(A \supset \Box A)) \not\subseteq s(A)$.

M12. $s(A) = s(\Box\Diamond A)$

M13. If $v(A) = T$ then $s(\Diamond A) = S$.

M14. $s(\Box(\Box A \supset A)) \subseteq s(\Box A)$

M15. $s(\Box A) \subseteq s(A)$ iff $s(A) = S$.

P. If $v(A) = T$ then $s(A) \neq \varnothing$.

Y. If $s(A) \neq \varnothing$ then $s(\Diamond A) = S$.

M6 can be replaced by: $s(A \rightarrow B) \subseteq \overline{s(A)} \cup s(B)$.

s is a **K**-*set-assignment* if there are t, C, and $S \subseteq C$ such that t: Wffs $\rightarrow$ Sub C, t satisfies M1–M5, and for all A, $s(A) = t(A) \cap S$. S is called the designated subset.

A **T**-*set-assignment* is a **K**-set-assignment where t also satisfies M6.

A **B**-*set-assignment* is a **K**-set-assignment where t also satisfies M6 and M12.

Complete set-assignment semantics in $L(\neg, \wedge, \Box)$ or $L(\neg, \rightarrow, \wedge)$
The classes of *modal semantics for implication* $<v, s>$ satisfying the following conditions are complete for the respective logics:

S4	M1–M7 (condition P may be added)
S5	M1–M9 (conditions P and Y together may be substituted for M9)
S4Grz	M1–M7, M10, and M11
T	s is a **T**-set-assignment
B	s is a **B**-set-assignment and $<v, s>$ satisfies M13
QT	s is a **K**-set-assignment
G*	M1–M5, M14, and M15
MSI	M1–M3

For each of **S4**, **S5**, **S4Grz**, **T**, **B**, and **QT** we may restrict S to be finite.

The classes of *weak* modal semantics for implication $<v, s>$ satisfying the following conditions are complete for the respective logics:

K	s is a **K**-set-assignment
G	M1–M5, M14, and M15 ($s(\Box A) \neq \varnothing$ may be added)

We may restrict S to be finite.

Alternate complete set-assignment semantics which have a simple presentation can replace those which don't. For **QT**, **T**, **B** and **K** the models are $<v, s, S, T>$ where $T \subseteq S$, and the truth-conditions are:

$\neg$ and $\wedge$ are evaluated classically

$v(A \rightarrow B) = T$ iff $(s(A) \cap T) \subseteq (s(B) \cap T)$ and (not both $v(A) = T$ and $v(B) = F$)]

The conditions on the models for the various logics are:

QT $\quad s$ satisfies M1–M5

T $\quad s$ satisfies M1–M6

B $\quad s$ satisfies M1–M6, M12, and M13*: if $v(A) = T$ then $s(A) \supseteq T$

For **K** the truth-conditions are different:

K $\quad s$ satisfies M1–M5, evaluates $\neg$ and $\wedge$ classically, and
$v(A \rightarrow B) = T$ iff $(s(A) \cap T) \subseteq (s(B) \cap T)$

For definitions of syntactic consequence relations and strong completeness theorems see §VI Appendix 1.C.

D. Intuitionistic Logics, Int and J

Int, Heyting's Intuitionist Propositional Calculus

Axiomatizations

Heyting's axiomatization, 1930

Int *in* $L(\neg, \rightarrow, \wedge, \vee)$

I. $A \rightarrow (A \wedge A)$

II. $(A \wedge B) \rightarrow (B \wedge A)$

III. $(A \rightarrow B) \rightarrow ((A \wedge C) \rightarrow (B \wedge C))$

IV. $((A \rightarrow B) \wedge (B \rightarrow C)) \rightarrow (A \rightarrow C)$

V. $A \rightarrow (B \rightarrow A)$

VI. $(A \wedge (A \rightarrow B)) \rightarrow B$

VII. $A \rightarrow (A \vee B)$

VIII. $(A \vee B) \rightarrow (B \vee A)$

IX. $((A \rightarrow C) \wedge (B \rightarrow C)) \rightarrow ((A \vee B) \rightarrow C)$

X. $\neg A \rightarrow (A \rightarrow B)$

XI. $((A \rightarrow B) \wedge (A \rightarrow \neg B)) \rightarrow \neg A$

rules $\quad \dfrac{A,\, A \rightarrow B}{B} \qquad \dfrac{A,\, B}{A \wedge B}$

Heyting used the rule of substitution rather than schemas.

Dummet's axiomatization, 1977, p.126

Int *in* L(¬, →, ∧, ∨)

1. $A \to (B \to A)$
2. $A \to (B \to (A \wedge B))$
3. $(A \wedge B) \to A$
4. $(A \wedge B) \to B$
5. $A \to (A \vee B)$
6. $B \to (A \vee B)$
7. $(A \vee B) \to ((A \to C) \to ((B \to C) \to C))$
8. $(A \to B) \to ((A \to (B \to C)) \to (A \to C))$
9. $(A \to B) \to ((A \to \neg B) \to \neg A)$
10. $A \to (\neg A \to B)$

rule $\dfrac{A, A \to B}{B}$

Segerberg's axiomatization, 1968

Int *in* L(→, ∧, ∨, ⊥)

$\neg A \equiv_{\text{Def}} A \to \bot$

1. $(A \wedge B) \to A$
2. $(A \wedge B) \to B$
3. $A \to (A \vee B)$
4. $B \to (A \vee B)$
5. $(A \to C) \to ((B \to C) \to ((A \vee B) \to C))$
6. $(A \to B) \to ((A \to C) \to (A \to (B \wedge C)))$
7. $(A \to (B \to C)) \to ((A \to B) \to (A \to C))$
8. $A \to (B \to A)$
9. $\bot \to A$

rule $\dfrac{A, A \to B}{B}$

Kripke Semantics (§VII.B.2)
In L(¬, →, ∧, ∨)

$\langle \mathsf{W}, \mathsf{R}, \mathsf{e} \rangle$ is a (Kripke) *model* if W is a nonempty set, R is a *reflexive, transitive* relation on W, and $\mathsf{e} : \text{PV} \to \text{Sub } \mathsf{W}$. e is called an *evaluation* and the model is *finite* if W is finite. The pair $\langle \mathsf{W}, \mathsf{R} \rangle$ is a *frame*. For $w \in \mathsf{W}$ define

$w \models A$ inductively, where $w \not\models A$ means 'not $w \models A$':

1. $w \models p$ iff for all z such that wRz, $z \in e(p)$
2. $w \models A \wedge B$ iff $w \models A$ and $w \models B$
3. $w \models A \vee B$ iff $w \models A$ or $w \models B$
4. $w \models \neg A$ iff for all z such that wRz, $z \not\models A$
5. $w \models A \to B$ iff for all z such that wRz, $z \not\models A$ or $z \models B$

Then $\langle W, R, e \rangle \models A$ iff for all $w \in W$, $w \models A$.

In L($\to, \wedge, \vee, \bot$):
Replace (4) above by: $w \not\models \bot$.

A *Kripke tree* is an anti-symmetric Kripke model with designated first element w, i.e., w has no predecessors under R and is related by R to all other elements of W.

PV and Wffs are not necessarily assumed to be completed infinite totalities. Assignments or evaluations such as e can be understood as meaning that we have a method such that given any variable p_i we can produce a subset of W.

The class of all Kripke models and the class of all Kripke tress of cardinality less than or equal to that of the real numbers are both *strongly complete* for **Int**.

The class of all *finite* Kripke models and the class of all *finite* Kripke trees are both *finitely strongly complete* for **Int** (for finite Γ, $\Gamma \vdash_{\mathbf{Int}} A$ iff $\Gamma \models A$) and can be proved so by intuitionistically acceptable means.

Set-Assignment Semantics (§VII.D)
A set-assignment **Int**-model is a triplet $<v, s, S>$, where v: PV $\to \{T, F\}$, s: Wffs $\to$ Subsets of S satisfies conditions Int 1–Int 10 below, and v is extended to all wffs by the *intuitionist truth-conditions*:

$\wedge$ and $\vee$ are evaluated classically

$\to$ is evaluated by the dual dependence table (as for **S4**):

$v(A \to B) = T$ iff $s(A) \subseteq s(B)$ and (not both $v(A) = T$ and $v(B) = F$)

and $\neg$ is evaluated by the table for *intuitionist negation*:

A	$s(A) = \varnothing$	$\neg A$
any value	fails	F
T	holds	F
F		T

That is, $v(\neg A) = T$ iff $v(A) = F$ and $s(A) = \varnothing$.
It is *finite* if S is finite.

Int 1. $s(A \wedge B) = s(A) \cap s(B)$

Int 2. $s(A \vee B) = s(A) \cup s(B)$

Int 3. $s(\neg A) \cup s(B) \subseteq s(A \rightarrow B)$

Int 4. $s(A) \cap s(A \rightarrow B) \subseteq s(B)$

Int 5. $s(A \rightarrow B) \subseteq s((A \wedge C) \rightarrow (B \wedge C))$

Int 6. $s(A \rightarrow B) \cap s(B \rightarrow C) \subseteq s(A \rightarrow C)$

Int 7. $s(A \rightarrow C) \cap s(B \rightarrow C) = s((A \vee B) \rightarrow C)$

Int 8. $s(A \rightarrow B) \cap s(A \rightarrow \neg B) = s(\neg A)$

Int 9. If $v(A) = T$ then $s(A) = S$.

Int 10. $s(A \wedge \neg A) = \varnothing$

The class of all **Int**-models where S has cardinality less than or equal to that of the real numbers is *strongly complete* for **Int**. The class of finite **Int**-models is *finitely strongly complete* for **Int** (for finite Γ, $\Gamma \vdash_{\mathbf{Int}} A$ iff $\Gamma \models A$) and can be proved so by intuitionistically acceptable means.

$\mathcal{I}$ is the class of set-assignment models which use the intuitionist truth-conditions and satisfy Int 1, Int 2, Int 4, Int 8, Int 10, and

Int K. If $\bigcap \{ s(C) : x \in s(C) \} \subseteq \overline{s(A)} \cup s(B)$, then $x \in s(A \rightarrow B)$.

$\mathcal{I}$ is strongly complete for **Int**; the class of finite models in $\mathcal{I}$ is finitely strongly complete for **Int** and can be proved so by intuitionistically acceptable means (Theorem VII.26).

J, Johansson's Minimal Intuitionist Calculus

Axiomatizations

J *in* $L(\neg, \rightarrow, \wedge, \vee)$

Delete from Heyting's axiomatization of **Int** axiom schema X:

$\neg A \rightarrow (A \rightarrow B)$

J *in* $L(\rightarrow, \wedge, \vee, \bot)$

Delete from Segerberg's axiomatization of **Int** axiom schema 9:

$\bot \rightarrow A$

Kripke Semantics (§VII.E.2)

A *Kripke model for* **J** in $L(\neg, \rightarrow, \wedge, \vee)$ is $\langle W, R, Q, e \rangle$ where:

$\langle W, R \rangle$ is transitive and reflexive

$Q \subseteq W$ is R-closed (i.e., if $w \in Q$ and wRz, then $z \in Q$)

$e : PV \rightarrow \mathrm{Sub}\, W$

Q is to be thought of as those states of information which are inconsistent.

Validity in such a model is defined as in the Kripke semantics for **Int** with the exception of the evaluation of negations. Clause (4) is replaced by:

$w \models \neg A$ iff for all z such that wRz, $z \nvDash A$ or $z \in Q$

In the language $L(\rightarrow, \wedge, \vee, \perp)$ clause (4) is replaced by: $w \nvDash \perp$.

Both the class of all Kripke models and all anti-symmetric Kripke models are strongly complete for **J**. The class of all finite anti-symmetric Kripke models is finitely strongly complete for **J**.

Set-Assignment Semantics (§VII.E.4)
A set-assignment **J**-model in $L(\neg, \rightarrow, \wedge, \vee)$ is a triplet $<v, s, S>$, where $v: PV \rightarrow \{T, F\}$, s: Wffs $\rightarrow$ Subsets of S satisfies conditions Int 1, Int 2, Int 4–9 and

Int 3′. $s(B) \subseteq s(A \rightarrow B)$

and v is extended to all wffs by the *minimal intuitionist truth-conditions*:

$\wedge$ and $\vee$ are evaluated classically

$\rightarrow$ is evaluated by the dual dependence table as for **Int**

and $\neg$ is evaluated by the *minimal intuitionist negation* table:

A	$s(A) \subseteq s(\neg A)$	$\neg A$
any value	fails	F
T	holds	F
F		T

That is, $v(\neg A) = T$ iff $v(A) = F$ and $s(A) \subseteq s(\neg A)$.

E. Many-Valued Logics, L_3, L_n, $L_{\aleph}$, K_3, G_3, G_n, $G_{\aleph}$ Paraconsistent J_3

The general form of many-valued semantics can be found in §VIII.B.

L_3, Łukasiewicz's 3-Valued Logic

Many-Valued Semantics

An L_3-*evaluation* is a map $e: PV \rightarrow \{0, \frac{1}{2}, 1\}$ which is extended to all wffs of $L(\neg, \rightarrow, \wedge, \vee)$ by the tables:

A	$\neg$A
1	0
$\frac{1}{2}$	$\frac{1}{2}$
0	1

A→B		B: 1	$\frac{1}{2}$	0
A	1	1	$\frac{1}{2}$	0
	$\frac{1}{2}$	1	1	$\frac{1}{2}$
	0	1	1	1

A∧B		B: 1	$\frac{1}{2}$	0
A	1	1	$\frac{1}{2}$	0
	$\frac{1}{2}$	$\frac{1}{2}$	$\frac{1}{2}$	0
	0	0	0	0

A∨B		B: 1	$\frac{1}{2}$	0
A	1	1	1	1
	$\frac{1}{2}$	1	$\frac{1}{2}$	$\frac{1}{2}$
	0	1	$\frac{1}{2}$	0

The *sole designated value* is 1, so that e⊨A means e(A) = 1; ⊨A means that e(A) = 1 for all **L**$_3$-evaluations. And Γ⊨$_{\mathbf{L}_3}$A means that for every **L**$_3$-evaluation **e**, if **e**(B) = 1 for every B in Γ, then **e**(A) = 1.

The primitives may be reduced by defining:

A∨B $\equiv_{\text{Def}}$ (A→B)→B A∧B $\equiv_{\text{Def}}$ $\neg$($\neg$A∨$\neg$B)

A↔B $\equiv_{\text{Def}}$ (A→B) ∧ (B→A)

Other significant connectives (abbreviations) and their tables are:

◇A $\equiv_{\text{Def}}$ $\neg$A→A □A $\equiv_{\text{Def}}$ $\neg$◇$\neg$A IA $\equiv_{\text{Def}}$ A↔$\neg$A

A	◇A
1	1
$\frac{1}{2}$	1
0	0

A	□A
1	1
$\frac{1}{2}$	0
0	0

A	IA
1	0
$\frac{1}{2}$	1
0	0

A $\to_3$ B $\equiv_{\text{Def}}$ A→(A→B)

A $\to_3$ B		B: 1	$\frac{1}{2}$	0
A	1	1	$\frac{1}{2}$	0
	$\frac{1}{2}$	1	1	1
	0	1	1	1

Axiomatizations

L$_3$ *in* L($\neg$, →) (*Wajsberg, 1931*)

L$_3$1. A→(B→A)

L$_3$2. (A→B) → ((B→C)→(A→C))

L$_3$3. ($\neg$A→$\neg$B) → (B→A)

L$_3$4. ((A→$\neg$A)→A) →A)

rule $\dfrac{\text{A, A} \to \text{B}}{\text{B}}$

$\mathbf{L_3}$ *in* L(¬, →) (§VIII.C.1.c)

Definitions of ∧, ∨, I, $\rightarrow_3$ are given above.

1. $A \rightarrow (B \rightarrow (A \wedge B))$
2. a. $(A \wedge B) \rightarrow A$
   b. $(A \wedge B) \rightarrow B$
3. $A \leftrightarrow \neg\neg A$
4. a. $\neg(A \rightarrow B) \rightarrow (A \wedge \neg B)$
   b. $(A \wedge \neg B) \rightarrow_3 \neg(A \rightarrow B)$
5. $(IA \wedge \neg B) \rightarrow_3 I(A \rightarrow B)$
6. $\neg A \rightarrow (A \rightarrow B)$
7. $B \rightarrow (A \rightarrow B)$
8. $(IA \wedge IB) \rightarrow (A \rightarrow B)$
9. $(B \wedge \neg B) \rightarrow_3 A$
10. $(B \wedge IB) \rightarrow_3 A$
11. $(\neg B \wedge IB) \rightarrow_3 A$
12. $B \rightarrow_3 (A \rightarrow_3 B)$
13. $(A \rightarrow_3 (B \rightarrow_3 C)) \rightarrow ((A \rightarrow_3 B) \rightarrow (A \rightarrow_3 C))$
14. $((\neg A \rightarrow_3 A) \wedge (IA \rightarrow_3 A)) \rightarrow A$

rule $\dfrac{A, A \rightarrow B}{B}$

Set-Assignment Semantics (§VIII.C.1.d)

An $\mathbf{L_3}$-model for L(¬, →) is a triplet $<v, s, S>$, where $v: PV \rightarrow \{T, F\}$, s: Wffs → Subsets of S, v is extended to all wffs by the intuitionist tables for ¬ and →:

$v(\neg A) = T$ iff $s(A) = \varnothing$ and $v(A) = F$

$v(A \rightarrow B) = T$ iff $s(A) \subseteq s(B)$ and (not both $v(A) = T$ and $v(B) = F$)

and s satisfies:

1. $s(\neg A) = \begin{cases} \overline{s(A)} & \text{if } s(A) = \varnothing \text{ or } s(A) = S \\ s(A) & \text{otherwise} \end{cases}$

2. $s(A \rightarrow B) = \begin{cases} S & \text{if } s(A) \subseteq s(B) \\ s(B) & \text{if } s(B) \subset s(A) \text{ and } s(A) = S \\ s(A) & \text{if } s(B) \subset s(A) \text{ and } s(A) \neq S \end{cases}$

3. $v(p) = T$ iff $s(p) = S$
4. If both $\varnothing \subset s(A) \subset S$ and $\varnothing \subset s(B) \subset S$, then $s(A) = s(B)$.

Condition 4 ensures that there are at most 3 possibilities for content sets: $\varnothing$, S, and some U such that $\varnothing \subset \mathsf{U} \subset \mathsf{S}$.

The defined connectives $\wedge, \vee$ are then evaluated classically, and

$$\mathsf{s}(A \vee B) = \mathsf{s}(A) \cup \mathsf{s}(B) \qquad \mathsf{s}(A \wedge B) = \mathsf{s}(A) \cap \mathsf{s}(B)$$

$\mathbf{L_3}$-models thus evaluate all four connectives $\{\neg, \rightarrow, \wedge, \vee\}$ by the intuitionist truth-conditions.

Alternate set-assignment semantics
These are not limited to using only three content sets for each model.

A *rich* $\mathbf{L_3}$*-model* evaluates the connectives by:

$\upsilon(\neg A) = \mathsf{T}$ iff $\mathsf{s}(A) = \varnothing$ and $\upsilon(A) = \mathsf{F}$ [as before]

$\upsilon(A \rightarrow B) = \mathsf{T}$ iff $(\mathsf{s}(A) \subseteq \mathsf{s}(B)$ or both $\varnothing \subset \mathsf{s}(A) \subset \mathsf{S}$ and $\varnothing \subset \mathsf{s}(B) \subset \mathsf{S})$ and (not both $\upsilon(A) = \mathsf{T}$ and $\upsilon(B) = \mathsf{F}$)

and satisfies conditions (1) and (3) for $\mathbf{L_3}$-models, while condition (2) is replaced by:

$$2.\ \mathsf{s}(A \rightarrow B) = \begin{cases} \mathsf{S} & \text{if } \mathsf{s}(A) \subseteq \mathsf{s}(B) \text{ or (both } \varnothing \subset \mathsf{s}(A) \subset \mathsf{S} \text{ and } \varnothing \subset \mathsf{s}(B) \subset \mathsf{S}) \\ \mathsf{s}(B) & \text{if } \mathsf{s}(B) \subset \mathsf{s}(A) \text{ and } \mathsf{s}(A) = \mathsf{S} \\ \mathsf{s}(A) & \text{if } \mathsf{s}(B) \subset \mathsf{s}(A) \text{ and } \mathsf{s}(A) \neq \mathsf{S} \end{cases}$$

Every $\mathbf{L_3}$-model is a rich $\mathbf{L_3}$-model.

The Lukasiewicz Logics $\mathbf{L_n}$, $\mathbf{L_{\aleph}}$, $\mathbf{L_{\aleph_0}}$

Many-Valued Semantics (§VIII.C.2.a)
An $\mathbf{L}$*-evaluation* is a map $\mathsf{e}: PV \rightarrow [0,1]$ which is extended to all wffs of $L(\neg, \rightarrow)$ by the following tables:

$$\mathsf{e}(\neg A) = 1 - \mathsf{e}(A)$$

$$\mathsf{e}(A \rightarrow B) = \begin{cases} 1 & \text{if } \mathsf{e}(A) \leq \mathsf{e}(B) \\ (1 - \mathsf{e}(A)) + \mathsf{e}(B) & \text{if } \mathsf{e}(B) < \mathsf{e}(A) \end{cases}$$

The connectives $\wedge$, $\vee$, and $\leftrightarrow$ are defined as for $\mathbf{L_3}$:

$A \vee B \equiv_{Def} (A \rightarrow B) \rightarrow B \qquad A \wedge B \equiv_{Def} \neg(\neg A \vee \neg B)$

$A \leftrightarrow B \equiv_{Def} (A \rightarrow B) \wedge (B \rightarrow A)$

These have the following tables:

$$\mathsf{e}(A \leftrightarrow B) = \begin{cases} 1 & \text{if } \mathsf{e}(A) = \mathsf{e}(B) \\ (1 - \mathsf{e}(A)) + \mathsf{e}(B) & \text{if } \mathsf{e}(A) > \mathsf{e}(B) \\ (1 - \mathsf{e}(B)) + \mathsf{e}(A) & \text{if } \mathsf{e}(B) > \mathsf{e}(A) \end{cases}$$

$e(A \vee B) = \max(e(A), e(B))$

$e(A \wedge B) = \min(e(A), e(B))$

For $n \geq 2$,

$\mathbf{L_n} = \{ A : e(A) = 1$ for every **L**-evaluation $e : PV \rightarrow \{ \frac{m}{n-1} : 0 \leq m \leq n-1 \} \}$

$\mathbf{L}_{\aleph_0} = \{ A : e(A) = 1$ for every **L**-evaluation e which takes only rational values in $[0,1] \}$

$\mathbf{L}_{\aleph} = \{ A : e(A) = 1$ for every **L**-evaluation $e \}$

When the values that e may take on PV are restricted in these definitions then the extension of e to all wffs obeys the same restriction.

$\mathbf{L}_{\aleph_0} = \mathbf{L}_{\aleph}$ (Theorem VIII.14)

Axiomatization

$\mathbf{L}_{\aleph}$ *(and* $\mathbf{L}_{\aleph_0}$*) in* $L(\neg, \rightarrow)$ *(Turquette, 1959)*

In Wajsberg's axiomatization of $\mathbf{L_3}$ replace $L_3 4$ by:

$L_{\aleph}$. $(A \rightarrow B) \vee (B \rightarrow A)$

Set-Assignment Semantics for both $\mathbf{L}_{\aleph}$ *and* $\mathbf{L}_{\aleph_0}$

An $\mathbf{L}_{\aleph}$-model $\langle v, s \rangle$ for $L(\neg, \rightarrow)$ uses the intuitionist tables for $\neg$ and $\rightarrow$ (p. 337; as for $\mathbf{L_3}$-models) and satisfies:

L1. $s(A \rightarrow B) = S$ iff $s(A) \subseteq s(B)$

L2. $s(\neg A) = S$ iff $s(A) = \varnothing$

L3. $s(B) \subseteq s(A \rightarrow B)$

L4. $s(A \rightarrow B) \subseteq s((B \rightarrow C) \rightarrow (A \rightarrow C))$

L5. If $s(A) \subseteq s(B)$ then $s(B \rightarrow C) \subseteq s(A \rightarrow C)$.

L6. $s(\neg A \rightarrow \neg B) \subseteq s(B \rightarrow A)$

L7. $s(\neg B) \subseteq s(\neg A)$ iff $s(A) \subseteq s(B)$

L8. $s(A) \subseteq s(B)$ or $s(B) \subseteq s(A)$

L9. $v(p) = T$ iff $s(p) = S$

and for the defined connectives

L10. $s(A \vee B) = s(A) \cup s(B)$

L11. $s(A \wedge B) = s(A) \cap s(B)$

The resulting tables for the defined connectives $\wedge$ and $\vee$ are then classical, and thus all four connectives use the intuitionist truth-conditions.

$\mathbf{K_3}$, Kleene's 3-Valued Logic

Many-Valued Semantics (§VIII.D.1)

A $\mathbf{K_3}$-*evaluation* is a map $e : PV \to \{T, F, U\}$ which is extended to all wffs of $L(\neg, \to, \wedge, \vee)$ by the tables:

A	¬A
T	F
U	U
F	T

A→B		B: T	U	F
A	T	T	U	F
	U	T	U	U
	F	T	T	T

A∧B		B: T	U	F
A	T	T	U	F
	U	U	U	F
	F	F	F	F

A∨B		B: T	U	F
A	T	T	T	T
	U	T	U	U
	F	T	U	F

These are what Kleene calls the *strong connectives.*

The *sole designated value* is T. There are no tautologies. The logic is solely a consequence relation:

$\Gamma \vDash_{\mathbf{K_3}} A$ iff for every $\mathbf{K_3}$-evaluation e, if $e \vDash \Gamma$ then $e(A) = T$.

If 1 is read for T, 0 for F, and $\frac{1}{2}$ for U, then these tables agree with those for Łukasiewicz's 3-valued logic $\mathbf{L_3}$ with one exception: if $e(A) = e(B) = U$ then Łukasiewicz assigns $e(A \to B) = T$, whereas Kleene assigns value U.

Set-Assignment Semantics

A $\mathbf{K_3}$-model for $L(\neg, \to, \wedge, \vee)$ is a triplet $<v, s, S>$, where $v : PV \to \{T, F\}$, s : Wffs → Subsets of S, v is extended to all wffs by the classical tables for $\wedge$ and $\vee$, and

$$v(A \to B) = T \text{ iff } v(\neg A) = T \text{ or } v(B) = T$$
$$v(\neg A) = T \text{ iff } v(A) = F \text{ and } s(A) = \varnothing$$

and s satisfies:

K1. $s(A) \subseteq s(B)$ or $s(B) \subseteq s(A)$

K2. $s(A \to B) = s(\neg A) \cup s(B)$

K3. $s(\neg A) = \begin{cases} s(A) & \text{if } \varnothing \subset s(A) \subset S \\ \overline{s(A)} & \text{otherwise} \end{cases}$

K4. $s(A \wedge B) = s(A) \cap s(B)$

K5. $s(A \vee B) = s(A) \cup s(B)$

K6. $v(p) = T$ iff $s(p) = S$

These semantics are strongly complete for $\mathbf{K_3}$.

The entire difference between these semantics and the classical ones lies in the table and set-assignments for negation.

Note that the content sets are only required to be linearly ordered under inclusion rather than being restricted to $\varnothing \subset \mathsf{U} \subset \mathsf{S}$ for some U as for $\mathbf{L_3}$-models. Indeed, the content sets need not even be linerarly ordered if K1 and K2 are replaced by:

$$\text{K7.}\quad s(A \rightarrow B) = \begin{cases} \mathsf{S} & \text{if } s(\neg A) = \mathsf{S} \text{ or } s(B) = \mathsf{S} \\ \varnothing & \text{if } s(A) = \mathsf{S} \text{ and } s(B) = \varnothing \\ s(B) & \text{otherwise} \end{cases}$$

The Gödel Logics $\mathbf{G_n}$ and $\mathbf{G_{\aleph}}$

Many-Valued Semantics (§VIII.F)

A **G**-*evaluation* is a map $e : PV \rightarrow [0,1]$ which is extended to all wffs of $L(\neg, \rightarrow, \wedge, \vee)$ by the following tables:

$$e(\neg A) = \begin{cases} 1 & \text{if } e(A) = 0 \\ 0 & \text{if } e(A) \neq 0 \end{cases}$$

$$e(A \rightarrow B) = \begin{cases} 1 & \text{if } e(A) \leq e(B) \\ e(B) & \text{otherwise} \end{cases}$$

$$e(A \wedge B) = \min(e(A), e(B)) \qquad e(A \vee B) = \max(e(A), e(B))$$

The table for $A \leftrightarrow B \equiv_{\text{Def}} (A \rightarrow B) \wedge (B \rightarrow A)$ is:

$$e(A \leftrightarrow B) = \begin{cases} 1 & \text{if } e(A) = e(B) \\ \min(e(A), e(B)) & \text{otherwise} \end{cases}$$

For $n \geq 2$ define

$\mathbf{G_n} = \{ A : e(A) = 1$ for every **G**-evaluation $e : PV \rightarrow \{ \frac{m}{n-1} : 0 \leq m \leq n-1 \} \}$

$\mathbf{G_{\aleph_0}} = \{ A : e(A) = 1$ for every **G**-evaluation e which takes only rational values in $[0,1] \}$

$\mathbf{G_{\aleph}} = \{ A : e(A) = 1$ for every **G**-evaluation $e \}$

$\mathbf{G_{\aleph_0}} = \mathbf{G_{\aleph}}$ (Theorem VIII.23)

The tables for $\mathbf{G_3}$ are:

A	¬A
1	0
$\frac{1}{2}$	0
0	1

	A → B	B: 1	$\frac{1}{2}$	0
A	1	1	$\frac{1}{2}$	0
	$\frac{1}{2}$	1	1	0
	0	1	1	1

$A \wedge B$		B: 1	$\frac{1}{2}$	0
A	1	1	$\frac{1}{2}$	0
	$\frac{1}{2}$	$\frac{1}{2}$	$\frac{1}{2}$	0
	0	0	0	0

$A \vee B$		B: 1	$\frac{1}{2}$	0
A	1	1	1	1
	$\frac{1}{2}$	1	$\frac{1}{2}$	$\frac{1}{2}$
	0	1	$\frac{1}{2}$	0

Axiomatization

$\mathbf{G}_{\aleph}$ *(and* $\mathbf{G}_{\aleph_0}$*) in* $L(\neg, \rightarrow, \wedge, \vee)$ *(Dummett, 1959)*

Add to any of the axiomatizations of **Int** (pp. 335–336):

$(A \rightarrow B) \vee (B \rightarrow A)$

Set-Assignment Semantics (§VIII.F)

For $\mathbf{G}_{\aleph}$ *and* $\mathbf{G}_{\aleph_0}$

A $\mathbf{G}_{\aleph}$-model for $L(\neg, \rightarrow, \wedge, \vee)$ is a triplet $<v, s, S>$, where $v: PV \rightarrow \{T, F\}$, $s: \text{Wffs} \rightarrow \text{Subsets of } S$, v is extended to all wffs by the intuitionist truth-conditions:

$\wedge$ and $\vee$ are classical

$v(\neg A) = T$ iff $s(A) = \varnothing$ and $v(A) = F$

$v(A \rightarrow B) = T$ iff $s(A) \subseteq s(B)$ and (not both $v(A) = T$ and $v(B) = F$)

and s satisfies:

G1. $s(\neg A) = \begin{cases} S & \text{if } s(A) = \varnothing \\ \varnothing & \text{otherwise} \end{cases}$

G2. $s(A \rightarrow B) = \begin{cases} S & \text{if } s(A) \subseteq s(B) \\ s(A) \cap s(B) & \text{otherwise} \end{cases}$

G3. $s(A \wedge B) = s(A) \cap s(B)$

G4. $s(A \vee B) = s(A) \cup s(B)$

G5. $v(p) = T$ iff $s(p) = S$

G6. $s(A) \subseteq s(B)$ or $s(B) \subseteq s(A)$

Alternatively, G6 can be added to the list of conditions for a set-assignment **Int**-model to obtain complete set-assignment semantics.

For $\mathbf{G}_3$

Replace G6 above by: there is some U such that $s: \text{Wffs} \rightarrow \{\varnothing, U, S\}$.

Paraconsistent $\mathbf{J}_3$

Many-Valued Semantics (§IX.B.2)

There are two distinct presentations of $\mathbf{J}_3$, in L(~, ∧, ◇) and in L(¬, →, ∧, ~). A $\mathbf{J}_3$-*evaluation* is a map $\mathbf{e}: \mathrm{PV} \to \{0, \frac{1}{2}, 1\}$ which is extended to all wffs of one of these languages by the appropriate tables below:

A	~A
1	0
$\frac{1}{2}$	$\frac{1}{2}$
0	1

A	¬A
1	0
$\frac{1}{2}$	0
0	1

A	◇A
1	1
$\frac{1}{2}$	1
0	0

		B		
A → B		1	$\frac{1}{2}$	0
A	1	1	$\frac{1}{2}$	0
	$\frac{1}{2}$	1	$\frac{1}{2}$	0
	0	1	1	1

		B		
A ∧ B		1	$\frac{1}{2}$	0
A	1	1	$\frac{1}{2}$	0
	$\frac{1}{2}$	$\frac{1}{2}$	$\frac{1}{2}$	0
	0	0	0	0

The *designated values are* 1 *and* $\frac{1}{2}$, so that $\mathbf{e} \vDash A$ means $\mathbf{e}(A) = 1$ or $\frac{1}{2}$. And $\vDash A$ means that $\mathbf{e} \vDash A$ for all $\mathbf{J}_3$-evaluations $\mathbf{e}$. Finally, $\Gamma \vDash A$ means that for every $\mathbf{J}_3$-evaluation $\mathbf{e}$, if $\mathbf{e} \vDash B$ for all $B \in \Gamma$, then $\mathbf{e} \vDash A$.

Four additional connectives (abbreviations) are important; they are defined in the axiomatizations below and the definitions are surveyed in §IX.B.3.

		B		
A ∨ B		1	$\frac{1}{2}$	0
A	1	1	1	1
	$\frac{1}{2}$	1	$\frac{1}{2}$	$\frac{1}{2}$
	0	1	$\frac{1}{2}$	0

		B		
A ↔ B		1	$\frac{1}{2}$	0
A	1	1	$\frac{1}{2}$	0
	$\frac{1}{2}$	$\frac{1}{2}$	$\frac{1}{2}$	0
	0	0	0	1

A	□A
1	1
$\frac{1}{2}$	0
0	0

A	©A
1	1
$\frac{1}{2}$	0
0	1

Axiomatizations

$\mathbf{J}_3$ *in* L(~, ∧, ◇) (§IX.E.1)

A ∨ B $\equiv_{\mathrm{Def}}$ ~(~A ∧ ~B)

A → B $\equiv_{\mathrm{Def}}$ ~◇A ∨ B

A ↔ B $\equiv_{\mathrm{Def}}$ (A → B) ∧ (B → A)

□A $\equiv_{\mathrm{Def}}$ ~◇~A

©A $\equiv_{\mathrm{Def}}$ ¬(◇A ∧ ◇~A)

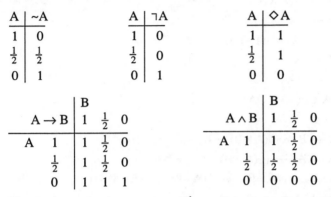

J_3 *in* L(~, ∧, ◇) (*continued*)

1. B→(A→B)
2. (A→(B→C)) → ((A→B)→(A→C))
3. (B→(A→C)) → ((A∧B)→C)
4. A→(B→(A∧B))
5. (A∧~A∧©A) → B
6. ((~A∧©A)→A) → A
7. ~~A ↔ A
8. ©A ↔ ©~A
9. ~◇A ↔ (~A∧©A)
10. ©(◇A)
11. [(~(A∧B) ∧ ©(A∧B)) ∧B] → (~A∧©A)
12. (~A∧©A) → [~(A∧B)∧©(A∧B)]
13. [(A∧B)∧©(A∧B)] ↔ [(A∧©A)∧(B∧©B)]

rule $\dfrac{A, A \to B}{B}$

J_3 *in* L(¬, →, ∧, ~) (§IX.E.2)

©A $\equiv_{Def}$ ¬[¬(A→(A∧¬A)) ∧ ¬(~A→ (A∧¬A))]

◇A $\equiv_{Def}$ ¬(A→(A∧¬A))

PC based on ¬, →, ∧

1. (~A∧©A) ↔ ¬A
2. ~~A ↔ A
3. ©(¬A)
4. [(A∧B)∧©(A∧B)] ↔ [(A∧©A)∧(B∧©B)]
5. (~A∧©A) → ©(A→B)
6. (B∧©B) → ©(A→B)

rule $\dfrac{A, A \to B}{B}$

Set-Assignment Semantics (§IX.F)

In L(~, ∧, ◇)

A **J_3**-model for L(~, ∧, ◇) is a triplet <υ,s,S>, where υ: PV→{T, F}, s: Wffs → Subsets of S, υ is extended to all wffs by

υ(A∧B) = T iff υ(A) = T and υ(B) = T

υ(~A) = T iff s(A) ≠ S

$v(\Diamond A) = T$ iff $v(A) = T$

and s, S satisfy:

1. $S \neq \varnothing$
2. $s(p) \neq \varnothing$ iff $v(p) = T$
3. $s(A) \subseteq s(B)$ or $s(B) \subseteq s(A)$
4. $s(A \wedge B) = s(A) \cap s(B)$
5. $s({\sim}A) = \begin{cases} s(A) & \text{if } \varnothing \subset s(A) \subset S \\ \overline{s(A)} & \text{otherwise} \end{cases}$
6. $s(\Diamond A) = \begin{cases} S & \text{if } s(A) \neq \varnothing \\ \varnothing & \text{if } s(A) = \varnothing \end{cases}$

The defined connectives are then evaluated:

$v(\neg A) = T$ iff $v(A) = F$

$v({\sim}A) = F$ iff $v(A) = T$ and $s(A) = S$

$v(A \rightarrow B) = T$ iff $v(A) = F$ or $v(B) = T$

$v(A \vee B) = T$ iff $v(A) = T$ or $v(B) = T$

In $L(\neg, \rightarrow, \wedge, \sim)$

$\neg$, $\rightarrow$, $\wedge$ are evaluated classically

$v({\sim}A) = T$ iff $s(A) \neq S$

and s, S satisfy:

1. $S \neq \varnothing$
2. $s(p) \neq \varnothing$ iff $v(p) = T$
3. $s(A) \subseteq s(B)$ or $s(B) \subseteq s(A)$
4. $s(A \wedge B) = s(A) \cap s(B)$
5. $s(A \rightarrow B) = \begin{cases} s(B) & \text{if } \varnothing \subset s(A) \subset S \\ \overline{s(A)} \cup s(B) & \text{otherwise} \end{cases}$
6. $s(\neg A) = \begin{cases} S & \text{if } s(A) = \varnothing \\ \varnothing & \text{otherwise} \end{cases}$
7. $s({\sim}A) = \begin{cases} s(A) & \text{if } \varnothing \subset s(A) \subset S \\ \overline{s(A)} & \text{otherwise} \end{cases}$

Alternate truth-default set-assignment semantics for $\mathbf{J_3}$ can be found in §IX.G.

Bibliography

I list only those works which are cited in the text or elsewhere in the bibliography. Page references are to the most recent English reference listed unless noted otherwise.

Quotation marks and logical notation in all quotations have been changed to conform with the conventions of this book (see p. 5 for the use of quotation marks).

ANDERSON, Alan R. and Nuel D. BELNAP, Jr.

1975 *Entailment*
Princeton Univ. Press.

ARRUDA, Ayda I.

1980 A survey of paraconsistent logic
Mathematical Logic in Latin America, ed. A. Arruda, R. Chuaqui, and N. C. A. da Costa, North-Holland.

198? Aspects of the historical development of paraconsistent logics
Typescript.

AUNE, Bruce

1976 Possibility
In *Edwards, 1967,* Vol.6, pp. 419–424.

BENNETT, Jonathan

1969 Entailment
Phil. Rev., vol. 78, pp. 197–235.

BERNAYS, Paul

1926 Axiomatische Untersuchung des Aussagen-Kalküls der *Principia mathematica*
Mathematische Zeitschrift, vol. 25, pp. 305–320.

BLOK, W. J. and KÖHLER, P.

1983 Algebraic semantics for quasi-classical modal logics
The Journal of Symbolic Logic, vol. 48, no. 4, pp. 941–963.

BLOK, W. J. and PIGOZZI, D.

1982 On the structure of varieties with equationally definable principal congruences I
Algebra Universalis, vol. 15, pp. 195–227.

1989 *Algebraizable Logics*
Memoirs of the American Mathematical Society, no. 396.

198? The deduction theorem in algebraic logic
Typescript.

BOCHENSKI, I. M.

1970 *A History of Formal Logic*
Chelsea. A revision and translation of the German *Formale Logik*, Verlag Karl Alber, Freiburg, 1956.

BOOLOS, George

1979 *The unprovability of consistency*
Cambridge Univ. Press.

1980 A Provability, truth, and modal logic
J. Phil. Logic, vol. 9, no. 1, pp. 1–7.

1980 B On systems of modal logic with provability interpretations
Theoria, vol. 46, no. 1, pp. 7–18

BROUWER, L. E. J.

1907 Over de grondslagen der wiskunde
Dissertation, Amsterdam. Translated as 'On the foundations of mathematics' in *Brouwer, 1975*, pp. 11–101

1908 De onbetrouwbaarheid der logische principes
Tijdschrift voor wijsbegeerte, vol. 2, pp. 152–158.
Translated as 'The unreliability of the logical principles' in *Brouwer, 1975*, pp. 107–111.

1912 Intuitionisme en formalisme
Inaugural address, Univ. of Amsterdam. Translated as 'Intuitionism and formalism' in *Bulletin of the American Math. Soc.*, vol. 20 (Nov. 1913), pp. 81–96, and reprinted in *Philosophy of Mathematics,* ed. P. Benacerraf and H. Putnam, Prentice-Hall, Englewood Cliffs N.J. 2nd edition, 1983, Cambridge Univ. Press, pp. 77–89.

1928 Intuitionistische Betrachtungen über den Formalismus
Sitzungsberichte der Preussischen Akademie der Wissenschaften, Phys.-math. Kl., pp. 48–52. Translated as 'Intuitionistic reflections on formalism' in *van Heijenoort, 1967,* pp. 490–492.

1975 *The collected works of L. E. J. Brouwer*
ed. A. Heyting, North-Holland.

CANTOR, Georg

1883 *Grundlagen einer allgemeinen Mannigfaltigkeitslehre*
Teubner, Leipzig.
The translation in the text comes from *Georg Cantor*, J. Dauben, Harvard Univ. Press, 1979, pp. 128–129.

CARNIELLI, Walter A.

1987 A Methods of proof for relatedness and dependence logic
Reports on Mathematical Logic, vol. 21, pp. 35–46.

1987 B Systematization of finite many-valued logics through the method of tableaux
The Journal of Symbolic Logic, vol. 52, pp. 473–493.

CHELLAS, Brian

1980 *Modal Logic*
Cambridge Univ. Press.

CHRISTENSEN, Niels Egmont

1973 Is there a "logic" or formal system based on the concept of a truth determinant?
Danish Yearbook of Philosophy, vol. 10, pp. 77–85.

CLEAVE, J. P.

1974 The notion of logical consequence in the logic of inexact predicates
Zeit. Math. Logik und Grundlagen, vol. 20, no. 4, pp. 307–324.

COPELAND, B. J.

1978 *Entailment, the Formalisation of Inference*
Doctor of Philosophy Thesis, Oxford.

1984 Horseshoe, hook, and relevance
Theoria, vol. L, pp. 148–164.

DA COSTA, Newton C.A.

1963 Calculs propositionnels pour les systèmes formels inconsistents
Comptes Rendus de l'Academie des Sciences de Paris, Série A, vol. 257, pp. 3790–3792.

1974 On the theory of inconsistent formal systems
Notre Dame Journal of Formal Logic, XV, no. 4, pp. 497–510.

See also D'OTTAVIANO and DA COSTA

DA COSTA, Newton C.A. and Diego MARCONI

198? An overview of paraconsistent logic in the 80's
To appear in *Logica Nova*, Akademie-Verlag.

DE MORGAN, Augustus

1847 *Formal Logic or the Calculus of Inferences Necessary and Probable*
London. Reprinted by Open Court, 1926.

DE SWART, H.

1977 An intuitionistically plausible interpretation of intuitionist logic
Journal of Symbolic Logic, vol. 42, no. 4, pp. 564–578.

D'OTTAVIANO, Itala M. L.

1985 A The completeness and compactness of a three-valued first-order logic
Revista Colombiana de Matemáticas, XIX, 1–2, pp. 31–42.

1985 B The model-extension theorems for $\mathbf{J}_3$-theories
Methods in Mathematical Logic, ed. C. A. Di Prisco, Lecture Notes in Mathematics, no. 1130, Springer-Verlag.

1987 Definability and quantifier elimination for $\mathbf{J}_3$-theories
Studia Logica, XLVI, 1, pp. 37–54.

D'OTTAVIANO, Itala M. L. and Newton C. A. da COSTA

1970 Sur un problème de Jaśkowski
C. R. Acad. Sc. Paris, 270, Série A, pp. 1349–1353.

DREBEN, Burton and Jean van HEIJENOORT

1986 Note to Gödel's dissertation
In *Gödel, 1986*, pp. 44–59.

DUMMETT, Michael

1959 A propositional calculus with denumerable matrix
The Journal of Symbolic Logic, vol. 24, pp. 97–106.

1973 The philosophical basis of intuitionistic logic
In *Logic Colloquium '73*, ed. H. E. Rose and J.C. Shepherdson, North-Holland, Amsterdam. Reprinted in Dummett, *Truth and Other Enigmas*, Harvard Univ. Press, 1978.

1977 *Elements of Intuitionism*
Clarendon Press, Oxford.

DUNN, J. Michael

1972 A modification of Parry's Analytic Implication
Notre Dame J. of Formal Logic, vol. 13, no. 2, pp. 195–205.

EDWARDS, Paul (ed.)
1967 *The Encyclopedia of Philosophy*
Macmillan and The Free Press.

EPSTEIN, Richard L.
1979 Relatedness and implication
Phil. Studies, vol. 36, no. 2, pp. 137–173.
1980 A (ed.) *Relatedness and Dependence in Propositional Logics*
Research Report of the Iowa State Univ. Logic Group.
1980 B Relatedness and dependence in propositional logics
Abstract, *The Journal of Symbolic Logic*, vol. 46, no. 1, p. 202.
1985 Truth is beauty
History and Phil. of Logic, vol. 6, pp. 117–125.
1987 The algebra of dependence logic
Reports on Mathematical Logic, vol. 21, pp. 19–34
1988 A general framework for semantics for propositional logics
In *Methods and Applications of Mathematical Logic*, Proceedings of the VII Latin American Symposium on Mathematical Logic, ed. W. Carnielli and L. P. de Alcantara, *Contemporary Mathematics*, American Math. Soc., no. 69.
198? A theory of truth based on a medieval solution to the liar paradox
Typescript.

EPSTEIN, Richard L. and Walter A. CARNIELLI
1989 *Computability*
Wadsworth & Brooks/Cole.

EPSTEIN, Richard L. and Roger D. MADDUX
1981 The algebraic nature of set assignments
In *Epstein, 1980 A*.

FINE, Kit
1979 Analytic implication
Notre Dame J. of Formal Logic, vol. 27, no. 2, pp. 169–179.

FITTING, M. C.
1969 *Intuitionistic Logic, Model Theory and Forcing*
North-Holland.

FREGE, Gottlob
1879 *Begriffschrift*
L. Nebert, Halle. Translated as *Begriffschrift, a formula language, modeled upon that of arithmetic, for pure thought*, in *van Heijenoort, 1967,* pp. 1–82.
1892 Über Sinn und Bedeutung
Zeit. für Philosophie und philosophische Kritik, vol. 100, pp. 25–50. Translated as 'On sense and reference' in *Translations from the Philosophical Writings of Gottlob Frege,* ed. M. Black and P. Geach, Basil Blackwell, 1970, pp. 56–78.
1918 Der Gedanke: eine logische Untersuchung
Beträge zur Philosophie des deutschen Idealismus, pp. 58–77.
Translated by A. and M. Quinton as 'The thought: a logical inquiry' in *Mind*, (new series) vol. 65, pp. 289–311, and reprinted in *Philosophical Logic,* ed. P. F. Strawson, Oxford Univ. Press, 1967, pp. 17–38.
1980 *Philosophical and Mathematical Correspondence*
Univ. of Chicago Press.

GENTZEN, Gerhard

1936 Die Widerspruchsfreiheit der reinen Zahlentheorie *Mathematische Annalen*, vol. 112, pp. 493–565.

GLIVENKO, V.

1929 Sur quelques points de la logique de M. Brouwer *Académie Royale de Belgique, Bulletins de la classe des sciences*, ser. 5, vol. 15, pp. 183–188.

GÖDEL, Kurt

1932 Zum intuitionistischen Aussagenkalkül *Akademie der Wissenschaften in Wien,* Math.-natur. Klasse, vol. 69, pp. 65–66. Translated as 'On the intuitionistic propositional calculus' in *Gödel, 1986*, pp. 223–225.

1933 A Zur intuitionistischen Arithmetik und Zahlentheorie *Ergebnisse eines mathematischen Kolloquiums*, vol. 4 (1931–32), pp. 34–38. Translated as 'On intutionistic arithmetic and number theory' in *The Undecidable*, ed. M. Davis, Raven Press, New York, 1965, pp. 75–81, and in *Gödel, 1986*, pp. 287–295.

1933 B Eine Interpretation des intuitionistischen Aussagenkalküls *Ergebnisse eines mathematischen Kolloquiums*, vol. 4 (1931–32), pp. 39–40. Translated as 'An interpretation of the intuitionistic sentential logic', in *The Philosophy of Mathematics*, ed. J. Hintikka, Oxford Univ. Press, 1969, pp. 128–129, and as 'An interpretation of the intuitionistic propositional calculus' in *Gödel, 1986*, pp. 301–303.

1933 C Über Unabhängigkeitsbeweise in Aussagenkalkül *Ergebnisse eines mathematischen Kolloquiums*, vol. 4 (for 1931–32), pp. 9–10. Translated as 'On independence proofs in the propositional calculus' in *Gödel, 1986*, pp. 269–271.

1986 *Collected Works, Volume 1* ed. Feferman et al., Oxford Univ. Press.

GOLDBLATT, Rob

1978 Arithmetical necessity, provability and intuitionistic logic *Theoria*, vol. 44, pp. 38–46

1979 *Topoi, the categorial analysis of logic* North-Holland.

GOLDFARB, Warren D.

1979 Logic in the twenties: the nature of the quantifier *The Journal of Symbolic Logic*, vol. 44, pp. 351–369.

GRZEGORCZYK, Andrzej

1967 Some relational systems and the associated topological spaces *Fundamenta mathematicae*, vol. 60, pp. 223–231.

HAACK, Susan

1974 *Deviant Logic* Cambridge Univ. Press.

1978 *Philosophy of Logics* Cambridge Univ. Press.

HANSON, William H.

1980 First-degree entailments and information *Notre Dame J. of Formal Logic*, vol. 21, no. 4, pp. 659–671.

HENKIN, Leon

1954 Boolean representation through propositional calculus
Fundamenta Mathematicae, vol. 41, pp. 89–96.

HEYTING, Arend

1930 Die formalen Regeln der intuitionistischen Logik
Sitzungsberichte der Preussischen (Berlin) Akademie der Wissenschaften, Phys.-Math. Kl, pp. 42–56. The quotations in the text are from *Bochenski, 1970*, pp. 293–294.

HUGHES, G. E. and M. J. CRESSWELL

1968 *An Introduction to Modal Logic*
Methuen. 2nd printing with corrections, 1971.

1984 *A Companion to Modal Logic*
Methuen and Co.

ISEMINGER, Gary

1986 Relatedness logic and entailment
The Journal of Non-classical Logic, vol. 3, no. 1, pp. 5–23.

JAŚKOWSKI, S.

1948 Rachunek zdań dla systemów dedukcyjnych sprzecznych
Studia Societatis Scientiarum Torunensis, Section A, vol. 1, no. 5, pp. 57–77. Translated as 'Propositional calculus for contradictory deductive systems', *Studia Logica*, XXIV, 1969, pp. 143–157.

JOHANSSON, Ingebrigt

1936 Der minimalkalkül, ein reduzierter intutionistischer Formalismus
Compositio mathematica, vol. 4, pp. 119–136. The translation in the text is by D. Steiner.

JÓNSSON, B. and Alfred TARSKI

1951 Boolean algebras with operators, Part I
Amer. J. Math., vol. 73, pp. 891–939.

KALMÁR, László

1935 Über die Axiomatisierbarkeit des Aussagenkalküls
Acta Scientiarum Mathematicarum, vol. 7, pp. 222–243.

KIELKOPF, Charles F.

1977 *Formal Sentential Entailment*
University Press of American, Washington D.C.

KLEENE, Stephen Cole

1952 *Introduction to Metamathematics*
North-Holland. Sixth reprint with corrections, 1971.

KNEALE, William and Martha

1962 *The Development of Logic*
Clarendon Press, Oxford.

KOLMOGOROFF, A. N.

1925 Sur le principe de tertium non datur
Matématicéskij Sbornik, vol. 32, pp. 646–667. Translated as 'On the principle of excluded middle', in *van Heijenoort*, pp. 416–437.

1932 Zur Deutung der intuitionistischen Logik
Mathematische Zeit., vol. 35, pp. 58–65

KÖRNER, Stephan
1976 *Philosophy of Logic*
Univ. of California Press.
KRAJEWSKI, Stanisław
1986 Relatedness logic
Reports on Mathematical Logic, vol. 20, pp. 7–14.
KRIPKE, Saul A.
1959 A completeness theorem in modal logic
The Journal of Symbolic Logic, vol. 24, pp. 1–14.
1965 Semantical analysis of intuitionistic logic, I
In *Formal Systems and Recursive Functions,* ed. J. N. Crossley and M. A. E. Dummett, North-Holland, Amsterdam, pp. 92–130.
1972 Naming and necessity
In *Semantics of Natural Language*, ed. D. Davidson and G. Harman, pp. 253–355.
1975 Outline of a theory of truth
J. of Philosophy, vol. 72, pp. 690–716.
LEIVANT, Daniel
1985 Syntactic translations and provably recursive functions
Journal of Symbolic Logic, vol. 50, no. 3, pp.682–688.
LEMMON, E. J.
1977 *An Introduction to Modal Logic*
In collaboration with Dana Scott, edited by K. Segerberg, *American Philosophical Quarterly,* Monograph 11, Basil Blackwell, Oxford.
LEWIS, C. I.
1912 Implication and the algebra of logic
Mind, vol. 21 (new series), pp. 522–531.
LEWIS, C. I. and C. H. LANGFORD
1932 *Symbolic Logic*
The Century Company. 2nd edition with corrections, Dover, 1959.
LEWIS, David K.
1973 *Counterfactuals*
Harvard Univ. Press.
ŁOS, Jerzy
1951 An algebraic proof of completeness for the two-valued propositional calculus
Colloquium Mathematicum, vol. 2, pp. 236–240.
ŁUKASIEWICZ, Jan
1920 O logice trójwartościowej
Ruch Filozoficzny, vol. 5, pp. 170–171. Translated as 'On three-valued logic' in *Łukasiewicz, 1970*, pp. 87–88, and in *McCall, 1967*, pp. 16–18.
1922 On determinism
Translation of the original Polish lecture, in *Łukasiewicz, 1970*, pp. 110–128, and in *McCall*, pp. 19–39.
1930 Philosophische Bemerkungen zu mehrwertigen Systemen des Aussagenkalküls

Comptes Rendus des Séances de la Société des Sciences et des Lettres de Varsovie, vol. 23, cl.iii, pp. 51–77. Translated as 'Philosophical remarks on many-valued systems of propositional logic' in *Łukasiewicz, 1970*, pp. 153–178, and in *McCall, 1967*, pp. 40–65.

1952 On the intuitionistic theory of deduction
Konikl. Nederl. Akademie van Wetenschappen, Proceedings, Series A, no. 3, pp. 202–212. Reprinted in *Łukasiewicz*, 1970, pp. 325–335.

1953 A system of modal logic
J. Computing Systems, vol. 1, pp. 111–149. Reprinted in *Łukasiewicz, 1970*, pp. 352–390.

1970 *Selected Works*
ed. L Borkowski, North-Holland.

ŁUKASIEWICZ, Jan and Alfred TARSKI

1930 Untersuchungen über den Aussagenkalkül
Comptes Rendus des Séances de la Société des Sciences et des Lettres de Varsovie, vol. 23, cl.iii, pp. 39–50. Translated as 'Investigations into the sentential calculus' in *Łukasiewicz, 1970*, pp. 131–152, and in *Tarski, 1956*, pp. 38–59. References in the text are to the latter.

MARCISZEWSKI, Witold (ed.)

1981 *Dictionary of Logic*
Martinus Nijhoff.

MATES, Benson

1953 *Stoic Logic*
Univ. of California Publications in Philosophy, Vol.26. Reprinted by the Univ. of California Press, 1961.

1986 *The Philosophy of Leibniz*
Oxford Univ. Press.

McCALL, Storrs (ed.)

1967 *Polish Logic*
Oxford Univ. Press.

McCARTY, Charles

1983 Intuitionism: an introduction to a seminar
Journal of Philosophical Logic, vol. 12, pp. 105–149.

McKINSEY, J. C. C.

1939 Proof of the independence of the primitive symbols of Heyting's calculus of propositions
Journal of Symbolic Logic, vol. 4, pp. 155–158.

McKINSEY, J. C. C. and Alfred TARSKI

1948 Some theorems about the sentential calculi of Lewis and Heyting
Journal of Symbolic Logic, vol. 13, no. 1, pp. 1–15.

MONTEIRO, A.

1967 Construction des algèbres de Łukasiewicz trivalentes dans les algèbres de Boole monadiques, I
Math. Japonicae, vol. 12, pp. 1–23.

PARRY, William Tuthill

1933 Ein Axiomensystem für eine neue Art von Implikation (analytische Implikation)
Ergebnisse eines mathematischen Kolloquiums, vol. 4, pp. 5–6.

1971 Comparison of entailment theories
Typescript of an address to the Association of Symbolic Logic, an abstract of which appears in *Journal of Symbolic Logic,* vol. 37 (1972), pp. 441–442.

197? Analytic implication: its history, justification, and varieties
Typescript of an address to the International Conference on Relevance Logics.

PERZANOWSKI, Jerzy

1973 The deduction theorem for the modal propositional calculi formalized after the manner of Lemmon, Part I
Reports on Mathematical Logic, vol. 1, pp. 1–12.

PIGOZZI, Donald. *See* BLOK and PIGOZZI

PORTE, Jean

1982 Fifty years of deduction theorems
In *Proceedings of the Herbrand Symposium Logic Colloquium '81,* ed. J. Stern, North-Holland, pp. 243–250.

POST, Emil L.

1921 Introduction to a general theory of elementary propositions
Amer. Journal of Math., vol. 43, pp. 163–185. Reprinted in *van Heijenoort, 1967,* pp. 264–283

PRIOR, Arthur

1948 Facts, propositions and entailment
Mind, (new series) vol. 57, pp. 62–68

1955 *Formal Logic*
2nd edition with corrections, 1963, Clarendon Press, Oxford.

1960 The autonomy of ethics
Australasian J. of Phil., vol. 38, pp. 199–206. Reprinted in *Prior, 1976,* pp. 88–96.

1964 Conjunction and contonktion revisited
Analysis vol. 24, pp. 191–195. Reprinted in *Prior, 1976,* pp. 159–164.

1967 Many-valued logic
In *Edwards, 1967,* vol. 5, pp. 1–5.

1976 *Papers in Logic and Ethics*
ed. P. T. Geach and A. J. P. Kenny, Univ. of Massachusetts Press.

PUTNAM, Hilary

1975 *Mind, Language and Reality*
Cambridge Univ. Press.

QUINE, Willard Van Orman

1950 *Methods of Logic*
4th edition, Harvard Univ. Press.

1970 *Philosophy of Logic*
Prentice-Hall.

RASIOWA, Helena

1974 *An Algebraic Approach to Non-classical Logics*
North-Holland.

RESCHER, Nicholas

1968 Many-valued logic
In *Topics in Philosophical Logic*, N. Rescher, D. Reidel.

1969 *Many-valued Logics*
McGraw-Hill.

ROSSER, J. Barkley
1953 *Logic for Mathematicians*
McGraw-Hill.
ROSSER, J. Barkley and Atwell R. TURQUETTE
1952 *Many-valued Logics*
North-Holland.
RUSSELL, Bertrand
See WHITEHEAD and RUSSELL
SCOTT, Theodore Kermit
1966 *John Buridan: Sophisms on Meaning and Truth*
Appleton-Century-Crofts, New York.
SEARLE, John R.
1970 *Speech Acts*
Cambridge Univ. Press.
1983 *Intentionality*
Cambridge Univ. Press.
SEGERBERG, Krister
1968 Propositional logics related to Heyting's and Johansson's
Theoria, vol. 34, pp. 26–61.
1971 *An Essay in Classical Modal Logic*
Filosofiska Studier, no. 13, Uppsala Univ.
SHAW-KWEI, MOH
1954 Logical paradoxes for many-valued logics
Journal of Symbolic Logic, vol. 19, no. 1, pp. 37–40.
SILVER, Charles
1980 A simple strong completeness proof for sentential logic
Notre Dame J. of Formal Logic, XXI, pp. 179–181.
SLUGA, Hans
1987 Semantic content and cognitive sense
In *Frege Synthesized*, ed. L Haaparanta and J. Hintikka, D. Reidel.
SMILEY, T. J.
1959 Entailment and deducibility
Proc. Aristotelian Soc., vol. 59, pp. 233–254.
1976 Comment on 'Does many-valued logic have any use?' by D. Scott, in
Körner, 1976, pp. 74–88.
SMULLYAN, Raymond
1978 *What is the name of this book?*
Prentice-Hall.
SPECKER, Ernst
1960 Die Logik Nicht Gleichzeitig Entscheidbarer Aussagen
Dialectica, vol. 14, pp. 239–246. Translated as, 'The logic of propositions which are not simultaneously decidable', in *The Logico-Algebraic Approach to Quantum Mechanics*, vol. 1, ed. C. A. Hooker, D. Reidel, 1975, pp. 135–140.
SURMA, Stanisław J.
1973 A (ed.) *Studies in the History of Mathematical Logic*
Polish Academy of Sciences, Warsaw.

1973 B A history of the significant methods of proving Post's theorem about the completeness of the classical propositional calculus
In *Surma, 1973 A*, pp. 19–32.

TARSKI, Alfred

1930 Über einige fundamentale Begriffe der Metamathematik
Comptes Rendus des séances de la Société des Sciences et des Lettres de Varsovie, cl. iii, vol. 23, pp. 22–29. Translated as 'On some fundamental concepts of metamathematics' in *Tarski, 1956*, pp. 30–37.

1936 The concept of truth in formalized languages
Reprinted in *Tarski, 1956*, pp. 152–278, where a detailed publication history of it is given.

1956 *Logic, Semantics, Metamathematics*
2nd edition with corrections, 1983, edited by J. Corcoran, Hackett Publ., Indianapolis.

See also JÓNSSON and TARSKI, ŁUKASIEWICZ and TARSKI, McKINSEY and TARSKI

TROELSTRA, A. S. and VAN DALEN, Dirk

1988 *Constructivism in Mathematics*
North-Holland.

TURQUETTE, Atwell R.

1959 Review of papers by Rose and Rosser, Meredith, and Chang
Journal of Symbolic Logic, vol. 24, pp. 248–249.

See also ROSSER and TURQUETTE

VAN FRAASSEN, Bas C.

1967 Meaning relations among predicates
Nous, vol. 1, no. 2, pp. 161–179.

VAN HEIJENOORT, Jean (ed.)

1967 *From Frege to Gödel: A source book in mathematical logic 1879–1931*
Harvard Univ. Press.

WAJSBERG, Mordechaj

1931 Aksjomatyzacja trójwartościowego rachunku zdan
Comptes Rendus des séances de la Société des Sciences et des Lettres de Varsovie, cl.iii, vol. 24, pp. 126–145. Translated as 'Axiomatization of the three-valued propositional calculus' in *McCall, 1967*, pp. 264–284.

WALTON, Douglas N.

1979 Relatedness in intensional action chains
Phil. Studies, vol. 36, no. 2, pp. 175–225.

1982 *Topical Relevance in Argumentation*
John Benjamins, Philadelphia.

1985 *Arguer's Position*
Greenwood Press, London.

WHITEHEAD, Alfred North and Bertrand RUSSELL

1910–13 *Principia Mathematica*
Cambridge Univ. Press.

WILLIAMSON, Colwyn

1968 Propositions and abstract propositions
In *Studies in Logical Theory*, ed. N. Rescher, *American Philosophical Quarterly*, Monograph no. 2, Basil Blackwell, Oxford.

WÓJCICKI, Ryszard
Theory of Logical Calculi
D. Reidel.

WRIGLEY, Michael
1980 Wittgenstein on inconsistency
Philosophy, vol. 55, pp. 471–484.

YABLO, Stephen
1985 Truth and reflection
J. Phil. Logic, vol. 14, pp. 297–349.

Glossary of Notation

General

§	indicates a section of this book
iff	'if and only if'
■	end of proof
⇐|, ⇒|	direction of proof

Formal Languages

$L(p_0, p_1, \ldots \neg, \to, \wedge, \vee)$	
	a formal language, *17, 29, 108*
$p_0, p_1, \ldots$	propositional variables of the formal language
$p, q, q_0, q_1, \ldots$	
	metavariables ranging over $\{p_0, p_1, \ldots\}$
PV	the collection of all propositional variables $p_0, p_1, \ldots$, *27, 93*
$A, B, C, A_0, A_1, \ldots$	
	metavariables ranging over wffs of the formal language or propositions, *16, 29*
Wffs	the collection of all wffs of the formal language, *29, 93, 108*
$\Gamma, \Sigma, \Delta, \ldots$	collections of propositions or wffs, *20, 31*
L, **M**	logics
$L_{\mathbf{L}}$	the formal language for logic **L**
L	the class of all formal languages for the general framework, *108*
L/α	language L with connective α deleted, *312*
γ	a formal connective
$\equiv_{\text{Def}}$	equivalent by definition, used for introducing defined connectives or abbreviations
$\to$-wff	a wff with principal connective $\to$, *30*
C(A)	A is a subformula of C, *31*
A(q)	q is a variable appearing in A, *31*
PV(A)	the collection of all propositional variables appearing in A, *31*
⟦A⟧	the Gödel number of A, *179*

Connectives and Abbreviations in the Formal Language

$\neg$	formalization of 'not', *8*	
$\wedge$	formalization of 'and', *8*	
$\vee$	formalization of 'or', *8*	
$\rightarrow$	formalization of 'if ... then ...', 'implies', etc., *8*	
$\leftrightarrow$	formalization of 'if and only if', defined as $(A \rightarrow B) \wedge (B \rightarrow A)$, *24*	
$\supset$	the material conditional, dependent on truth-values only; defined as $\neg(A \wedge \neg B)$, *57, 151*	
$\equiv$	material equivalence, defined as $(A \supset B) \wedge (B \supset A)$, *162*	
$\Box$	necessity operator in classical modal logics, $\neg A \rightarrow A$ or $A \rightarrow (A \rightarrow A)$, *154, 161* in $\mathbf{L_3}$, $\neg \Diamond \neg A$, *237* in $\mathbf{J_3}$, $\neg(A \rightarrow (A \wedge \neg A))$, *270*	
$\Diamond$	possibility operator in classical modal logics, $\neg \Box \neg A$ or $\neg(A \rightarrow \neg A)$, *154, 155* in $\mathbf{L_3}$, $\neg A \rightarrow A$, *237* in $\mathbf{J_3}$, $\sim \Box \sim A$, *267*	
$⥽$	formalization of 'implies' in classical modal logics; in this text '$\rightarrow$' is used, *147*	
$\sim$	weak negation in $\mathbf{J_3}$, *266*	
$\bot$	falsity, a propositional constant, *109, 225*	
$\mid$	Sheffer stroke, 'nand', *34, 79*	
$\downarrow$	formalization of 'neither ... nor ...', *34*	
$\bigwedge$	conjunction of all the wffs in the collection, *137, 272*	
$\bigvee$	disjunction of all the wffs in the collection, *247, 254, 299*	
R(A, B)	relatedness abbreviation, true in a model of **R** or **S** iff R(A, B), *77–78* of **S** iff $s(A) \cap s(B) \neq \varnothing$, *73*	abbreviations for $A \rightarrow (B \rightarrow B)$
D(A, B)	dependence abbreviation true in a model of **D** iff $s(A) \supseteq s(B)$, *125* of **Dual D** iff $s(A) \subseteq s(B)$, *134*	abbreviations for $A \rightarrow (B \rightarrow B)$
E(A, B)	equivalence of contents abbreviation for **Eq**, true in a model iff $s(A) = s(B)$, *136*	abbreviations for $A \rightarrow (B \rightarrow B)$
M(A, B)	abbreviation in certain classical modal logics for $(A \vee B) \rightarrow B$, true in model iff $s(A) \subseteq s(B)$, *159*	
$\neg^n$	$\neg$ repeated n times, *202*	
$\rightarrow_3$	abbreviation for $A \rightarrow (A \rightarrow B)$ for Deduction Theorem for $\mathbf{L_3}$, *238*	
I	indeterminacy abbreviation for $\mathbf{L_3}$, $A \leftrightarrow \neg A$, *236*	
©	abbreviation for $\mathbf{J_3}$ indicating that the wff has a classical (absolute) truth-value, $\neg(\Diamond A \wedge \Diamond \sim A)$, *269*	
D_n	wff used to determine whether a logic has n-valued semantics, *254*	

Semantic Terms

T	true, *13*
F	false, *13*
U	unknown, undefined, *250*
M	a model, *31, 92–93, 109–110*
M*	the model correlated to M by translation *, *304–305*
v	a valuation, *92, 93, 109, 110*
	a model for classical logic (**PC**), *18, 31*
<v,s>	a set-assignment model, *90–92, 109*
s	a set-assignment, *109*
S, U, T, C	content sets
Sub S	the collection of subsets of S, *93*
∅	the empty set
s(A)	the content set assigned to A
$\overline{s(A)}$	the complement of s(A)
B, C, A, N	relations governing the tables for →, ∧, ∨, ¬ respectively, *90–92*
<v,R>	a model for **S**, *72*; a model for **R**, *74*
<v,D>	a model for **D**, *122*
⟨W, R, e, w⟩	a Kripke possible-worlds model for a classical modal logic, *151*
	a Kripke model for the intuitionistic logic **Int**, or a Kripke tree, *200*
W	a collection of possible worlds, *151*
	a collection of states of information, *201*
R	an accessibility relation, *151*
e	an evaluation for: a classical modal logic, *151–152*
	the intuitionistic logic **Int**, *200–201*
	Johansson's minimal calculus **J**, *225*
	a many-valued logic, *233*
w	a possible world, *151*
⟨W, R⟩	a frame, *152, 200*
⟨W, R, Q, e⟩	a model for Johansson's minimal calculus **J**, *225*
Q	a collection of inconsistent states of information, *225*

Syntactic Consequence Relations

$\vdash A$	A is a theorem, *41*
$\Sigma \vdash A$	A is a syntactic consequence of Σ, *41*
$A \vdash B$	B is a syntactic consequence of {A}
$\Sigma, A \vdash B$	abbreviation for $\Sigma \cup \{A\} \vdash B$
Th(Σ)	the theory of Σ (the collection of all syntactic consequences of Σ) *42*
$\vdash_{\mathbf{L}}^{\Box}$	a modal logic consequence relation using the rule of necessitation, *192*
$\vdash_{\mathbf{J_3}, \Diamond}$	the consequence relation for $\mathbf{J_3}$ where ◇ is primitive, *276*
$\vdash_{\mathbf{J_3}, \neg}$	the consequence relation for $\mathbf{J_3}$ where ¬ is primitive, *279*

Semantic Consequence Relations

⊨A	A is valid, A is a tautology, *20, 96*
Γ⊨B	B is a semantic consequence of Γ (every model which validates every wff in Γ also validates B), *20, 96*
Γ⊨Δ	every wff in Δ is a semantic consequence of Γ, *20*
A⊨B	B is a semantic consequence of A, *19, 31, 96*
v⊨A	A is true in the classical model v, i.e., v(A) = T
<v,s>⊨A	A is true in model <v, s>, i.e., v(A) = T
<v,B>⊨A	A is true in model <v, B>, i.e., v(A) = T
M⊨A	A is true in model M, *109*
w⊨A	A is true in (at) world w, *151*
	the state of information w verifies A, *200–201*
⟨W, R, e⟩⊨A	A is true in model ⟨W, R, e⟩, *151, 200*
⟨W, R⟩	frame ⟨W, R⟩ validates A, *152*
$\models^{(j)}$, ⊨*	special definitions of validity for **G***, *182–183*
M	a formal semantic structure, *109*
	a class of models, *304*
SA	the class of all formal set-assignment semantic structures for all formal languages in L, *109*
RB	the class of all formal relation based semantic structures for all formal languages in L, *110*
C(*SA*)	the collection of all consequence relations for *SA*, *110*
C(*RB*)	the collection of all consequence relations for *RB*, *110*

Translations

L↪**M**	there is a translation from logic **L** to logic **M**, *290*
L↠**M**	there is a grammatical translation from logic **L** to logic **M**, *291*
A*	the translation by * of A
Γ*	the collection of translations of all wffs in Γ
M*	the model correlated to M by translation *, *304–305*

Index

Italic page numbers indicate a definition, theorem, or quotation.
All page numbers greater than 323 refer to the Summary of Logics.

Nijhoff International Philosophy Series

1. Rotenstreich, N.: Philosophy, History and Politics. Studies in Contemporary English Philosophy of History. 1976. ISBN 90-247-1743-4.
2. Srzednicki, J.T.J.: Elements of Social and Political Philosophy. 1976. ISBN 90-247-1744-2.
3. Tatarkiewicz, W.: Analysis of Happiness. 1976. ISBN 90-247-1807-4.
4. Twardowski, K.: On the Content and Object of Presentations. A Psychological Investigation. Translated and with an Introduction by R. Grossman. 1977. ISBN 90-247-1926-7.
5. Tatarkiewicz, W.: A History of Six Ideas. An Essay in Aesthetics. 1980. ISBN 90-247-2233-0.
6. Noonan, H.W.: Objects and Identity. An Examination of the Relative Identity Thesis and Its Consequences. 1980. ISBN 90-247-2292-6.
7. Crocker, L.: Positive Liberty. An Essay in Normative Political Philosophy. 1980. ISBN 90-247-2291-8.
8. Brentano, F.: The Theory of Categories. 1981. ISBN 90-247-2302-7.
9. Marciszewski, W. (ed.): Dictionary of Logic as Applied in the Study of Language. Concepts, Methods, Theories. 1981. ISBN 90-247-2123-7.
10. Ruzsa, I.: Modal Logic with Descriptions. 1981. ISBN 90-247-2473-2.
11. Hoffman, P.: The Anatomy of Idealism. Passivity and Activity in Kant, Hegel and Marx. 1982. ISBN 90-247-2708-1.
12. Gram, M.S.: Direct Realism. A Study of Perception. 1983. ISBN 90-247-2870-3.
13. Srzednicki, J.T.J., Rickey, V.F. and Czelakowski, J. (eds.): Leśniewski's Systems. Ontology and Mereology. 1984. ISBN 90-247-2879-7.
14. Smith, J.W.: Reductionism and Cultural Being. A Philosophical Critique of Sociobiological Reductionism and Physicalist Scientific Unificationism. 1984. ISBN 90-247-2884-3.
15. Zumbach, C.: The Transcendent Science. Kant's Conception of Biological Methodology. 1984. ISBN 90-247-2904-1.
16. Notturno, M.A.: Objectivity, Rationality and the Third Realm. Justification and the Grounds of Psychologism. A Study of Frege and Popper. 1985. ISBN 90-247-2956-4.
17. Dilman, I. (ed.): Philosophy and Life. Essays on John Wisdom. 1984. ISBN 90-247-2996-3.
18. Russell, J.J.: Analysis and Dialectic. Studies in the Logic of Foundation Problems. 1984. ISBN 90-247-2990-4.
19. Currie, G. and Musgrave, A. (eds.): Popper and the Human Sciences. 1985. ISBN 90-247-2998-X.
20. Broad, C.D.: Ethics. Edited by C. Lewy. 1985. ISBN 90-247-3088-0.
21. Seargent, D.A.J.: Plurality and Continuity. An Essay in G.F. Stout's Theory of Universals. 1985. ISBN 90-247-3185-2.
22. Atwell, J.E.: Ends and Principles in Kant's Moral Thought. 1986. ISBN 90-247-3167-4.
23. Agassi, J. and Jarvie, I.Ch. (eds.): Rationality. The Critical View. 1987. ISBN 90-247-3275-1.
24. Srzednicki, J.T.J. and Stachniak, Z. (eds.): S. Leśniewski's Lecture Notes in Logic. 1988. ISBN 90-247-3416-9.
25. Taylor, B.M. (ed.): Michael Dummett. Contributions to Philosophy. 1987. ISBN 90-247-3463-0.
26. Bar-On, A.Z.: The Categories and Principle of Coherence. Whitehead's Theory of Categories in Historical Perspective. 1987. ISBN 90-247-3478-9.
27. Dziemidok, B. and McCormick, P. (eds.): On the Aesthetics of Roman Ingarden. Interpretations and Assessments. 1989. ISBN 0-7923-0071-8
28. Srzednicki, J.T.J. (ed.): Stephan Körner. Philosophical Analysis and Reconstruction. 1987. ISBN 90-247-3543-2.
29. Brentano, F.: On the Existence of God. Lectures given at the Universities of Würzburg and Vienna (1868–1891). 1987. ISBN 90-247-3538-6.
30. Augustynek, Z.: Time. Past, Present and Future. *Forthcoming*.
31. Pawlowski, T.: Aesthetic Values. 1989. ISBN 0-7923-0418-7.
32. Ruse, M. (ed.): What the Philosophy of Biology Is. Essays Dedicated to David Hull. 1989. ISBN 90–247–3778–8.
33. Young, J.: Willing and Unwilling: A Study in the Philosophy of Arthur Schopenhauer. 1987. ISBN 90-247-3556-4.
34. Lavine, T.Z. and Tejera, V. (eds.): History and Anti-History in Philosophy. 1989. ISBN 0-7923-0455-1.
35. Epstein, R.L.: The Semantic Foundations of Logic. Volume 1: Propositional Logics. 1990. ISBN 0-7923-0622-8.

Nijhoff International Philosophy Series

36. Geach, P. (ed.): Logic and Ethics. *Forthcoming.*
37. Winterbourne, A.: The Ideal and the Real. 1988. ISBN 90-247-3774-5.
38. Szaniawski, K. (ed.): The Vienna Circle and the Lvov-Warsaw School. 1989. ISBN 90-247-3798-2.
39. Priest, G.: In Contradiction. A Study of the Transconsistent. 1987. ISBN 90-247-3630-7.